W9-CID-474

Probability and Mathematical Statistics (Continued)

PURI, VILAPLANA, and WERTZ • New Perspectives in Theoretical and Applied Statistics
RANDLES and WOLFE • Introduction to the Theory of Nonparametric Statistics
RAO • Linear Statistical Inference and Its Applications, *Second Edition*
RAO • Real and Stochastic Analysis
RAO and SEDRANSK • W.G. Cochran's Impact on Statistics
RAO • Asymptotic Theory of Statistical Inference
ROHATGI • An Introduction to Probability Theory and Mathematical Statistics
ROHATGI • Statistical Inference
ROSS • Stochastic Processes
RUBINSTEIN • Simulation and The Monte Carlo Method
SCHEFFE • The Analysis of Variance
SEBER • Linear Regression Analysis
SEBER • Multivariate Observations
SEN • Sequential Nonparametrics: Invariance Principles and Statistical Inference
SERFLING • Approximation Theorems of Mathematical Statistics
SHORACK and WELLNER • Empirical Processes with Applications to Statistics
TJUR • Probability Based on Radon Measures

Applied Probability and Statistics

ABRAHAM and LEDOLTER • Statistical Methods for Forecasting
AGRESTI • Analysis of Ordinal Categorical Data
AICKIN • Linear Statistical Analysis of Discrete Data
ANDERSON, AUQUIER, HAUCK, OAKES, VANDAELE, and WEISBERG • Statistical Methods for Comparative Studies
ARTHANARI and DODGE • Mathematical Programming in Statistics
ASMUSSEN • Applied Probability and Queues
BAILEY • The Elements of Stochastic Processes with Applications to the Natural Sciences
BAILEY • Mathematics, Statistics and Systems for Health
BARNETT • Interpreting Multivariate Data
BARNETT and LEWIS • Outliers in Statistical Data, *Second Edition*
BARTHOLOMEW • Stochastic Models for Social Processes, *Third Edition*
BARTHOLOMEW and FORBES • Statistical Techniques for Manpower Planning
BECK and ARNOLD • Parameter Estimation in Engineering and Science
BELSLEY, KUH, and WELSCH • Regression Diagnostics: Identifying Influential Data and Sources of Collinearity
BHAT • Elements of Applied Stochastic Processes, *Second Edition*
BLOOMFIELD • Fourier Analysis of Time Series: An Introduction
BOX • R. A. Fisher, The Life of a Scientist
BOX and DRAPER • Empirical Model-Building and Response Surfaces
BOX and DRAPER • Evolutionary Operation: A Statistical Method for Process Improvement
BOX, HUNTER, and HUNTER • Statistics for Experimenters: An Introduction to Design, Data Analysis, and Model Building
BROWN and HOLLANDER • Statistics: A Biomedical Introduction
BUNKE and BUNKE • Statistical Inference in Linear Models, Volume I
CHAMBERS • Computational Methods for Data Analysis
CHATTERJEE and PRICE • Regression Analysis by Example
CHOW • Econometric Analysis by Control Methods
CLARKE and DISNEY • Probability and Random Processes: A First Course with Applications, *Second Edition*
COCHRAN • Sampling Techniques, *Third Edition*
COCHRAN and COX • Experimental Designs, *Second Edition*

Applied Probability and Statistics (Continued)

CONOVER • Practical Nonparametric Statistics, *Second Edition*
CONOVER and IMAN • Introduction to Modern Business Statistics
CORNELL • Experiments with Mixtures: Designs, Models and The Analysis of Mixture Data
COX • Planning of Experiments
COX • A Handbook of Introductory Statistical Methods
DANIEL • Biostatistics: A Foundation for Analysis in the Health Sciences, *Fourth Edition*
DANIEL • Applications of Statistics to Industrial Experimentation
DANIEL and WOOD • Fitting Equations to Data: Computer Analysis of Multifactor Data, *Second Edition*
DAVID • Order Statistics, *Second Edition*
DAVISON • Multidimensional Scaling
DEGROOT, FIENBERG and KADANE • Statistics and the Law
DEMING • Sample Design in Business Research
DILLON and GOLDSTEIN • Multivariate Analysis: Methods and Applications
DODGE • Analysis of Experiments with Missing Data
DODGE and ROMIG • Sampling Inspection Tables, *Second Edition*
DOWDY and WEARDEN • Statistics for Research
DRAPER and SMITH • Applied Regression Analysis, *Second Edition*
DUNN • Basic Statistics: A Primer for the Biomedical Sciences, *Second Edition*
DUNN and CLARK • Applied Statistics: Analysis of Variance and Regression, *Second Edition*
ELANDT-JOHNSON and JOHNSON • Survival Models and Data Analysis
FLEISS • Statistical Methods for Rates and Proportions, *Second Edition*
FLEISS • The Design and Analysis of Clinical Experiments
FOX • Linear Statistical Models and Related Methods
FRANKEN, KÖNIG, ARNDT, and SCHMIDT • Queues and Point Processes
GALLANT • Nonlinear Statistical Models
GIBBONS, OLKIN, and SOBEL • Selecting and Ordering Populations: A New Statistical Methodology
GNANADESIKAN • Methods for Statistical Data Analysis of Multivariate Observations
GREENBERG and WEBSTER • Advanced Econometrics: A Bridge to the Literature
GROSS and HARRIS • Fundamentals of Queueing Theory, *Second Edition*
GUPTA and PANCHAPAKESAN • Multiple Decision Procedures: Theory and Methodology of Selecting and Ranking Populations
GUTTMAN, WILKS, and HUNTER • Introductory Engineering Statistics, *Third Edition*
HAHN and SHAPIRO • Statistical Models in Engineering
HALD • Statistical Tables and Formulas
HALD • Statistical Theory with Engineering Applications
HAND • Discrimination and Classification
HOAGLIN, MOSTELLER and TUKEY • Exploring Data Tables, Trends and Shapes
HOAGLIN, MOSTELLER, and TUKEY • Understanding Robust and Exploratory Data Analysis
HOEL • Elementary Statistics, *Fourth Edition*
HOEL and JESSEN • Basic Statistics for Business and Economics, *Third Edition*
HOGG and KLUGMAN • Loss Distributions
HOLLANDER and WOLFE • Nonparametric Statistical Methods
IMAN and CONOVER • Modern Business Statistics
JAGERS • Branching Processes with Biological Applications
JESSEN • Statistical Survey Techniques
JOHNSON • Multivariate Statistical Simulation

(continued on back)

Eswar Phadia.

Statistical Analysis with Missing Data

Statistical Analysis with Missing Data

RODERICK J. A. LITTLE
University of California at Los Angeles

DONALD B. RUBIN
Harvard University

JOHN WILEY & SONS
New York • Chichester • Brisbane • Toronto • Singapore

Copyright © 1987 by John Wiley & Sons, Inc.

All rights reserved. Published simultaneously in Canada. Reproduction or translation of any part of this work beyond that permitted by Section 107 or 108 of the 1976 United States Copyright Act without the permission of the copyright owner is unlawful. Requests for permission or further information should be addressed to the Permissions Department, John Wiley & Sons, Inc.

Library of Congress Cataloging in Publication Data:

Little, Roderick J. A.
Statistical analysis with missing data.
(Wiley series in probability and mathematical statistics. Applied probability and statistics)
Includes indexes.
1. Mathematical statistics. 2. Missing observations (Statistics) I. Rubin, Donald B. II. Title. III. Series.

QA276.L57 1986 519.5 86-19075
ISBN 0-471-80254-9

Printed in the United States of America

10 9

To Robin, David, Andrew,
Kathryn, Scott, and Paul

Preface

The literature on the statistical analysis of data with missing values has flourished since the early 1970s, spurred by advances in computer technology that made previously laborious numerical calculations a simple matter. This book aims to survey current methodology for handling missing-data problems and present a likelihood-based theory for analysis with missing data that systematizes these methods and provides a basis for future advances. Part I of the book discusses historical approaches to missing-value problems in three important areas of statistics: analysis of variance of planned experiments, survey sampling, and multivariate analysis. These methods, although not without value, tend to have an ad hoc character, being solutions worked out by practitioners with limited research into their theoretical properties. Part II presents a systematic approach to the analysis of data with missing values, where inferences are based on likelihoods derived from formal statistical models for the data and the missing-data mechanisms. Applications of the approach are presented in a variety of contexts, including regression, factor analysis, contingency table analysis, time series, and sample survey inference. Many of the historical methods in Part I can be derived as examples (or approximations) of this likelihood-based approach.

The book is intended for the applied statistician, and hence emphasizes examples over the precise statement of regularity conditions or proofs of theorems. Nevertheless, readers are expected to be familiar with basic principles of inferences based on likelihoods, briefly reviewed in Chapter 5. The book also requires an understanding of standard models of complete-data analysis—the normal linear model, multinomial models for counted data—and the properties of standard statistical distributions, especially the multivariate normal distribution. Particular chapters require familiarity in areas of statistical activity—analysis of variance for experimental designs (Chapter 2), survey sampling (Chapters 3 and 12), loglinear models for contingency tables

(Chapter 9). Specific examples also introduce other statistical topics, such as factor analysis and time series (Chapter 8). The discussion of these examples is self-contained and does not require specialized knowledge, but such knowledge will, of course, enhance the reader's appreciation of the main statistical issues. We have managed to cover about three-quarters of the material in the book in a 40-hour graduate statistics course.

Despite recent advances in the analysis of data with missing values, weaknesses in the literature remain and are reflected in the book. Specifically, most of the book concerns the derivation of point estimates of parameters and approximate standard errors, with interval estimation and testing based on large-sample theory. Tests and interval estimates from small samples with missing data have had very limited development, although a Bayesian approach is outlined in Section 6.3.2 for a particular problem. Also, methods are based on fairly standard statistical models such as the multivariate normal and multinomial models. Very little work has been done on diagnostic tests of the validity of such models when data are incomplete, or on the robustness of estimates derived from them. We hope that our systematic presentation of missing-data methods will stimulate more work in these areas. We also hope that the book will stimulate the development of widely available software for missing-data methods, an area where, at this time, the practitioner is liable to be frustrated.

Many individuals are due thanks for their help in producing this book. NSF and NIMH (through grants NSF-SES-83-11428, NSF-SES-84-11804, and NIMH-MH-37188) helped support some aspects of the research reported here, Mark Schluchter performed the calculations needed in Section 8.5, Leisa Weld and T. E. Raghunathan carefully read the final manuscript and made helpful suggestions, and our students in Biomathematics M232 at UCLA and Statistics 220r at Harvard University also made helpful suggestions. Finally we thank Judy Siesen for typing and retyping our many drafts, and Bea Shube for her kind support and encouragement.

In closing, we wish to add that we have found that many statistical problems can be usefully viewed as missing-value problems even when the data set is fully recorded, and moreover, that missing-data research can be an excellent springboard for learning about statistics in general. We hope our readers will agree with us and find the book stimulating.

R. J. A. LITTLE
D. B. RUBIN

Los Angeles, California
Cambridge, Massachusetts
January 1987

Contents

Statistical Analysis
with Missing Data

PART I

Overview and Historical Approaches

CHAPTER 1

Introduction

1.1. MISSING DATA

Standard statistical methods have been developed to analyze rectangular data sets. The rows of the data matrix represent units, also called cases, observations, or subjects depending on context, and the columns represent variables measured for each unit. The entries in a data matrix are real numbers, representing the values of essentially continuous variables, such as age and income, or representing categories of response, which may be ordered (e.g., level of education) or unordered (e.g., race and sex). This book concerns the analysis of such a data matrix when some of the values in the matrix are not observed. For example, respondents in a household survey may refuse to report income. In an industrial experiment some results are missing because of mechanical breakdowns unrelated to the experimental process. In an opinion survey some individuals may be unable to express a preference for one candidate over another. In the first two examples it is natural to treat the values that are not observed as missing, in the sense that there are true underlying values that would have been observed if survey techniques had been better or the industrial equipment had been better maintained. In the third example, however, it is less clear that a well-defined candidate preference has been masked by the nonresponse; thus it is less natural to treat the unobserved values as missing. Instead, the lack of a response is essentially an additional point in the sample space of the variable being measured, which identifies a "don't know" stratum of the population.

Most statistical software packages allow the identification of nonresponse strata by creating one or more special codes for those entries of the data matrix that are not observed. More than one code might be used to identify particular types of nonresponse, such as "don't know," or "refuse to answer," or "out

Table 1.1a Data Matrix for Children in a Survey Summarized by the Pattern of Missing Data: 1 = Observed, 0 = Missing

	Variables					
Pattern	Age	Gender	Weight 1	Weight 2	Weight 3	Number of Children with Pattern
A	1	1	1	1	1	1770
B	1	1	1	1	0	631
C	1	1	1	0	1	184
D	1	1	0	1	1	645
E	1	1	1	0	0	756
F	1	1	0	1	0	370
G	1	1	0	0	1	500

of legitimate range." Statistical packages typically exclude units that have missing value codes for any of the variables involved in an analysis. This strategy is generally inappropriate, since the investigator is usually interested in making inferences about the entire target population, rather than the portion of the target population that would provide responses to all relevant variables in the analysis. Our aim is to describe a collection of techniques that are more generally appropriate. Some of these techniques have already found their way into computer packages (e.g., the BMDPAM program in Dixon, 1983), and it is quite likely that many more will do so in the next few years.

EXAMPLE 1.1. *Nonresponse in a Binary Outcome Measured at Three Times Points.* Woolson and Clarke (1984) analyze data from the Muscative Coronary Risk Factor Study, a longitudinal study of coronary risk factors in schoolchildren. Table 1.1a summarizes the data matrix by the pattern of missing data. Five variables (gender, age, and obesity for three rounds of the survey) are recorded for 4856 cases—gender and age are completely recorded, but the three obesity variables are sometimes missing with six patterns of missingness. Since age is recorded in five categories and the obesity variables are binary, the data can be displayed as counts in a contingency table. Table 1.1b displays the data in this form, with missingness of obesity treated as a third category of the variable, where O = obese, N = not obese, and M = missing. Thus the pattern MON denotes missing at the first round, obese at the second round, and not obese at the third round, and other patterns are defined similarly.

Woolson and Clarke analyze these data by fitting multinomial distributions over the $3^3 - 1 = 26$ response categories for each column in Table 1.1. That is, missingness is regarded as defining strata of the population. We suspect that for these data it makes good sense to regard the nonrespondents

Table 1.1b Number of Children Classified by Population and Relative Weight Category in Three Rounds of a Survey

Response	Males					Females				
	Age Group					Age Group				
Category[a]	5–7	7–9	9–11	11–13	13–15	5–7	7–9	9–11	11–13	13–15
NNN	90	150	152	119	101	75	154	148	129	91
NNO	9	15	11	7	4	8	14	6	8	9
NON	3	8	8	8	2	2	13	10	7	5
NOO	7	8	10	3	7	4	19	8	9	3
ONN	0	8	7	13	8	2	2	12	6	6
ONO	1	9	7	4	0	2	6	0	2	0
OON	1	7	9	11	6	1	6	8	7	6
OOO	8	20	25	16	15	8	21	27	14	15
NNM	16	38	48	42	82	20	25	36	36	83
NOM	5	3	6	4	9	0	3	0	9	15
ONM	0	1	2	4	8	0	1	7	4	6
OOM	0	11	14	13	12	4	11	17	13	23
NMN	9	16	13	14	6	7	16	8	31	5
NMO	3	6	5	2	1	2	3	1	4	0
OMN	0	1	0	1	0	0	0	1	2	0
OMO	0	3	3	4	1	1	4	4	6	1
MNN	129	42	36	18	13	109	47	39	19	11
MNO	18	2	5	3	1	22	4	6	1	1
MON	6	3	4	3	2	7	1	7	2	2
MOO	13	13	3	1	2	24	8	13	2	3
NMM	32	45	59	82	95	23	47	53	58	89
OMM	5	7	17	24	23	5	7	16	37	32
MNM	33	33	31	23	34	27	23	25	21	43
MOM	11	4	9	6	12	5	5	9	1	15
MMN	70	55	40	37	15	65	39	23	23	14
MMO	24	14	9	14	3	19	13	8	10	5

Source: Woolson and Clarke (1984).

[a] NNN indicates not obese in 1977, 1979, and 1981; O indicates obese and M indicates missing in a given year.

as having a true underlying value for the obesity variable. Hence we would argue for treating the nonresponse categories as missing values and estimating the joint distribution of the three dichotomous outcome variables from the partially missing data. Appropriate methods for handling such categorical data with missing values are described in Chapter 9. The methods involve quite straightforward modifications of existing algorithms for categorical data analysis currently available in statistical software packages.

1.2. A BROAD TAXONOMY OF METHODS WITH PARTIALLY MISSING DATA

The literature on the analysis of partially missing data is comparatively recent; review papers include Afifi and Elashoff (1966), Hartley and Hocking (1971), Orchard and Woodbury (1972), Dempster, Laird, and Rubin (1977), and Little (1982). Methods proposed in this literature can be roughly grouped into the following categories (not mutually exclusive):

1. *Procedures Based on Completely Recorded Units.* When some variables are not recorded for some of the units, a simple expedient mentioned in Section 1.1 is simply to discard incompletely recorded units and to analyze only the units with complete data (e.g., Nie et al., 1975). This strategy is discussed in Chapter 3. It is generally easy to carry out and may be satisfactory with small amounts of missing data. It can lead to serious biases, however, and it is not usually very efficient.

2. *Imputation-Based Procedures.* The missing values are filled in and the resultant completed data are analyzed by standard methods. Commonly used procedures for imputation include *hot deck* imputation, where recorded units in the sample are substituted; *mean* imputation, where means from sets of recorded values are substituted; and *regression* imputation, where the missing variables for a unit are estimated by predicted values from the regression on the known variables for that unit. Applications of imputation methods to designed experiments, multivariate analysis, and sample surveys are described in Chapter 2, 3, and 4, respectively. For valid inferences to result, modifications to the standard analyses are required to allow for the differing status of the real and the imputed values. These modifications are relatively simple for a generalization discussed in Chapter 12 that produces multiple imputations for each missing value.

3. *Weighting Procedures.* Randomization inferences from sample survey data without nonresponse are commonly based on *design weights*, which are inversely proportional to the probability of selection. For example, let y_i be the value of a variable Y for unit i in the population. Then the population mean is often estimated by

$$\sum \pi_i^{-1} y_i / \sum \pi_i^{-1} \tag{1}$$

where the sums are over sampled units, π_i is the probability of inclusion in the sample for unit i, and π_i^{-1} is the design weight for unit i. Weighting procedures modify the weights in an attempt to adjust for nonresponse. The estimator (1) is replaced by

$$\sum (\pi_i \hat{p}_i)^{-1} y_i / \sum (\pi_i \hat{p}_i)^{-1}, \tag{2}$$

where the sums are now over sampled units that respond, and $\hat{p}_i$ is an estimate of the probability of response for unit i, usually the proportion of responding units in a subclass of the sample. Weighting is related to mean imputation; for example, if the design weights are constant in subclasses of the sample, then both imputing the subclass mean for missing units in each subclass and weighting responding units by the proportion responding in each subclass, lead to the same estimates of population means, although not the same estimates of sampling variances unless adjustments are made to the data with means imputed. Weighting methods are described further in Chapter 4.

4. *Model-Based Procedures.* A broad class of procedures is generated by defining a model for the partially missing data and basing inferences on the likelihood under that model, with parameters estimated by procedures such as maximum likelihood. Advantages of this approach are flexibility; the avoidance of ad hoc methods, in that model assumptions underlying the resulting methods can be displayed and evaluated; and the availability of large sample estimates of variance based on second derivatives of the loglikelihood, which take into account incompleteness in the data. Model-based procedures are the main focus of this book, and are developed in Chapters 5–12, which comprise Part II.

1.3. MISSING-DATA PATTERNS

Some methods of analysis, described in Chapter 6, are intended for particular patterns of missing data and use only standard complete-data analyses. Other methods, such as the EM algorithm described in Chapters 7–9, are applicable to more general missing-data patterns, but usually involve more computing than methods designed for special patterns. Thus it pays to sort the data to see if they can be arranged in an orderly pattern.

EXAMPLE 1.2. *A Special Pattern of Missing Data.* The data pattern in Table 1.2, where 1 indicates observed and 0 indicates missing, was obtained from the results of a panel study of students in 10 Illinois schools, analyzed by Marini, Olsen, and Rubin (1980). The first block of variables was recorded for all individuals at the start of the study, and hence is completely observed. The second block consists of variables measured for all respondents in the follow-up study, fifteen years later. Of all respondents to the original survey, 79% responded to the follow-up, and thus the subset of variables in block 2 is regarded as 79% observed. Block 1 variables are consequently *more observed* than block 2 variables. The data for the fifteen-year follow-up survey were collected in several phases, and for economic reasons the group of variables forming the third block were recorded for a subset of those responding to block 2 variables. Thus, block 2 variables are more observed than block 3

Table 1.2 Patterns of Missing Data across Four Blocks of Variables

Pattern	Adolescent Variables, Block 1	Variables Measured for All Follow-Up Respondents, Block 2	Variables Measured Only for Initial Follow-Up Respondents, Block 3	Parent Variables, Block 4	Number of Cases	Percentage of Cases
A	1	1	1	1	1594	36.6
B	1	1[a]	1[a]	0	648	14.9
C	1	1	0	1[b]	722	16.6
D	1	1[a]	0	0	469	10.8
E	1	0	0	1[b]	499	11.5
F	1	0	0	0	420	9.6
					4352	100.0

Source: Marini, Olsen, and Rubin (1980).

[a] Observations falling outside monotone pattern 2 (block 1 more observed than block 4; block 4 more observed than block 2; block 2 more observed than block 3).

[b] Observations falling outside monotone pattern 1 (block 1 more observed than block 2; block 2 more observed than block 3; block 3 more observed than block 4).

variables. Blocks 1, 2, and 3 form a *monotone* pattern of missing data. The fourth block of variables consists of a small number of items measured by a questionnaire mailed to the parents of all students in the original adolescent sample. Of these parents, 65% responded. The four blocks of variables do not form a monotone pattern. However, by sacrificing a relatively small amount of data, monotone patterns can be obtained. The authors analyze two monotone data sets. First, the values of block 4 variables for patterns C and E (marked with the letter *b*) are omitted, leaving a monotone pattern with block 1 more observed than block 2, which is more observed than block 3, which is more observed than block 4. Second, the values of block 2 variables for Patterns B and D and the values of block 3 variables for pattern B (marked with the letter *a*) are omitted, leaving a monotone pattern with block 1 more observed than block 4, which is more observed than block 2, which is more observed than block 3. In contrast to this example, the creation of a monotone pattern for the data of Example 1.1 would involve the substantial loss of information.

1.4. MECHANISMS THAT LEAD TO MISSING DATA

Knowledge, or absence of knowledge, of the mechanisms that led to certain values being missing is a key element in choosing an appropriate analysis and in interpreting the results. Sometimes the mechanism is under the control of the statistician. For example, we might view survey sampling as leading to missing data, where some variables (sample design variables) are recorded for all units in the population and survey variables are "missing" for units that are not selected. The mechanism leading to missing data is then simply the process of sample selection. If units are selected by probability sampling, then the mechanism is under the control of the sampler (provided the design is successfully implemented) and may be called "ignorable." If the sampling frame is deficient or some units do not respond, however, then the mechanism leading to the observed data is not so well understood. Any analysis of the data is then dependent on assumptions about the missing-data mechanism, which should be made explicit.

The technique of *double sampling* in survey methodology provides another instance where a pattern of missing data is obtained under the control of the sampler. A large sample is selected, and certain basic characteristics are recorded. Then a random subsample is selected from the original sample, and more variables are measured. The resulting data form a monotone pattern. Regression methods used to analyze such data can be viewed as incomplete data methods, although they are not usually viewed in this way.

The case of *censoring* illustrates a situation where the mechanism leading to missing data may not be under the control of the statistician, but is understood. The data consist of times to the occurrence of an event (e.g., death of an experimental animal, birth of a child, failure of a light bulb). For some units in the sample, time to occurrence is censored because the event had not occurred before the termination of the experiment. If the time to censoring is recorded, then we have the partial information that the failure time exceeds the time to censoring. The analysis of the data needs to take account of this information to avoid biased results.

In many data analyses the mechanism leading to missing data does not enter explicitly; in such cases, an assumption is being made that the mechanism is ignorable. It is possible, however, to include the mechanism in the statistical model by including a distribution for response indicator variables that take the value 1 if an item is recorded and the value 0 otherwise. The mechanism that leads to missing values generally cannot be ignored. For example, nonresponse in an income survey may be related to unobserved incomes, and thus is nonignorable. These ideas are developed in Chapter 5, when outlining likelihood theory in the presence of nonresponse. Chapter 11 is devoted to nonignorable missing-data mechanisms.

1.5. UNIVARIATE SAMPLES WITH MISSING VALUES

Perhaps the simplest data structure is a univariate random sample for which some units are missing. Let y_i denote the value of a variable Y for unit i, and suppose that for a simple random sample of size n, $y_1, \ldots, y_m$ are recorded and $y_{m+1}, \ldots, y_n$ are missing. An obvious consequence of the missing values in this example is the reduction in sample size from n to m. We might contemplate carrying out the same analyses on the reduced sample as we intended for the size-n sample. For example, if we assume the values are normally distributed and wish to make inferences about the mean, we might estimate the mean by the sample mean of the responding units and assign the estimate of variance s^2/m, where s^2 is the sample variance of the responding units. When we do this we are in effect ignoring the mechanism that caused certain values to be missing.

For a univariate sample the missing-data mechanism is ignorable if the missing values are *missing at random* in the sense that the observed units are a random subsample of the sampled units. If the probability that y_i is observed depends on the value of y_i, then the missing-data mechanism is nonignorable, and analyses on the reduced sample that do not allow for this are subject to bias.

EXAMPLE 1.3. *Stochastic Censoring of a Univariate Normal Sample.* The data in Figure 1.1 illustrate the importance of understanding the process that creates missing data. Figure 1.1*a* presents a stem and leaf plot (i.e., a histogram with individual values retained) of $n = 100$ standard normal deviates. The population mean (zero) for this sample is estimated by the sample mean, which has the value -0.03. Figure 1.1*b* presents a subsample of data obtained from the original sample in Figure 1.1*a* by deleting units independently with probability 0.5. The probability of deletion does not depend on the value of y; consequently, the resulting sample of size $m = 52$ is a random subsample of the original values whose sample mean, -0.11, can be used to estimate the population mean of y without bias.

Figures 1.1*c* and 1.1*d* illustrate nonignorable missing-data mechanisms. In Figure 1.1*c* negative values from the original sample have been retained and positive values have been deleted. Thus letting R_i be the response indicator,

$$\Pr(R_i = 1 | y_i) = \Pr(y_i \text{ observed} | y_i) = \begin{cases} 1, & y_i < 0, \\ 0, & y_i \geqslant 0. \end{cases}$$

The probability of response depends on y, and hence the response mechanism is nonignorable. Standard complete-data analyses that ignore the missing-data mechanism with such data are in general biased. In particular, the sample mean, -0.89, obviously underestimates the population mean of y. The missing-data mechanism here is called censoring, with observed values *censored from above*, or *right censored*, at the value zero.

Figure 1.1*d* illustrates a form of *stochastic censoring*, where the probability that y_i is observed lies between zero and one. Specifically, the probability that

Stem	(a)	(b)	(c)	(d)
−3.5	7		7	7
−3				
−2.5	8		8	8
−2				
−1.5	57889	578	57889	57889
−1	001111222233	1112233	001111222233	001111222233
−0.5	5556666778888899999	566788899999	5556666778888899999	555666778888899999
−0	0112222223344	011234	0112222223344	0112222234
+0	0011222222233344444	0122222234		012224
+0.5	56777778899	677789		
+1	0011113444	11144		
+1.5	56778	6		
+2	023	02		
+2.5				
+3	3			
	Sample mean = −0.03	Sample mean = −0.11	Sample mean = −0.89	Sample mean = −0.81
	(a) Uncensored Sample, $n = 100$	(b) Ignorable Censoring Mechanism, $\Pr(y_i \text{ observed}) = 0.5$, $m = 52$	(c) Pure Censoring from above at 0.0, $m = 51$	(d) Nonignorable Censoring Mechanism, $\Pr(y_i \text{ observed}) = \Phi(-2.05 y_i)$, $m = 53$

Figure 1.1. Stem and leaf displays of distribution of standard normal sample with stochastic censoring.

y_i is observed is $\Phi(-2.05y_i)$, where $\Phi(\cdot)$ is the cumulative standard normal distribution function. The probability decreases as y_i increases, and thus most of the observed values of y are negative. The missing-data mechanism is again nonignorable, and the sample mean is again a systematic underestimate of the population mean.

Now suppose we are faced with an incomplete sample as in Figure 1.1*c* or 1.1*d*, and we wish to estimate the population mean. If the censoring mechanism is *known*, then methods are available that correct for the selection bias of the sample mean. These methods are commonly based on the *method of maximum likelihood*. If, however, the censoring mechanism is *unknown*, the problem is much more difficult. The principal evidence that the response mechanism is not ignorable lies in the fact that the observed samples are *asymmetric*, which contradicts the assumption that the original data have a (symmetric) normal distribution. If we are confident that the uncensored sample has a symmetric distribution, we can use this information to adjust for selection bias using, for example, the method of maximum likelihood. On the other hand, if the statistician has little knowledge about the form of the uncensored distribution, he cannot say whether the data are a censored sample from a symmetric distribution or a random subsample from an asymmetric distribution. In the former case the sample mean is biased for the population mean; in the latter case it is not.

EXAMPLE 1.4. *An Application of Example 1.3 to Historical Heights.* Wachter and Trussell (1982) present an interesting illustration of this problem, involving the estimation of historical heights. The distribution of heights in historical populations is of considerable interest, because of the information it provides about nutrition, and hence indirectly about living standards. Most of the recorded information concerns the heights of recruits for the Armed Services. The samples are subject to censoring, since minimal height standards were often in operation and were enforced with variable strictness, depending on the demand for and supply of recruits. Thus a typical observed distribution of heights might take the form of the nonshaded histogram in Figure 1.2, adapted from Wachter and Trussell (1982). The shaded area in the figure represents the heights of men excluded from the recruit sample and is drawn under the assumption that heights are normally distributed in the uncensored population. Wachter and Trussell discuss methods for estimating the mean and variance of the uncensored distribution under this crucial normal assumption. In this example there is considerable external evidence that heights in unrestricted populations *are* nearly normal, so the inferences have some validity. In many other problems involving missing data such information is not available or is highly tenuous in nature. As discussed in Chapter 11, the sensitivity of answers from an incomplete sample to unjustifiable or tenu-

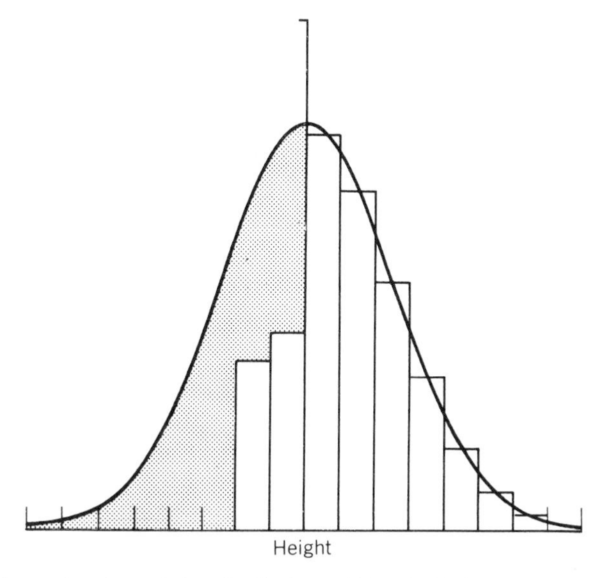

Figure 1.2. Observed and population distributions of historical heights. Population distribution is normal, observed distribution is represented by the histogram, and the shaded area represents missing data.

ous assumptions is a basic problem in the analysis of data subject to unknown missing-data mechanisms, such as can occur in survey data subject to nonresponse.

1.6. MORE THAN ONE VARIABLE, BUT ONLY ONE SUBJECT TO NONRESPONSE

Suppose now we add a variable X to the variable Y in the previous example, where X is not subject to nonresponse, that is, is recorded for all n units in the sample. The monotone data pattern of Figure 1.3 is then obtained. A wide variety of situations lead to the pattern in this figure. In the sample survey context, the variable Y may denote a questionnaire item subject to non-response, such as household income, and X may denote an item recorded for all units in the sample, for example, a survey design variable such as locality or a completely recorded survey item such as age. In an experimental situation the variable X may be a completely recorded covariate or it may be a fixed variable controlled by the experimenter, such as treatment indicator in a randomized design. The data on Y may be incomplete because of uncontrolled

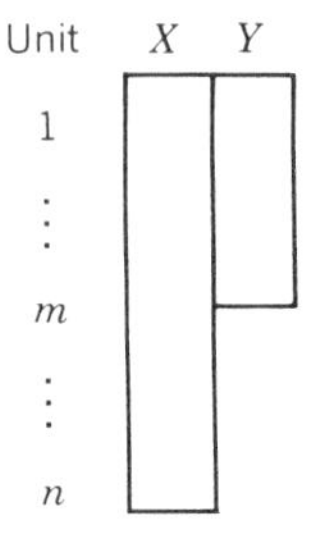

Figure 1.3. Monotone pattern with X more observed than Y.

events in the course of the data collection, such as nonresponse, wild values that have been discarded, or errors in recording the data. Alternatively, they may be missing by design, as in a calibration experiment where X is a cheap measurement recorded for a large sample and Y is an expensive measurement recorded for a subsample.

The variables X and Y may be modeled as continuous (interval scaled) variables or as categorical. The case where X and Y are bivariate normally distributed continuous variables has received considerable attention in the literature, as discussed in Chapter 6 in the context of monotone patterns of incomplete data. If X is categorical and Y is continuous, the data have a one-way analysis of variance structure, with values of Y missing within cells defined by values of X. This case is discussed in detail in Chapter 2. If X and Y are both categorical, then the complete units can be arranged as a two-way contingency table, with cells defined by each pair of values of X and Y. The units measured on Y alone then form a partially classified supplemental margin. Methods of analysis for partially classified contingency tables like this are discussed in Chapter 9.

For data with this pattern it is useful to classify the missing-data mechanism according to whether the probability of response (1) depends on Y and possibly X as well, (2) depends on X but not on Y, or (3) is independent of X and Y. Rubin (1976) proposes the following terminology already introduced in the simple case with one Y variable. If case 3 applies, we say that the missing data are missing at random (MAR) and the observed data are observed at random (OAR) or, more simply, that the missing data are missing completely at random (MCAR). In this case the observed values of Y form a random subsample of the sampled values of Y, as in Figure 1.1*a*. If case 2 applies, we say that the missing data are missing at random (MAR). In this case, the observed values of Y are not necessarily a random subsample of the sampled values, but they are a random sample of the sampled values within subclasses defined by values of X. If case 1 applies, the data are neither MAR nor OAR.

In cases 2 and 3 the missing-data mechanism is ignorable for likelihood-based inferences. In case 3 this mechanism is ignorable for both sampling-based and likelihood-based inferences. In case 1 the mechanism is nonignorable. These classifications may be clearer with a specific example.

EXAMPLE 1.5. *Two Continuous Variables, One Subject to Nonresponse.* Suppose that X = age and Y = income. If the probability that income is recorded is the same for all individuals, regardless of their age or income, then the data are MAR and OAR (i.e., MCAR). If the probability that income is recorded varies according to the age of the respondent but does not vary according to the income of respondents within an age group, then the data are MAR but not OAR (and so not MCAR). If the probability that income is recorded varies according to income within each age group, then the data are neither MAR nor OAR. This latter nonignorable case is hardest to deal with analytically, which is unfortunate, since it is also the most likely case in this application.

The significance of these assumptions about the missing-data mechanism depends somewhat on the objective of the analysis. For example, if interest lies in the marginal distribution of X, then the data on Y, and the mechanism leading to missing Y values, are usually irrelevant ("usually" because one can construct examples where this is not the case, but such examples are typically of theoretical rather than practical interest). If interest lies in the conditional distribution of Y given X, as, for example, when we are studying how the distribution of income varies according to age, then the analysis based on the m completely observed units may be satisfactory if the data are MAR. On the other hand, if interest is in the marginal distribution of Y, or summary measures such as the mean of Y, then an analysis based on the m complete units is generally biased unless the data are MCAR. With X and Y observed on the same n units, the data on X are not usually useful in estimating the mean of Y; however, for the pattern in Figure 1.3 the data on X are useful for this purpose, both in increasing the efficiency with which the mean of Y is estimated and in reducing the effects of selection bias if the data are not MCAR. This point is of great importance and will be illustrated in subsequent chapters.

The task of estimating the joint distribution of X and Y from data with the monotone pattern of Figure 1.3 assuming MAR is often simplified by exploiting a factorization of the joint distribution, as discussed in some detail in Chapter 6. Let $f(X, Y)$ denote the density of the joint distribution of X and Y. Recall that this density can be factored in the form

$$f(X, Y) = f(X)f(Y|X), \tag{1.1}$$

where $f(X)$ denotes the marginal distribution of X and $f(Y|X)$ denotes the conditional distribution of Y given X. Here and subsequently, functions are distinguished by their arguments, as in (1.1). Inference about the marginal distribution of age can be based on the n sampled values of age. Inferences about the conditional distribution of income given age can be based on the m units with X and Y observed. The results of these analyses can then be combined to estimate the joint distribution of age and income or the conditional distribution of age given income. Estimation of the conditional distribution of income given age is often a form of regression analysis, and the strategy of factoring the distribution relates to the idea of *imputing* the missing values of income by regressing income on age and then calculating predictions from the regression equation. The analysis of the data in Figure 1.3 can thus be viewed as a classical regression–prediction problem.

EXAMPLE 1.6. $p + 1$ *Variables, One Subject to Nonresponse.* In many problems the values of more than one variable are available for the n sampled units. The data can be represented schematically as in Figure 1.3, where X is now a matrix with p columns. The first incomplete-data problem to receive systematic attention in the statistics literature has the pattern of Figure 1.3, namely, the problem of missing data in designed experiments. In the context of agricultural trials this is often called the missing-plot problem. Interest is in the relationship between a dependent variable Y, such as yield of crop, on a set of factors, such as variety, type of fertilizer, and temperature. The n data points are chosen so that the design matrix is easily inverted, as with complete or fractional replications of factorial designs. The missing-data problem arises when, at the conclusion of the experiment, the values of the outcome variable Y for $n - m$ units are missing, perhaps because no values were possible, as when particular plots were not amenable to seeding, or because values were recorded and then discarded or lost.

Standard analyses of the incomplete data assume that the data are MAR, that is, the probability that y_i is missing may vary across the design variables, but for given values of x_i, the ith row of X, the probability that y_i is missing does not depend on the value of y_i. In practical situations the plausibility of this assumption needs to be checked. The analysis aims to exploit the "near-balance" of the resulting data set to simplify computations. For example, one tactic is to substitute estimates of the missing values of Y and then to carry out the analysis assuming the data to be complete. Questions needing attention then address the choice of appropriate values to impute and how to modify the subsequent analyses to allow for the substitutions. This problem is discussed in Chapter 2.

Problems of item nonresponse in sample surveys often have the pattern of Figure 1.3, with Y representing a variable subject to nonresponse and the X variables representing background items that are completely observed. Here the emphasis is more commonly on estimating the marginal distribution of Y, in contrast to the analysis of variance problem, where interest concerns the conditional distribution of Y given X. Nevertheless there are similarities between the two problems.

Essentially all the methods in the survey literature for handling the problem of nonresponse assume that the data are MAR, although in many practical problems this assumption is highly questionable. In addition to discussing randomization-based methods in Chapter 4, in Chapter 12 we outline some recent literature that relaxes this assumption and uses likelihood-based methods.

1.7. MULTIVARIATE MISSING DATA

The missing-data structures discussed in Sections 1.5 and 1.6 are univariate in the sense that the missing values are confined to a single outcome variable. We now review some missing-data structures that are multivariate.

Many multivariate statistical analyses, including least squares regression, factor analysis, and discriminant analysis are based on an initial reduction of the data to the sample mean vector and sample covariance matrix of the variables. The question of how to estimate these quantities from incomplete data is, therefore, an important one. Early literature, discussed selectively in Chapter 3, proposed ad hoc solutions. A more systematic likelihood-based approach, which is the focus of Part II, is introduced in Chapter 5 and applied to a variety of situations in the following chapters.

EXAMPLE 1.7. *Estimating the Mean and Covariance Matrix with Monotone Missing Data.* Suppose the data can be arranged in a monotone pattern. A simple approach to estimating the mean and covariance matrix is to restrict the analysis to the units with all variables observed. However, this method of analysis discards a considerable amount of data. Also, for many examples, including the data summarized in Table 1.2, the completely observed units are not a random sample of the original sample (i.e., the data are not MCAR), and the resulting estimates are biased. A more successful strategy is to assume the data have a multivariate normal distribution and to estimate the means and covariance matrix by maximum likelihood. In Chapter 6 we show that for monotone data this task is not as difficult as one might suppose, because estimation is simplified by a factorization of the joint distribution of the variables, as in equation (1.1), allowing maximum likelihood estimates to be found from a sequence of regressions.

EXAMPLE 1.8. *Estimating the Mean and Covariance Matrix with General Missing-Data Patterns.* Many data sets with missing values do not exhibit convenient monotone patterns or close approximations such as are displayed in Table 1.2. Methods for estimating the mean and covariance matrix of a set of variables have also been developed that can be applied to *any* pattern of missing values. As in the previous example, these methods are often based on maximum likelihood estimation, assuming the variables are multivariate normally distributed, and estimation involves iterative algorithms.

The estimation–maximization (EM) algorithm developed in Chapter 7 is an important general technique for finding maximum likelihood estimates from incomplete data. It is applied to the case of multivariate normal data in Chapter 8. The resulting algorithm is particularly instructive, because it is closely related to an iterative version of a method that imputes estimates of the missing values by regression. Thus even in this complex problem, a link can be established between efficient model-based methods and more traditionally pragmatic approaches based on substituting reasonable estimates for missing values. Chapter 8 also presents more esoteric uses of the EM algorithm to handle problems such as variance components models, factor analysis, and time series, which can be viewed as missing-data problems for multivariate normal data with specialized mean and covariance structures.

EXAMPLE 1.9. *Estimation When Some Variables Are Categorical.* The reduction of the data to a vector of means and a covariance matrix is generally not appropriate when the variables are categorical. In that case the data can be arranged as a contingency table with partially classified margins as in Example 1.1. Methods for analyzing such data are discussed in Chapter 9.

Chapter 10 considers multivariate data where some of the variables are continuous and some are categorical. Again, a problem not usually thought of as related to missing data, the estimation of finite mixtures, is considered from a missing-data perspective.

EXAMPLE 1.10. *Estimation When The Data May Not Be Missing At Random.* Essentially all the literature on multivariate incomplete data assumes that the data are MAR, and much of it also assumes that the data are OAR. Chapter 11 deals explicitly with the case when the data are not MAR. The final chapter, Chapter 12, presents the likelihood approach to survey nonresponse including situations when the data are and are not MAR.

REFERENCES

Afifi, A. A., and Elashoff, R. M. (1966). Missing observations in multivariate statistics I: Review of the literature, *J. Am. Statist. Assoc.* **61**, 595–604.

Dempster, A. P., Laird, N. M., and Rubin, D. B. (1977). Maximum likelihood from incomplete data via the EM algorithm (with discussion), *J. Roy. Statist. Soc.* **B39**, 1–38.

Dixon, W. J. (Ed.) (1983). *BMDP Statistical Software*, 1983 revised printing, University of California Press: Berkeley.

Hartley, H. O., and Hocking, R. R. (1971). The analysis of incomplete data, *Biometrics* **27**, 783–808.

Little, R. J. A. (1982). Models for nonresponse in sample surveys, *J. Am. Statist. Assoc.* **77**, 237–250.

Marini, M. M., Olsen, A. R., and Rubin, D. B. (1980). Maximum likelihood estimation in panel studies with missing data. *Sociological Methodology* 1980, San Francisco: Jossey Bass.

Nie, N. H., Hull, C. H., Jenkins, J. G., Steinbrenner, K., and Bent, D. H. (1975). *SPSS, 2nd ed.* McGraw-Hill, New York.

Orchard, T., and Woodbury, M. A. (1972). A missing information principle: Theory and applications. *Proc. 6th Berkeley Symposium on Math. Statist. and Prob.* **1**, 697–715.

Rubin, D. B. (1976). Inference and missing data, *Biometrika* **63**, 581–592.

Wachter, K. W., and Trussell, J. (1982). Estimating historical heights, *J. Am. Statist. Assoc.* **77**, 279–301.

Woolson, R. F., and Clarke, W. R. (1984). Analysis of categorical incomplete longitudinal data, *J. Roy. Statist. Soc.* **A147**, 87–99.

PROBLEMS

1. Find the monotone pattern for the data of Example 1.1 that involves minimal deletion of observed values. Can you think of better criterea for deleting values than this one?
2. List methods for handling missing values in an area of statistical application of interest to you, based on experience or relevant literature.
3. What assumptions about the missing-data mechanism are implied by the statistical analyses used to analyze the data in Problem 2? Do these assumptions appear realistic?
4. What impact does the occurrence of missing values have on (a) estimates and (b) tests and confidence intervals for the analyses in Problem 2? For example, are estimates consistent for underlying population quantities, and do tests have the stated significance levels?
5. Let $Y = (y_{ij})$ be a data matrix and $R = (R_{ij})$ be the corresponding missing-data indicator matrix, where usually $R_{ij} = 1$ indicates observed and $R_{ij} = 0$ indicates missing.
   (a) Propose situations where two values of R_{ij} are not sufficient.
   (b) Nearly always it is assumed that R is fully observed. Describe a realistic case when it may make sense to regard part of R itself as missing.
   (c) Suppose $R_{ij} = 0$ or 1. When attention is focused only on the units that fully respond, the conditional distribution of y_i given $R_i = (1, \ldots, 1)$ is being estimated, where y_i and R_i are the ith rows of Y and

R, respectively. Propose situations where it makes sense to define the conditional distribution of y_i given other missing-data patterns. Propose situations where it makes no sense to define these other distributions.

(d) Express the marginal distribution of y_i in terms of the conditional distributions of y_i given the various missing-data patterns and their probabilities.

CHAPTER 2

Missing Data in Experiments

2.1. INTRODUCTION

Controlled experiments are generally carefully designed to allow revealing statistical analyses to be made using straightforward computations. In particular, corresponding to a standard experimental design, there is a standard least squares analysis, which yields estimates of parameters, standard errors for contrasts of parameters, and the analysis of variance (ANOVA) table. The estimates, standard errors, and ANOVA table corresponding to most designed experiments are easily computed because of balance in the design. For example, with two factors being studied, the analysis is particularly simple when the same number of observations are taken at each combination of factor levels; Cochran and Cox (1957), Davies (1960), Kempthorne (1952), Winer (1962), and more recent textbooks on experimental design catalogue many examples of specialized analyses.

Since the levels of the factors in an experiment are fixed by the experimenter, missing values, if they occur, do so far more frequently in the outcome variable, Y, than in the factors, X. Consequently, we restrict attention to missing values in Y. When such data are missing, the balance present in the original design is destroyed. As a result, even assuming MCAR, the proper least squares analysis becomes more complicated. An intuitively attractive approach is to fill in the missing values to restore the balance and then proceed with the standard analysis.

The advantages of filling in the missing values in an experiment rather than trying to analyze the actual observed data include the following: (1) It is easier to specify the data structure using the terminology of experimental design (e.g., as a balanced incomplete block), (2) it is easier to compute necessary statistical summaries, and (3) it is easier to interpret the results of analyses since standard displays and summaries can be used. Ideally, we would hope that simple rules could be devised for filling in the missing values in such a way that the resultant complete-data analyses would be correct. In fact, much progress toward this goal can be made.

Assuming that the missingness is unrelated to the values of the outcome variable (i.e., under MAR), there exists a variety of methods for filling in values that yield correct estimates of all estimable effect parameters. Furthermore, it is easy to correct the residual (error) mean square, standard errors, and sums of squares that have one degree of freedom. Unfortunately, it is more complicated to provide correct sums of squares with more than one degree of freedom but it can be done.

Methods that fill in one value per missing value strictly apply only to analyses based on one fixed-effect linear model with one error term. Examples involving the fitting of more than one fixed-effect linear model include hierarchical models, which attribute sums of squares to effects in a particular order by fitting a sequence of nested fixed-effect models; split-plot and repeated measures designs, which use different error terms for testing different effects; and random and mixed effect models, which treat some parameters as random variables. For analyses employing more than one fixed-effect model, in general a different set of missing values have to be filled in for each fixed-effect model. Further discussion is given, for example, in Anderson (1946) and Jarrett (1978). See also Section 8.5 of this book.

2.2. THE EXACT LEAST SQUARES SOLUTION WITH COMPLETE DATA

Let X be an $n \times p$ matrix whose ith row, $x_i = (x_{i1}, \ldots, x_{ip})$, provides the values of the fixed factors for the ith unit. For example, in a 2×2 design with two observations per cell and the levels of the factors labeled 0, 1:

$$X = \begin{bmatrix} 100 \\ 100 \\ 101 \\ 101 \\ 110 \\ 110 \\ 111 \\ 111 \end{bmatrix}$$

The outcome variable $Y = (y_1, \ldots, y_n)^{\mathrm{T}}$ is assumed to satisfy the linear model

$$Y = X\beta + e \tag{2.1}$$

where $e = (e_1, \ldots, e_n)^T$, and the e_i are independent and identically distributed with zero mean and common variance σ^2; β is the parameter to be estimated, a $p \times 1$ vector. The least squares estimate of β is

$$\hat{\beta} = (X^T X)^{-1}(X^T Y) \tag{2.2}$$

if $X^T X$ is full rank and is undefined otherwise. If $X^T X$ is full rank, $\hat{\beta}$ is the minimum variance unbiased estimate of β. If the e_i are normally distributed, $\hat{\beta}$ is also the maximum likelihood estimate of β and is normally distributed with mean β and variance $\sigma^2(X^T X)^{-1}$.

The best unbiased estimate of σ^2 is

$$\hat{\sigma}^2 = \sum_{i=1}^{n} \frac{(y_i - \hat{y}_i)^2}{n - p} \tag{2.3}$$

where $\hat{y}_i = x_i\hat{\beta}$; if the e_i are normal, $(n - p)\hat{\sigma}^2/\sigma^2$ is distributed as χ^2 on $n - p$ degrees of freedom. The best unbiased estimate of the variance–covariance matrix of $(\hat{\beta} - \beta)$ is given by

$$V = \hat{\sigma}^2(X^T X)^{-1}. \tag{2.4}$$

If the e_i are normal, $(\hat{\beta}_i - \beta_i)/\sqrt{v_{ii}}$ (where v_{ii} is the ith diagonal element of V) has a t distribution on $n - p$ degrees of freedom; $(\hat{\beta} - \beta)$ has a scaled multivariate t distribution with scale $V^{1/2}$.

Tests of the hypothesis that some set of linear combinations of β are all zero are carried out by calculating the sum of squares attributable to the set of linear combinations. More precisely, suppose C is a $p \times w$ matrix specifying the w linear combinations of β that are to be tested. Then the sum of squares attributable to the w linear combinations is

$$S = (C^T\hat{\beta})^T[C^T(X^T X)^{-1}C]^{-1}(C^T\hat{\beta}). \tag{2.5}$$

The test that $C^T\beta = 0$ is made by comparing S/w to $\hat{\sigma}^2$:

$$F = \frac{S/w}{\hat{\sigma}^2}. \tag{2.6}$$

If the e_i are normal, then F in (2.6) is the likelihood ratio test for $C^T\beta = 0$; if in addition $C^T\beta = 0$, then F is distributed as Snedecor's F distribution with w and $n - p$ degrees of freedom, since S/σ^2 and $(n - p)\hat{\sigma}^2$ are independent χ^2 random variables with w and $n - p$ degrees of freedom, respectively. Proofs of the preceding results can be found in standard books on regression analysis, such as Draper and Smith (1981) and Weisberg (1980). Also, as these references point out, careful interpretation of these tests may be required in nonorthogonal designs; for example, this test of a collection of A effects in a model with

A effects, B effects, and interactive effects addresses A adjusted for B and interactions.

Standard experimental designs are chosen to make estimation and testing easy and precise. In particular, the matrix X^TX is usually easy to invert, with the result that $\hat{\beta}$, $\hat{\sigma}^2$, V, and the sum of squares attributable to specific collections of linear combinations of β, such as treatment effects and block effects, are easy to calculate. In fact, these summaries of the data are usually calculated by simple averaging of the observations and their squares. This simplicity can be important in experiments with several factors and many parameters to be estimated, because then X^TX can be of large dimension. The inversion of large matrices was especially cumbersome before the days of modern computing equipment, but still can be troublesome in some computing environments when p is very large (e.g., >50).

2.3. THE CORRECT LEAST SQUARES ANALYSIS WITH MISSING DATA

We assume that X represents an experimental design such that if all of Y were observed, the analysis of the data would be conducted using existing standard formulas and computer programs. The question of interest is how to use these complete-data formulas and computer programs when part of Y is missing.

Assuming that the reason for the occurrence of missing data in Y does not depend on any Y values (i.e., under MAR), the correct analysis of the observed data is easy to describe. Simply ignore the rows of X corresponding to missing y_i and carry out the calculations indicated in Section 2.2 using the remaining rows of X, which correspond to observed y_i. The problem with this suggestion is that the specialized formulas and computing routines used with complete Y cannot be used, since the original balance is no longer present. As a result, (1) the data structure is more cumbersome to specify, since the entire design matrix must be defined rather than simply providing a description using terminology for standard experimental designs; (2) the computational demands with respect to time and storage allocation can be much more substantial, since general formulas such as those given in (2.2) are being used rather than specialized formulas appropriate for particular designs; and (3) ease of statistical interpretation may be hindered, since the displays and summaries associated with specialized designs are not produced.

Nevertheless, the equations given in Section 2.2 applied to the m units with y_i observed define the correct least squares estimates, standard errors, sum of squares and F tests when faced with missing data. We will use $\hat{\beta}_*$, $\hat{\sigma}_*^2$, V_*, and S_* to refer to the quantities in Eqs. (2.2)–(2.5) as calculated from the m units with y_i observed. The remainder of this chapter describes how to obtain these summaries essentially using only the procedures needed for complete data, which use the special structure in X to simplify computions.

2.4. FILLING IN LEAST SQUARES ESTIMATES

2.4.1. Yates's Method

The classical and standard approach to missing data in ANOVA is due in general to Yates (1933): (1) fill in the least squares estimates of all missing values $\hat{y}_i = x_i\hat{\beta}_*$ where $\hat{\beta}_*$ is defined by (2.2) applied to the m rows of (Y, X) that have y_i observed, and (2) use complete-data methods of analysis. This approach of filling in least squares estimates may at first seem to be circular and of little practical help since it appears to require knowledge of $\hat{\beta}_*$ to estimate the missing y_i as $x_i\hat{\beta}_*$ before $\hat{\beta}_*$ can be calculated! It turns out, perhaps surprisingly, that it can be relatively easy to calculate $\hat{y}_i = x_i\hat{\beta}_*$ for the missing y_i before calculating $\hat{\beta}_*$ directly, at least if only a few values are missing.

The rationale for Yates's procedure is that it yields (1) the correct least squares estimates of β, $\hat{\beta}_*$, and (2) the correct residual sum of squares; that is, the resultant estimate of $\sigma^2(n - p)$ will be correct and equal $\hat{\sigma}_*^2(m - p)$. It is quite easy to prove these two facts. Let $\hat{y}_i = x_i\hat{\beta}_*$, $i = 1, \ldots, m_0$ denote the least squares estimate of the m_0 missing values, which for notational simplicity are the first m_0 observations. Complete-data methods applied to the filled-in data minimize the quantity

$$\mathrm{SS}(\beta) = \sum_{i=1}^{m_0} (\hat{y}_i - x_i\beta)^2 + \sum_{i=m_0+1}^{n} (y_i - x_i\beta)^2$$

with respect to β. By definition, $\beta = \hat{\beta}_*$ minimizes the second summation in $\mathrm{SS}(\beta)$; but also by definition, $\beta = \hat{\beta}_*$ minimizes the first summation in $\mathrm{SS}(\beta)$, setting it equal to zero. Consequently, with least squares estimates of missing values filled in, (1) $\mathrm{SS}(\beta)$ is minimized at $\beta = \hat{\beta}_*$ and (2) $\mathrm{SS}(\hat{\beta}_*)$ equals the minimal residual sum of squares over the m observed values of y_i. Hence (1) the correct least squares estimate of β, $\hat{\beta}_*$, equals the least squares estimate of β found by the complete-data ANOVA program, and (2) the correct least squares estimate of σ^2, $\hat{\sigma}_*^2$ is found from the complete-data ANOVA estimate of σ^2, $\hat{\sigma}^2$ by

$$\hat{\sigma}_*^2 = \hat{\sigma}^2 \frac{(n - p)}{(m - p)}.$$

The analysis with missing y_i set equal to $\hat{y}_i$ is not perfect: It yields an estimated variance–covariance matrix of $\hat{\beta}$ that is too small and sums of squares attributable to collections of linear combinations of β that are too big, although for small fractions of missing data these biases are often relatively minor. We now consider methods for calculating the values $\hat{y}_i$.

2.4.2. Using a Formula for the Missing Values

One approach is to use a formula for the missing values, fill them in, and then proceed. In the first application of this idea, Allan and Wishart (1930) provided formulas for the least squares estimate of one missing value in a randomized block design and of one missing value in a Latin square design. For example, in a randomized block with T treatments and B blocks, the least squares estimate of a missing value in treatment t and block b is

$$\frac{Ty_+^{(t)} + By_+^{(b)} - y_+}{(T-1)(B-1)}$$

where $y_+^{(t)}$ and $y_+^{(b)}$ are the sum of the observed values of Y for treatment t and block b, respectively, and y_+ is the sum of all observed values of Y. Wilkinson (1958a) extended this work by giving tables providing formulas for many designs and many patterns of missing values.

2.4.3. Iterating to Find the Missing Values

Hartley (1956) proposed a general noniterative method for estimating one missing value that he suggested should be used iteratively for more than one. The method for one missing value involves substituting three different trial values for the missing value, with the residual sum of squares calculated for each trial value. Since the residual sum of squares is quadratic in the missing value, the minimizing value of the one missing value can then be found. This method is not as attractive as alternative methods.

Healy and Westmacott (1956) described a popular iterative technique that is sometimes attributed to Yates and even sometimes to Fisher. With this method, (1) trial values are substituted for all missing values, (2) the complete-data analysis is performed, (3) predicted values are obtained for the missing values, (4) these predicted values are substituted for the missing values, (5) a new complete-data analysis is performed, and so on, until the missing values do not change appreciably, or equivalently, until the residual sum of squares essentially stops decreasing.

We shall show in Example 8.5 that the Healy and Westmacott method is an example of an EM algorithm, introduced here in Chapter 7, and each iteration decreases the residual sum of squares (or equivalently, increases the likelihood under the corresponding normal linear model). In some cases, convergence can be slow and special acceleration techniques have been suggested (Pearce, 1965, p. 111; Preece, 1971). Although these can improve the rate of convergence in some examples, they can also destroy the monotone decrease of the residual sum of squares in other examples (see Jarrett, 1978, for a summary of conditions).

2.4.4. ANCOVA with Missing-Value Covariates

A general noniterative method due to Bartlett (1937) is to fill in guesses for the missing values, and then perform an analysis of covariance (ANCOVA) with a missing-value covariate for each missing value. The ith missing-value covariate is defined to be the indicator for the ith missing value, that is, zero everywhere except for the ith missing value where it equals one. The coefficient of the ith missing-value covariate, when subtracted from the initial guess of the ith missing value, yields the least squares estimate of the ith missing value. Furthermore, the residual mean square and all contrast sums of squares adjusted for the missing-value covariates are their correct values. We prove these results in Section 2.5.

Although this method is quite attractive in some ways, it often cannot be directly implemented because specialized ANOVA routines may not have the capability to handle multiple covariates. It turns out, however, that Bartlett's method can be applied using only the existing complete-data ANOVA routine and a routine to invert an $m_0 \times m_0$ symmetric matrix. The next section proves that Bartlett's method leads to the correct least squares analysis. The subsequent section concerns how to obtain this analysis using only the complete-data ANOVA routine.

2.5. BARTLETT'S ANCOVA METHOD

2.5.1. Useful Properties of Bartlett's Method

Bartlett's ANCOVA method has the following useful properties. First, it is noniterative and thus avoids questions of convergence. Second, if there is a singular pattern of missing values (i.e., a pattern such that some parameters are inestimable as when all values under one treatment are missing), the method will warn the user, whereas iterative methods will produce an answer, possibly a quite inappropriate one. A third advantage is, as mentioned earlier, that the method produces not only the correct estimates and correct residual sum of squares, but correct standard errors, sums of squares, and F tests as well.

2.5.2. Notation

Suppose each missing y_i is filled in with some initial guess in order to create a complete vector of values for Y. Call the initial guesses $\tilde{y}_i$, $i = 1, \ldots, m_0$. Also let Z be the $n \times m_0$ matrix of m_0 missing-value covariates: the first row of Z, z_1, equals $(1, 0, \ldots, 0)$, ..., the m_0th row of Z equals $(0, \ldots, 0, 1)$, and all z_i with $i > m_0$ equal $(0, \ldots, 0)$, since they correspond to observed y_i. The analysis of covariance uses both X and Z to predict Y.

Analogous to (2.1), the model for Y is now

$$Y = X\beta + Z\gamma + e, \tag{2.7}$$

where γ is a column vector of m_0 regression coefficients for the missing-value covariates. The residual sum of squares to be minimized over (β, γ) is

$$\mathrm{SS}(\beta, \gamma) = \sum_{i=1}^{m_0} (\tilde{y}_i - x_i\beta - z_i\gamma)^2 + \sum_{i=m_0+1}^{n} (y_i - x_i\beta - z_i\gamma)^2.$$

By the definition of Z,

$$\mathrm{SS}(\beta, \gamma) = \sum_{i=1}^{m_0} (\tilde{y}_i - x_i\beta - \gamma_i)^2 + \sum_{i=m_0+1}^{n} (y_i - x_i\beta)^2. \tag{2.8}$$

2.5.3. The ANCOVA Estimates of Parameters and Missing Y Values

As before, let $\hat{\beta}_*$ equal the correct least squares estimate of β obtained by applying (2.2) to the observed values, that is, to the last $m = n - m_0$ rows of (Y, X); this minimizes the second summation in (2.8). But with $\beta = \hat{\beta}_*$, setting γ equal to $(\hat{\gamma}_1, \ldots, \hat{\gamma}_{m_0})^{\mathrm{T}}$ where

$$\hat{\gamma}_i = \tilde{y}_i - x_i\hat{\beta}_*, \qquad i = 1, \ldots, m_0 \tag{2.9}$$

minimizes the first summation in (2.8) by making it identically zero, so that

$$\mathrm{SS}(\hat{\beta}_*, \hat{\gamma}) = \sum_{i=m_0+1}^{n} (y_i - x_i\hat{\beta}_*)^2. \tag{2.10}$$

Thus $(\hat{\beta}_*, \hat{\gamma})$ minimizes $\mathrm{SS}(\beta, \gamma)$ and gives the least squares estimate of (β, γ) obtained from the ANCOVA model in (2.7). Equation (2.9) also implies that the correct least squares estimate of the missing y_i, that is, $\hat{y}_i = x_i\hat{\beta}_*$, is given by $\tilde{y}_i - \hat{\gamma}_i$, or in words:

(Correct least squares predicted value for ith missing value)

equals (initial guess for ith missing value)

minus (coefficient of ith missing value covariate). (2.11)

Bartlett's original description of this method set all $\tilde{y}_i$ equal to zero, but setting all $\tilde{y}_i$ equal to the grand mean of all observations is computationally more attractive and yields the correct total sum of squares about the grand mean.

2.5.4. ANCOVA Estimates of the Residual Sums of Squares and the Variance–Covariance Matrix of $\hat{\beta}$

Equation (2.10) establishes that the residual sum of squares from the ANCOVA is the correct residual sum of squares; the ANCOVA degrees of freedom corresponding to this residual sum of squares is $n - p - m_0 = m - p$, which is also correct. Consequently, the residual mean square is correct and equal to $\hat{\sigma}_*^2$. If the variance–covariance matrix of $\hat{\beta}_*$ from the ANCOVA equals V_* obtained by applying (2.4) to the m units with y_i observed, then all standard errors, sums of squares, and tests of significance will also be correct. The estimated variance–covariance matrix of $\hat{\beta}_*$ from the ANCOVA is the estimated residual mean square, $\hat{\sigma}_*^2$, times the upper left $p \times p$ submatrix of $[(X, Z)^{\mathrm{T}}(X, Z)]^{-1}$, say, U. Since the estimated residual mean square is correct, we need only show that U^{-1} is the sum of cross-products of X for the units with y_i observed. From standard results on matrices,

$$U = [X^{\mathrm{T}}X - (X^{\mathrm{T}}Z)(Z^{\mathrm{T}}Z)^{-1}(Z^{\mathrm{T}}X)]^{-1}. \tag{2.12}$$

By the definition of z_i,

$$X^{\mathrm{T}}Z = \sum_{i=1}^{m_0} x_i^{\mathrm{T}} z_i \tag{2.13}$$

and

$$Z^{\mathrm{T}}Z = \sum_{i=1}^{m_0} z_i^{\mathrm{T}} z_i = I. \tag{2.14}$$

From (2.13) and (2.14)

$$(X^{\mathrm{T}}Z)(Z^{\mathrm{T}}Z)^{-1}(Z^{\mathrm{T}}X) = \left(\sum_{i=1}^{m_0} x_i^{\mathrm{T}} z_i\right)\left(\sum_{j=1}^{m_0} z_j^{\mathrm{T}} x_j\right). \tag{2.15}$$

But

$$z_i z_j^{\mathrm{T}} = \begin{cases} 1 & \text{if } i = j, \\ 0 & \text{otherwise,} \end{cases}$$

whence (2.15) equals

$$\sum_{i=1}^{m_0} x_i^{\mathrm{T}} x_i$$

and from (2.12)

$$U = \left[\sum_{i=m_0+1}^{n} x_i^{\mathrm{T}} x_i \right]^{-1}$$

so that $\hat{\sigma}_*^2 U = V_*$, the variance–covariance matrix of $\hat{\beta}_*$ found by ignoring the missing observations, as required to complete the proof that the ANCOVA produces least squares values for all summaries.

2.6. LEAST SQUARES ESTIMATES OF MISSING VALUES BY ANCOVA USING ONLY COMPLETE-DATA METHODS

The preceding theory relating incomplete ANOVA to a complete ANCOVA analysis would be merely of academic interest if the ANCOVA needed special software to implement. We now describe how to implement the missing-value covariate method to calculate least squares estimates of m_0 missing values using only complete-data ANOVA routines and a routine to invert an $m_0 \times m_0$ symmetric matrix (the sweep operator, described in Section 6.5, can be used for this purpose). In Section 2.7 the analysis is extended to produce correct standard errors and sums of squares for one degree of freedom hypotheses. The argument here will appeal to the ANCOVA results; a direct algebraic proof appears in Rubin (1972).

By ANCOVA theory, the vector $\hat{\gamma}$ can be written as

$$\hat{\gamma} = B^{-1} \rho, \tag{2.16}$$

where B is the $m_0 \times m_0$ cross-products matrix for the residuals of the m_0 missing-value covariates after adjusting for the design matrix X, and ρ is the $m_0 \times 1$ vector of cross products of Y and the residuals of the missing-value covariates after adjusting for the design matrix. If B is singular, the pattern of missing data is such that an attempt is being made to estimate inestimable parameters, such as the effect of a treatment when all observations on that treatment are missing. The method requires (1) the calculation of B and ρ using the complete-data ANOVA routine, (2) the inversion of B to obtain $\hat{\gamma}$ from (2.16), and (3) the calculation of the missing values from (2.11).

To find B and ρ, begin by performing a complete-data ANOVA on the first missing-value covariate, that is, using the first column of Z, which is all zeros except for a one where the first missing value occurs, as the dependent variable rather than Y. The residuals from this analysis for the m_0 missing values comprise the first row of B. Repeat the complete-data ANOVA on the ith missing-value covariate, $i = 1, \ldots, m_0$, which is all zeros except for a one where the ith missing value occurs, and let the ith row of B equal the residuals for the m_0 missing values from this analysis. The vector ρ is calculated by performing the complete-data ANOVA on the real Y data with initial guesses $\tilde{y}_i$ filled in for y_i, $i = 1, \ldots, m_0$; the residuals for the m_0 missing values comprise the vector ρ.

These procedures work for the following reasons. The ij entry of B is

$$b_{ij} = \sum_{k=1}^{n} (z_{ik} - \hat{z}_{ik})(z_{jk} - \hat{z}_{jk}),$$

where z_{ik} and $\hat{z}_{ik}$ (z_{jk} and $\hat{z}_{jk}$) are observed and fitted values for observation k from the ANOVA of the ith (jth) missing value covariate on X. Now $\sum_{k=1}^{n} (z_{ik} - \hat{z}_{ik})x_{lk} = 0$ for all X variables in the design matrix, by elementary properties of least squares estimates. Hence $\sum_{k=1}^{n} (z_{ik} - \hat{z}_{ik})\hat{z}_{jk} = 0$, since $\hat{z}_{jk}$ is a fixed linear combination of X variables $\{x_{lk}: l = 1, \ldots, p\}$ for observation k. Consequently,

$$b_{ij} = \sum_{k=1}^{n} (z_{ik} - \hat{z}_{ik})z_{jk} = z_{ij} - \hat{z}_{ij},$$

the residual for the ith missing data covariate for the jth missing value, since $z_{jk} = 1$ when $j = k$ and 0 otherwise. Similarly, the jth component of ρ is the sum over all n observations of the residual for Y (with initial values filled in) times the residual for the jth missing-value covariate. By an argument completely analogous to that just given, this is simply the residual for the jth missing value.

EXAMPLE 2.1. *Estimating Missing Values In A Randomized Block.* The following example of a randomized block design is taken from Cochran and Cox (1957, p. 111) and Rubin (1972, 1976b). Suppose that the two observations, u_1 and u_2, are missing as presented in Table 2.1. We formulate model (2.1) using a seven-dimensional parameter β consisting of five parameters for the means of the five treatments and two parameters for the block effects; the

Table 2.1 Strength Index of Cotton in a Randomized Block Experiment

Treatments (pounds potassium oxide per acre)	Blocks			
	1	2	3	Totals
36	u_1	8.00	7.93	15.93
54	8.14	8.15	7.87	24.16
72	7.76	u_2	7.74	15.50
108	7.17	7.57	7.80	22.54
144	7.46	7.68	7.21	22.35
	30.53	31.40	38.55	100.48

residual mean square is formed from the treatment by block interaction, with $(5-1)\times(3-1)=8$ degrees of freedom when no data are missing.

Inserting the grand mean $\bar{y}=7.7292$ for both missing values, we find the residual in the u_1 cell to be -0.0798 and in the u_2 cell to be -0.1105. Thus, $\rho=-(0.0798, 0.1105)^{\mathrm{T}}$. Also, we obtain the correct total sum of squares, $\mathrm{TSS}_*=1.1679$.

Inserting one for u_1 and zeros everywhere else, we find that the residual in the u_1 cell is 0.5333 and that in the u_2 cell is 0.0667. Similarly, inserting one for u_2 and zeros everywhere else we find the residual in the u_1 cell to be 0.0667 and that in the u_2 cell to be 0.5333. Hence

$$B=\begin{bmatrix}0.5333 & 0.0667\\ 0.0667 & 0.5333\end{bmatrix} \quad \text{and} \quad B^{-1}=\begin{bmatrix}1.9048 & -0.2381\\ -0.2381 & 1.9048\end{bmatrix}.$$

The least squares estimates of the missing values are

$$(\bar{y}, \bar{y}) - B^{-1}\rho = (7.8549, 7.9206)^{\mathrm{T}}.$$

Thus the least squares estimate of the u_1 cell is 7.8549 and that of the u_2 cell is 7.9206. The least squares estimates for the missing cells given by Cochran and Cox were found by an iterative method and agree with the values found here.

Estimated parameters based on the analysis of the filled-in data will be the correct least squares values. For example, the correct estimates of the treatment means are simply the treatment averages of observed and filled-in data (7.9283, 8.0533, 7.8069, 7.5133, 7.4500). Furthermore the correct residual sum of squares, and thus the correct residual mean square $\hat{\sigma}_*^2$, is obtained when the number of missing values m_0 is subtracted from the residual degrees of freedom, $n-p$. However, sums of squares and standard errors generally will be incorrect.

2.7. CORRECT LEAST SQUARES ESTIMATES OF STANDARD ERRORS AND ONE DEGREE OF FREEDOM SUMS OF SQUARES

A simple extension of the technique of Section 2.6 yields the correct estimates of standard errors and one degree of freedom sums of squares.

Let $\lambda=C^{\mathrm{T}}\beta$, where C is a vector of p constants, be a linear combination of β with estimate $\hat{\lambda}=C^{\mathrm{T}}\hat{\beta}$ from the ANOVA of least squares filled-in data. Since least squares estimates of the missing values have been filled in, $\hat{\beta}=\hat{\beta}_*$ and so $\hat{\lambda}=\hat{\lambda}_*$, the correct least squares estimate of λ. The standard error of $\hat{\lambda}$ obtained from the complete-data ANOVA is

$$\mathrm{SE} = \hat{\sigma}\sqrt{C^{\mathrm{T}}(X^{\mathrm{T}}X)^{-1}C} \tag{2.17}$$

and the sum of squares attributable to λ from this analysis is

$$\mathrm{SS} = \hat{\lambda}^2/C^{\mathrm{T}}(X^{\mathrm{T}}X)^{-1}C. \tag{2.18}$$

The correct standard error of $\hat{\lambda} = \hat{\lambda}_*$ is, from Section 2.5.4,

$$\mathrm{SE}_* = \hat{\sigma}_*\sqrt{C^{\mathrm{T}}UC} \tag{2.19}$$

and the correct sum of squares attributable to λ is

$$\mathrm{SS}_* = \hat{\lambda}_*^2/C^{\mathrm{T}}UC. \tag{2.20}$$

Let H be the $(m_0 \times 1)$ vector of complete-data ANOVA estimates of λ taking each of the m_0 missing-data covariates as the dependent variable rather than Y; that is, in matrix terms

$$H^{\mathrm{T}} = C^{\mathrm{T}}(X^{\mathrm{T}}X)^{-1}X^{\mathrm{T}}Z. \tag{2.21}$$

Clearly, H can be calculated at the same time B is being calculated: The ith component in H and the ith row in B are obtained from the complete-data ANOVA of the ith missing-data covariate. Standard ANCOVA theory or matrix algebra using results in Section 2.5.4 shows that

$$C^{\mathrm{T}}UC = C^{\mathrm{T}}(X^{\mathrm{T}}X)^{-1}C + H^{\mathrm{T}}B^{-1}H. \tag{2.22}$$

Equations (2.17), (2.19), (2.21), (2.22), and the fact that

$$\hat{\sigma}_*^2 = \hat{\sigma}^2(n-p)/(m-p)$$

imply that SE_* can be simply expressed in terms of output from the complete-data ANOVA:

$$\mathrm{SE}_* = \sqrt{\frac{n-p}{m-p}(\mathrm{SE}^2 + \hat{\sigma}^2 H^{\mathrm{T}}B^{-1}H)}. \tag{2.23}$$

Similarly, (2.18), (2.20) with $\lambda = \lambda_*$, (2.21), and (2.22) imply that SS_* can be simply expressed in terms of output from the complete-data ANOVA:

$$\mathrm{SS}_* = \mathrm{SS}/[1 + (\mathrm{SS}/\hat{\lambda}^2)H^{\mathrm{T}}B^{-1}H]. \tag{2.24}$$

EXAMPLE 2.2. *Adjusting Standard Errors For Filled-In Missing Values (Example 2.1 continued).* To apply the method just described, $m_0 + 2$ complete-data ANOVA's are required: an initial one on initial filled-in Y data, one for each of the m_0 missing-data covariates, and a final ANOVA on the least squares filled-in Y data. Following Rubin (1976a), we consider the data in Table 2.1 and the linear combination of parameters that corresponds to contrasting treatment 1 and treatment 2; in terms of the parameterization of

Example 2.1, $C^{\mathrm{T}} = (1, -1, 0, 0, 0, 0, 0)$ and $X^{\mathrm{T}}X$ is a 7×7 block diagonal matrix whose upper left 5×5 submatrix is diagonal with all elements equal to 3. Thus $\hat{\lambda}$ is simply the mean of the three observations of treatment 1 minus the mean of the three observations of treatment 2 with associated complete-data standard error $\hat{\sigma}\sqrt{(2/3)}$ and sum of squares $3\hat{\lambda}^2/2$.

As in Example 2.1, for the initial ANOVA, estimate both missing values by the grand mean to obtain residuals ρ: $-(0.0798, 0.1105)$ and the correct total sum of squares, $\mathrm{TSS}_* = 1.1679$.

For $i = 1, 2, \ldots, m_0$, fill in 1 for the ith missing value and set all other values to zero and analyze the resultant missing-data covariate by the complete-data ANOVA program: r_i is the vector of residuals corresponding to the m_0 missing values and h_i is the estimate of the linear combination of parameters being tested. The resultant $B = (r_1^{\mathrm{T}}, r_2^{\mathrm{T}})$ for our example is given in Example 2.1, the resultant $H^{\mathrm{T}} = (h_1, h_2)$ is $(0.3333, 0.0000)$, and consequently, the resultant $H^{\mathrm{T}}B^{-1}H$ is 0.2116.

Now fill in the least squares estimates of the missing values, as found in Example 2.1, $(7.8549, 7.9206)$, and compute the ANOVA on the filled-in data. The resultant estimate of λ is $\hat{\lambda} = -0.1250$, with $\hat{\sigma}^2 = 0.0368$, $\mathrm{SE} = 0.1567$, and $\mathrm{SS} = 0.0235$. From (2.23), the correct standard error of $\hat{\lambda}$ is

$$\mathrm{SE}_* = \sqrt{(8/6)(0.0246 + 0.0368 \times 0.2116)} = 0.2077$$

and from (2.24) the correct sum of squares attributable to λ is

$$\mathrm{SS}_* = 0.0235/[1 + 1.5 \times 0.2116] = 0.0178.$$

2.8. CORRECT LEAST SQUARES SUMS OF SQUARES WITH MORE THAN ONE DEGREE OF FREEDOM

A generalization of the technique of Section 2.7 yields the correct sum of squares with more than one degree of freedom. The technique presented here is due to Rubin (1976b); related earlier work includes Tocher (1952) and Wilkinson (1958b), and later work includes Jarrett (1978).

Let $\lambda = C^{\mathrm{T}}\beta$, where C is a $p \times w$ matrix of constants, be w linear combinations of β for which the sum of squares is desired, and let $\hat{\lambda}_* = C^{\mathrm{T}}\hat{\beta}_*$ be the correct least squares estimate of λ. When least squares estimates of the missing values have been filled in, $\hat{\beta} = \hat{\beta}_*$ and thus $\hat{\lambda} = \hat{\lambda}_*$. We suppose for simplicity that the w linear combinations have been chosen to be orthonormal with complete data in the sense that

$$C^{\mathrm{T}}(X^{\mathrm{T}}X)^{-1}C = I. \tag{2.25}$$

That is, with complete data, the variance–covariance matrix of $\hat{\lambda}$ is $\sigma^2 I$. Thus the sum of squares attributable to λ from the complete-data ANOVA is

$$\text{SS} = \hat{\lambda}^{\text{T}}\hat{\lambda}. \tag{2.26}$$

The correct sum of squares to attribute to λ is

$$\text{SS}_* = \hat{\lambda}_*^{\text{T}}(C^{\text{T}}UC)^{-1}\hat{\lambda}_*. \tag{2.27}$$

Letting H be the $m_0 \times w$ matrix of complete-data ANOVA estimates of λ for the m_0 missing-data covariates, standard ANCOVA theory or matrix algebra using results in Section 2.5.4 shows that (2.22) holds in general; hence, since the components of $\hat{\lambda}$ are orthonormal and $\hat{\lambda} = \hat{\lambda}_*$ with least squares estimates for missing values,

$$\text{SS}_* = \hat{\lambda}^{\text{T}}(I + H^{\text{T}}B^{-1}H)^{-1}\hat{\lambda} \tag{2.28}$$

or, using Woodbury's identity (Rao, 1965, p. 29) and (2.26),

$$\text{SS}_* = \text{SS} - (H\hat{\lambda})^{\text{T}}(HH^{\text{T}} + B)^{-1}(H\hat{\lambda}). \tag{2.29}$$

Equation (2.28) involves the inversion of a $w \times w$ symmetric matrix, whereas (2.29) involves the inversion of an $m_0 \times m_0$ matrix. Consequently, (2.28) is preferable when $w < m_0$.

EXAMPLE 2.3. *Adjusting Sums of Squares for the Filled-In Values (Example 2.2 continued).* The treatment sum of squares has four degrees of freedom, which we span with the following orthonormal contrasts of five treatment means:

$$\sqrt{\tfrac{3}{20}}(4, -1, -1, -1, -1, 0, 0),$$
$$\sqrt{\tfrac{1}{4}}(0, 3, -1, -1, -1, 0, 0),$$
$$\sqrt{\tfrac{1}{2}}(0, 0, 2, -1, -1, 0, 0),$$
$$\sqrt{\tfrac{3}{2}}(0, 0, 0, 1, -1, 0, 0).$$

Note that with complete data the linear combinations have variance–covariance matrix $\sigma^2 I$.

The values of the four contrasts obtained from the complete-data ANOVA of the first missing-data covariate gives the first row of H, and the complete-data ANOVA of the second missing-data covariate gives the second row of H:

$$H = \begin{bmatrix} 0.5164 & 0.0000 & 0.0000 & 0.0000 \\ -0.1291 & -0.1667 & 0.4714 & 0.0000 \end{bmatrix}.$$

Thus H is calculated at the same time B is calculated.

From the final complete-data ANOVA of the data with least squares

Table 2.2 The Corrected Analysis of Variance on the Filled-in Data

Source of Variation	d.f.	SS	MS	F
Blocks, unadjusted	2	0.0977		
Treatments, adjusted for blocks	4	0.7755	0.1939	3.9486
Error	6	0.2947	0.0491	
Total	12	1.1679		

Treatment means: (7.9283, 8.0533, 7.8069, 7.5133, 7.4500)

Contrast: Treatment 1 − treatment 2 = −0.1250
Standard error = 0.2077

estimates filled in, we have SS = 0.8191, $\hat{\lambda}^{\mathrm{T}} = (0.3446, 0.6949, 0.4600, 0.0775)$. From (2.29), $\mathrm{SS}_* = 0.7755$.

A summary of the resulting ANOVA for this example appears in Table 2.2, where the blocks sum of squares (unadjusted for treatments) has been found by subtracting the corrected treatment and error sums of squares (0.7755 and 0.2947) from the corrected total sum of squares (1.1679) found in Example 2.1.

REFERENCES

Allan, F. G., and Wishart, J. (1930). A method of estimating the yield of a missing plot in field experiments, *J. Agric. Sci.* **20**, 399–406.

Anderson, R. L. (1946). Missing plot techniques, *Biometrics* **2**, 41–47.

Bartlett, M. S. (1937). Some examples of statistical methods of research in agriculture and applied botany, *J. Roy. Statist. Soc.* **B4**, 137–170.

Cochran, W. G., and Cox, G. (1957). *Experimental Design.* London: Wiley.

Davies, O. L. (1960). *The Design and Analysis of Industrial Experiments.* New York: Hafner.

Draper, N. R., and Smith, H. (1981). *Applied Regression Analysis.* New York: Wiley.

Hartley, H. O. (1956). Programming analysis of variance for general purpose computers, *Biometrics* **12**, 110–122.

Healy, M. J. R., and Westmacott, M. (1956). Missing values in experiments analyzed on automatic computers, *Appl. Statist.* **5**, 203–206.

Jarrett, R. G. (1978). The analysis of designed experiments with missing observations, *Appl. Statist.* **27**, 38–46.

Kempthorne, O. (1952). *The Design and Analysis of Experiments.* New York: Wiley.

Pearce, S. C. (1965). *Biological Statistics: An Introduction.* New York: McGraw-Hill.

Preece, D. A. (1971). Iterative procedures for missing values in experiments, *Technometrics* **13**, 743–753.

Rao, C. R. (1965). *Linear Statistical Inference.* New York: Wiley.

Rubin, D. B. (1972). A non-iterative algorithm for least squares estimation of missing values in any analysis of variance design, *Appl. Statist.* **21**, 136–141.

Rubin, D. B. (1976a). Inference and missing data (with discussion), *Biometrika* **63**, 581–592.

Rubin, D. B. (1976b). Non-iterative least squares estimates, standard errors and F-tests for any analysis of variance with missing data, *J. Roy. Statist. Soc.* **B38**, 270–274.

Snedecor, G. W., and Cochran, W. G. (1967). *Statistical Methods*, Ames: Iowa State University Press.

Tocher, K. D. (1952). The design and analysis of block experiments, *J. Roy. Statist. Soc.* **B14**, 45–100.

Weisberg, S. (1980). *Applied Linear Regression.* New York: Wiley.

Winer, B. J. (1962). *Statistical Principles in Experimental Design.* New York: McGraw-Hill.

Wilkinson, G. N. (1958a). Estimation of missing values for the analysis of incomplete data, *Biometrics*, **14**, 257–286.

Wilkinson, G. N. (1958b). The analysis of variance and derivation of standard errors for incomplete data, *Biometrics*, **14**, 360–384.

Yates, F. (1933). The analysis of replicated experiments when the field results are incomplete, *Emp. J. Exp. Agric.* **1**, 129–142.

PROBLEMS

1. Review the literature on missing values in ANOVA from Allan and Wishart (1930) through Jarrett (1978).
2. Prove that $\hat{\beta}$ in (2.2) is (a) the least squares estimate of β, (b) the minimum variance unbiased estimate, and (c) the maximum likelihood estimate under normality. Which of these properties does $\hat{\sigma}^2$ enjoy, and why?
3. Outline the distributional results leading to (2.6) being distributed as F.
4. Summarize the argument that Bartlett's ANCOVA method leads to correct least squares estimates of missing values.
5. Prove that (2.12) follows from the definition of U^{-1}.
6. Provide intermediate steps leading to (2.13), (2.14), and (2.15).
7. Using the notation and results of Section 2.5.4, justify (2.16) and the method for calculating B and ρ that follows it.
8. Carry out the computations leading to the results of Example 2.1.
9. Justify (2.17)–(2.20).
10. Show (2.22) and then (2.23) and (2.24).
11. Carry out the computations leading to the results of Example 2.2.
12. Carry out the computations leading to the results of Example 2.3.
13. Carry out a standard ANOVA for the following data, where three values have been deleted from a (5×5) Latin square (Snedecor and Cochran, 1967, p. 313).

Yields (Grams) of Plots of Millet Arranged in a Latin Square[a]

	Column				
Row	1	2	3	4	5
1	B: —	E: 230	A: 279	C: 287	D: 202
2	D: 245	A: 283	E: 245	B: 280	C: 260
3	E: 182	B: —	C: 280	D: 246	A: 250
4	A: —	C: 204	D: 227	E: 193	B: 259
5	C: 231	D: 271	B: 266	A: 334	E: 338

[a] Spacings (in.): A, 2; B, 4; C, 6; D, 8; E, 10.

CHAPTER 3

Quick Methods for Multivariate Data with Missing Values

3.1. INTRODUCTION

In Chapter 2 we discussed the analysis of data with missing values confined to a single outcome variable Y, which is related to completely observed predictor variables by a linear model. In this chapter we discuss three quick approaches to the more general problem with values missing for more than one of a set of variables: complete-case analysis, available-case analysis, and simple forms of imputation. A review of the earlier literature on missing data that includes some of the methods discussed here is given by Afifi and Elashoff (1966). Although the methods appear in statistical computing software and are widely used, we do not generally recommend any of them except in special cases where the amount of missing data is limited. The procedures in Part II provide sounder solutions in more general circumstances.

Consider a rectangular $(n \times K)$ data matrix $Y = (y_{ij})$, where y_{ij} is the value of a variable Y_j for observation i, $i = 1, \ldots, n$, $j = 1, \ldots, K$. In the absence of missing values, many multivariate statistical analyses involve initial reduction of the data to the vector of sample means $\bar{y} = (\bar{y}_1, \ldots, \bar{y}_K)$ and the sample covariance matrix $S = (s_{jk})$, where $s_{jk} = (n-1)^{-1} \sum_{i=1}^{n} (y_{ij} - \bar{y}_j)(y_{ik} - \bar{y}_k)$. Thus an important problem concerns the estimation of $\bar{y}$ and S when some of the values y_{ij} are missing. We shall discuss quick missing-data procedures with an emphasis on this problem.

As we stressed in Chapter 1, the performance of any missing-data method depends heavily on the mechanisms that lead to missing values. Except when stated otherwise, the methods in this chapter are only appropriate under the strong assumption that the data are *missing completely at random* (MCAR), which means that missingness is not related to the variables under study. In general, the methods are not appropriate when the data are not MCAR but are missing at random (MAR), which means that missingness is related to the observed data but not to the missing data. In contrast, the likelihood-based methods in Part II are appropriate under this weaker MAR assumption about the missing-data mechanism. This advantage is of great importance in many practical problems.

3.2. COMPLETE-CASE ANALYSIS

Analyses using *complete cases* confine attention to cases where all K variables are present. Advantages of this approach are (1) simplicity, since standard complete-data statistical analyses can be applied without modifications, and (2) comparability of univariate statistics, since these are all calculated on a common sample base of cases. Disadvantages stem from the potential loss of information in discarding incomplete cases. The loss in sample size can be considerable, particularly if K is large. For example, if $K = 20$ and each variable is missing or observed independently according to a Bernoulli process with a 10% chance of missingness, then the expected proportion of complete cases is $0.9^{20} = 0.12$. That is, only about $12/0.9 = 13$ percent of the observed data values will be retained.

A crucial concern is whether the selection of complete cases leads to biases in sample estimates. Under the MCAR assumption of Section 1.4, the complete cases are effectively a random subsample of the original cases, and then discarding data in incomplete cases does not bias estimates. Commonly, however, the completely recorded cases differ in important ways from the original sample. For example, in a panel survey, individuals who are lost to follow-up are often different from those who remain in the study. In such situations, complete-case analysis can yield seriously biased results.

The nature of these biases depends on the missingness mechanisms leading to incomplete cases and the nature of the analysis involved. To take a simple example, suppose $K = 2$ and $Y_1 =$ age and $Y_2 =$ income are two variables measured in a survey. Suppose that either Y_1 or Y_2 may be missing and that missingness is related to Y_2 but not to Y_1. Specifically, $\Pr(y_{i1}$ missing, y_{i2} present $| y_{i1}, y_{i2}) = \phi_{01}(y_{i2})$; $\Pr(y_{i1}$ present, y_{i2} missing $| y_{i1}, y_{i2}) = \phi_{10}(y_{i2})$, and $\Pr(y_{i1}$ present, y_{i2} present $| y_{i1}, y_{i2}) = 1 - \phi_{01}(y_{i2}) - \phi_{10}(y_{i2})$, where ϕ_{10} and ϕ_{01} are functions of y_{i2} but not y_{i1}. Suppose that the form of ϕ_{10} and ϕ_{01} are such that high- and low-income cases are more likely to be incomplete than those for middle-income individuals. Then marginal distributions of age and income based on complete cases are distorted by the overrepresentation of middle-income earners in the depleted sample. Estimation of the correlation between Y_1 and Y_2 and of parameters of the linear regression of Y_2 on Y_1 from complete cases is also subject to bias. The linear regression of Y_1 on Y_2, however, is not subject to selection bias, since the selection is a function of the independent variable Y_2 and not the dependent variable Y_1. Even milder restrictions on the relationship between missingness and the measured variables apply to certain other analyses. For example, if Y_1 and Y_2 are dichotomous and inference concerns the odds ratio in the 2×2 table of counts classified by Y_1 and Y_2, then complete-case analysis does not yield selection bias if the logarithm of the probability of response is an additive function of Y_1 and Y_2 (Kleinbaum, Morgenstern and Kupper, 1981).

The discarded information from incomplete cases can be used to study whether the complete cases are plausibly a random subsample of the original sample, that is, whether MCAR is a reasonable assumption. A simple procedure is to compare the distribution of a particular variable Y_j based on complete cases with the distribution of Y_j based on incomplete cases for which Y_j is recorded. Sample sizes often restrict the comparison to summary statistics such as means, as in the BMDP8D program in Dixon (1983). Significant differences indicate that the MCAR assumption is invalid and the complete-case analysis yields biased estimates. Such tests are useful but have limited power when the sample of incomplete cases is small. Also the tests can offer no direct evidence on the validity of the MAR assumption.

A strategy for adjusting for the bias in the selection of complete cases is to assign them case weights for use in subsequent analyses. This strategy is found most commonly in sample surveys, and is used particularly to handle the problem of unit nonresponse, where all the survey items are missing for cases in the sample that did not participate. Information available for respondents and nonrespondents, such as their geographic location, can be used to assign weights to the respondents that at least partially adjust for nonresponse bias. We discuss the formation of suitable weights in Chapter 4.

3.3. AVAILABLE-CASE METHODS

Complete-case analysis is potentially wasteful for univariate analyses such as estimation of means and marginal frequency distributions, since values of a particular variable are discarded when they belong to cases that are missing other variables. A natural alternative procedure for univariate analyses is to include all cases where the variable of interest is present, an option we will term *available-case* analysis. The method uses all the available values; its disadvantage is that the sample base changes from variable to variable according to the pattern of missing data. As many processors of large data sets know, this variability in the sample base creates practical problems, particularly when tables are computed for various conceptual sample bases (e.g., all women, ever-married women, currently married women in a demographic fertility survey). The analyst wishes to associate a fixed sample size to each base as a check that the tables are correctly defined. The changes in the sample bases in available-case analysis prevent such simple checks. They also yield problems of comparability across variables if missingness is a function of the variables under study, that is, if the data are not MCAR.

Estimates of means and variances can be calculated under MCAR, using the available-case procedure we have described, but modifications are required to estimate measures of covariation such as covariances or correlations. A natural extension is to *pairwise* available-case methods, where measures of covariation for Y_j and Y_k are based on cases i for which both y_{ij} and y_{ik} are present. In particular, one might compute pairwise covariances:

$$s_{jk}^{(jk)} = \sum_{(jk)} (y_{ij} - \bar{y}_j^{(jk)})(y_{ik} - \bar{y}_k^{(jk)})/(n^{(jk)} - 1), \tag{3.1}$$

where $n^{(jk)}$ is the number of cases with both Y_j and Y_k observed, and the means $\bar{y}_j^{(jk)}$, $\bar{y}_k^{(jk)}$ and the summation in (3.1) are calculated over those $n^{(jk)}$ cases. Let $s_{jj}^{(j)}$, $s_{kk}^{(k)}$ denote sample variances of y_j and y_k from available cases. Combining these with $s_{jk}^{(jk)}$ yields the following estimate of correlation:

$$r_{jk}^{*} = s_{jk}^{(jk)}\big/\sqrt{s_{jj}^{(j)} s_{kk}^{(k)}}. \tag{3.2}$$

A criticism of (3.2) is that, unlike the population correlation being estimated, r_{jk}^{*} can lie outside the range $(-1, 1)$. This difficulty is avoided by computing pairwise correlations, where variances are estimated from the same sample base as the covariance:

$$r_{jk}^{(jk)} = s_{jk}^{(jk)}\big/\sqrt{s_{jj}^{(jk)} s_{kk}^{(jk)}}. \tag{3.3}$$

This estimate is discussed by Matthai (1951). It corresponds to the covariance estimate

$$s_{jk}^{*} = r_{jk}^{(jk)}\sqrt{s_{jj}^{(j)} s_{kk}^{(k)}}. \tag{3.4}$$

Still more estimates can be constructed by replacing the means in (3.1)–(3.4) by estimates from all available cases. Applying this idea to (3.1) yields

$$\tilde{s}_{jk}^{(jk)} = \sum_{(jk)} (y_{ij} - \bar{y}_j^{(j)})(y_{ik} - \bar{y}_k^{(k)})/(n^{(jk)} - 1), \tag{3.5}$$

which is called the ALLVALUE estimate of covariance in the BMDP8D program in Dixon (1983), also discussed in Wilks (1932).

Pairwise available-case estimates such as (3.1)–(3.5) attempt to recover some of the information in partially recorded units that is lost by complete-case analysis. Under MCAR, Eq. (3.1)–(3.5) yield consistent estimates of the covariances and correlations being estimated. When considered collectively, however, the estimates have deficiencies that can severely limit their utility in practical problems.

We have noted that (3.2) can yield correlations outside the acceptable range. On the other hand, (3.3) yields correlations that always lie between ± 1. For $K \geqslant 3$ variables, both (3.2) and (3.3) can yield estimated correlation matrices that are not positive definite. To take an extreme artificial example, consider the following data with 12 observations on three variables (—denotes missing):

Y_1	1	2	3	4	1	2	3	4	—	—	—	—
Y_2	1	2	3	4	—	—	—	—	1	2	3	4
Y_3	—	—	—	—	1	2	3	4	4	3	2	1

Equation (3.3) yields $r_{12}^{(12)} = 1$, $r_{13}^{(13)} = 1$, $r_{23}^{(23)} = -1$. These estimates are clearly unsatisfactory, since $\text{Corr}(Y_1, Y_2) = \text{Corr}(Y_1, Y_3) = 1$ implies $\text{Corr}(Y_2, Y_3) = 1$, not -1. In the same way, covariance matrices based on (3.1) or (3.4) are not necessarily positive definite. Since many analyses based on the covariance matrix, including multiple regression, require a positive-definite matrix, ad hoc modifications are required for these methods when this condition is not satisfied.

Since available-case methods apparently make use of all the data, one might expect them to dominate complete-case methods, a conclusion supported in simulations by Kim and Curry (1977) when the data are MCAR and correlations are modest. Other simulations, however, indicate superiority for complete-case analysis when correlations are large (Haitovsky, 1968; Azen and Van Guilder, 1971). Neither method, however, is generally satisfactory.

3.4. FILLING IN THE MISSING VALUES

3.4.1. Introduction

Both complete-case and available-case analysis make no use of cases with Y_j missing when estimating either the marginal distribution of Y_j or measures of covariation between Y_j and other variables. Suppose a case with Y_j missing contains the value of another variable Y_k that is highly correlated with Y_j. It is tempting to predict the missing value of Y_j from Y_k and then to include the filled-in (or *imputed*) value in analyses involving Y_j.

Imputation is a general and flexible method for handling missing-data problems. However, it is not without pitfalls. In the words of Dempster and Rubin (1983):

> The idea of imputation is both seductive and dangerous. It is seductive because it can lull the user into the pleasurable state of believing that the data are complete after all, and it is dangerous because it lumps together situations where the problem is sufficiently minor that it can be legitimately handled in this way and situations where standard estimators applied to the real and imputed data have substantial biases.

In this section we discuss some simple imputation methods and the problems of estimating the mean and covariance matrix of $Y_1, \ldots, Y_k$ from imputed data.

3.4.2. Imputing Unconditional Means

A particularly simple form of imputation is to estimate missing values y_{ij} by $\bar{y}_j^{(j)}$, the mean of the recorded values of Y_j. The average of the observed and imputed values is then clearly $\bar{y}_j^{(j)}$, the estimate from available-case analysis. The variance of the observed and imputed values is $[(n^{(j)} - 1)/(n - 1)]s_{jj}^{(j)}$, where $s_{jj}^{(j)}$ is the estimated variance from available cases. Under MCAR, $s_{jj}^{(j)}$ is a consistent estimate of the true variance, so the sample variance from the filled-in data set underestimates the variance by a factor of $(n^{(j)} - 1)/(n - 1)$. This underestimation is a natural consequence of imputing missing values at the center of the distribution. The sample covariance of Y_j and Y_k from the filled-in data is $[(n^{(jk)} - 1)/(n - 1)]\tilde{s}_{jk}^{(jk)}$, where $\tilde{s}_{jk}^{(jk)}$ is given by equation (3.5). Since under MCAR $\tilde{s}_{jk}^{(jk)}$ is a consistent estimate of the covariance, the estimate from filled-in data underestimates the magnitude of the covariance by a factor $(n^{(jk)} - 1)/(n - 1)$. Thus, although the covariance matrix from the filled-in data is positive semidefinite, the variances and covariances are systematically underestimated. Obvious adjustment factors, namely, $(n - 1)/(n^{(j)} - 1)$ for the variance of Y_j and $(n - 1)/(n^{(jk)} - 1)$ for the covariance of Y_j and Y_k, simply yield the estimates of (3.5), which as noted in Section 3 are not generally satisfactory.

3.4.3. Imputing Conditional Means: Buck's Method

A more promising form of imputation is to substitute means that are conditioned on the variables recorded in an incomplete case. If the variables $Y_1, \ldots, Y_K$ are multivariate normally distributed with mean μ and covariance matrix Σ, then the missing variables in a particular case have linear regressions on the observed variables, with regression coefficients that are well-known functions of μ and Σ. The method proposed by Buck (1960) first estimates μ and Σ from the sample mean and covariance matrix based on the complete cases, and then uses these estimates to calculate the linear regressions of the missing variables on the present variables, case by case. Substituting the observed values of the present variables for a case in the regressions yields

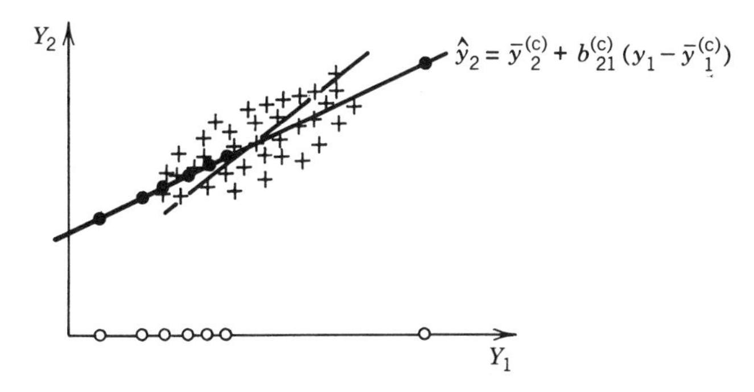

Figure 3.1. Buck's method for $K = 2$ variables.

predictions for the missing values in that case. The computation of different linear regressions for each pattern of missing data may appear formidable, but in fact is relatively simple using the sweep operator discussed in Section 6.5.

Buck's method is illustrated for $K = 2$ variables in Figure 3.1. The points marked as pluses represent cases with Y_1 and Y_2 both observed. These points are used to calculate the least squares regression line of Y_2 on Y_1, say $\hat{y}_2 = \bar{y}_2^{(c)} + b_{21}^{(c)}(y_1 - \bar{y}_1^{(c)})$, where the superscript c signifies complete cases. Cases with Y_1 observed but Y_2 missing are represented by circles on the Y_1 axis. Buck's procedure replaces them by the dots lying on the regression line. Cases with Y_2 observed and Y_1 missing would be imputed on the regression line of Y_1 on Y_2, the other line in the diagram.

The averages of observed and imputed values from this procedure are consistent estimates of the means under MCAR and mild assumptions about the moments of the distribution (Buck, 1960). They are also consistent when the missing-data mechanism depends on variables that are observed, although additional assumptions are required in this case. In particular, suppose that for the data in Figure 3.1 missingness of Y_2 depends on the values of Y_1, so that MAR holds, even though the distribution of Y_1 for complete and incomplete cases is different. Buck's method projects the incomplete cases to the regression line, a process that makes the assumption that the regression of Y_2 on Y_1 is linear. This assumption is particularly tenuous if the imputation involves extrapolation beyond the range of the complete data, as occurs for the incomplete cases with the two smallest and the single largest Y_1 values in Figure 3.1.

The filled-in data from Buck's method yield reasonable estimates of means, particularly if the normality assumptions are plausible. The sample covariance matrix from the filled-in data underestimates the sizes of variances and covariances, although the extent of underestimation is less than that obtained when unconditional means are substituted. Consider, for example, the sample variance of Y_2 from the filled-in data in Figure 3.1. Expressing the variance of Y_2 as the sum of the variance of the mean of Y_2 given Y_1 and the expected variance of Y_2 given Y_1, we obtain

$$\sigma_{22} = \beta_{21}^2 \sigma_{11} + \sigma_{22 \cdot 1},$$

where $\beta_{21} = \sigma_{12}/\sigma_{11}$ is the slope of the regression of Y_2 on Y_1, $\beta_{21}^2 \sigma_{11}$ is the component of variance explained by the regression of Y_2 on Y_1, and $\sigma_{22 \cdot 1}$ is the residual variance. The variability of the imputed values of Y_2 incorporates the component $\beta_{21}^2 \sigma_{11}$ of the variance but effectively sets the component $\sigma_{22 \cdot 1}$ to zero, since the imputations lie exactly on the regression line. It follows that the sample variance of Y_2, based on both the filled-in values and the real values, is biased and underestimates σ_{22} by the quantity $(n - n^{(2)})(n - 1)^{-1}\sigma_{22 \cdot 1}$, where $(n - n^{(2)})$ is the number of missing Y_2 values. Note that the extent of underestimation is small when Y_1 is a good predictor of Y_2 in the sense that $\sigma_{22 \cdot 1}$ is small compared with σ_{22}, but the bias does not go to zero as n becomes large unless that fraction of missing values goes to zero; thus this estimate of σ_{22} is generally inconsistent.

More generally, the sample variance of Y_j from data filled in by Buck's method underestimates σ_{jj} by the quantity $(n - 1)^{-1} \sum \sigma_{jj \cdot \text{obs}, i}$, where $\sigma_{jj \cdot \text{obs}, i}$ is the residual variance from regressing Y_j on the variables present (or observed) in case i if y_{ij} is missing and is zero if y_{ij} is observed. The sample covariance of Y_j and Y_k has a bias of $(n - 1)^{-1} \sum \sigma_{jk \cdot \text{obs}, i}$, where $\sigma_{jk \cdot \text{obs}, i}$ is the residual covariance of Y_j and Y_k from the multivariate regression of Y_j, Y_k on the variables observed in case i if both y_{ij} and y_{ik} are missing and zero otherwise. A consistent estimate of Σ can be constructed under the MCAR assumption by substituting consistent estimates of $\sigma_{jj \cdot \text{obs}, i}$ and $\sigma_{jk \cdot \text{obs}, i}$ (such as estimates based on the sample covariance matrix of the complete observations, sample sizes permitting) in the expressions for bias and then adding the resulting quantities to the sample covariance matrix of the filled-in data. This method is closely related to one iteration of the maximum likelihood procedure presented in Section 8.2, and unlike the adjustment for imputing the unconditional mean, does not reduce to the available-case estimates of Section 3.3.

Although the regression-based imputations in Buck's method appear to require that the variables $Y_1, \ldots, Y_k$ are interval scaled, the method can be applied to categorical variables by replacing each of them by a set of dummy variables, numbering one less than the number of categories. If a categorical variable is completely observed, then the dummy variables always appear as independent variables in the regressions for Buck's method and no problems arise. If it is sometimes missing, then the set of dummy variables also appears as dependent variables in linear regressions. The imputations from the regressions are linear estimates of the probability of falling into the categories represented by each of the dummy variables. Problems can arise from the fact that linear regression is used to predict these probabilities; for example, the predicted values can lie outside the range (0, 1). Thus Buck's method has limitations when some of the missing variables are categorical.

3.4.4. Other Approaches

If we assume MCAR and ignore sampling variability of the estimates of μ and Σ based on complete cases, then the conditional means imputed in Section 3.4.3 are the best point estimates of the missing values in the sense of minimizing the expected squared error. We have seen, however, that adjustments are required to the sample variances of the filled-in data to yield consistent estimates even under these assumptions. More generally, the marginal distributions of the completed data are distorted by mean imputation. This distortion is particularly disturbing when the tails of the distribution or standard errors of estimates are being studied. For example, an imputation method that imputes conditional means for missing incomes must tend to underestimate the percentage of cases in poverty.

These considerations suggest an alternative strategy where imputations are selected randomly from a distribution of plausible values, rather than from the center of this distribution. One way of achieving this is to add a suitable perturbation to the conditional mean. Methods of this type are often used in survey applications; consequently, discussion is deferred until Chapters 4 and 12.

To summarize, it is hard to recommend any of the simple methods discussed since (1) their performance is unreliable; (2) they often require ad hoc adjustments to yield satisfactory estimates, and (3) it is not easy to distinguish situations when the methods work from situations when they fail. Furthermore, the methods fail to provide simple correct answers when measures of the precision of estimates are required, as for interval estimation. This was seen in the special case of univariate nonresponse in Chapter 2.

The main focus of this book (Chapters 5–12) is a unified collection of procedures for dealing with missing data based on statistical models for the missing data mechanisms and the data. The methods based on this theory are trustworthy in the sense that under clearly stated assumptions they have optimal statistical properties, at least in large samples. The methods do not require ad hoc adjustments to work for both point and interval estimation, and the situations where the methods work are clearly delineated by the specifications of the models on which they are based. In practice, we rarely know the correct model specification, and various specifications may be tried.

REFERENCES

Afifi, A. A., and Elashoff, R. M. (1966). Missing observations in multivariate statistics I: Review of the literature, *J. Am. Statist. Associ.* **61**, 595–604.

Azen, S., and Van Guilder, M. (1981). Conclusions regarding algorithms for handling incomplete data, *Proceedings of the Statistical Computing Section, American Statistical Association 1981*, 53–56.

Buck, S. F. (1960). A method of estimation of missing values in multivariate data suitable for use with an electronic computer, *J. Roy. Statist. Soci.* **B22**, 302–306.

Dempster, A. P., and Rubin, D. B. (1983). Overview, in *Incomplete Data in Sample Surveys, Vol. II: Theory and Annotated Bibliography* (W. G. Madow, I. Olkin, and D. B. Rubin, Eds.). New York: Academic Press, 3–10.

Dixon, W. J. (ed.), (1983). *BMDP Statistical Software*, 1983 revised printing, Berkeley: University of California Press.

Haitovsky, Y. (1968). Missing data in regression analysis, *J. Roy. Statist. Soci.* **B30**, 67–81.

Kim, J. O., and Curry, J. (1977). The treatment of missing data in multivariate analysis," *Social. Meth. Res.* **6**, 215–240.

Kleinbaum, D. G., Morgenstern, H., and Kupper, L. L. (1981). Selection bias in epidemiological studies. *American Journal of Epidemidogy*, **113**, 452–463.

Matthai, A. (1951). Estimation of parameters from incomplete data with application to design of sample surveys. *Sankhya*, **2**, 145–152.

Wilks, S. S. (1932). Moments and distribution of estimates of population parameters from fragmentary samples, *Annals of Mathematical Statistics* **3**, 163–195.

PROBLEMS

1. List some standard multivariate statistical analyses that are based on the sample means, variances, and correlations.
2. Show that if missingness depends only on Y_2, and Y_1 has a linear regression on Y_2, then the sample regression of Y_1 on Y_2 based on complete cases yields unbiased estimates of the regression parameters.
3. Show that for dichotomous Y_1 and Y_2, the odds ratio based on complete cases is a consistent estimate of the population odds ratio if the log probability of response takes the form $\phi_1(Y_1) + \phi_2(Y_2)$.
4. Construct a data set where the estimated correlation (3.2) lies outside the range $(-1, 1)$.
5. (a) Why does the estimated correlation (3.3) always lie in the range $(-1, 1)$? (b) Suppose the means $\bar{y}_j^{(jk)}$ and $\bar{y}_k^{(jk)}$ in (3.2) are replaced by estimates from all available cases. Is the resulting estimated correlation (3.3) always in the range $(-1, 1)$? Either prove that it is or provide a counterexample.

6. Consider the relative merits of complete-case analysis and available-case analysis for estimating (a) means, (b) correlations, and (c) regression coefficients when the data are not MCAR.
7. Review the results of Haitovsky (1968), Kim and Curry (1977), and Azen and Van Guilder (1981). Describe situations where complete-cases analysis is more sensible than available-case analysis, and vice versa.
8. Prove the statement in Section 3.4.4 that, assuming the data are MCAR and ignoring sampling variability of estimates, conditional mean imputation minimizes the expected squared prediction error of imputations.
9. Consider a bivariate sample with $n = 45$, 20 complete cases, 15 cases with only Y_1 recorded, and 10 cases with only Y_2 recorded. The data are filled in using unconditional means, as in Section 3.4.2. Assuming MCAR, determine the percentage bias of estimates of the following quantities computed from the filled-in data: (a) the variance of Y_1 (σ_{11}); (b) the covariance of Y_1 and Y_2 (σ_{12}); (c) the slope of the regression of Y_2 on Y_1 (σ_{12}/σ_{11}). For (c) you can ignore bias terms of order $1/n$.
10. Repeat the previous example when the missing values are filled in by Buck's method and compare the answers.
11. Describe the circumstances where Buck's method clearly dominates both complete-case analysis and available-case analysis.
12. Suppose data are an incomplete random sample on Y_1 and Y_2, where Y_1 given θ is $N(\mu_1, \sigma_1^2)$ and Y_2 given Y_1 and θ is $N(\beta_0 + \beta_1 Y_1 + \beta_2 Y_1^2, \sigma_{2\cdot 1}^2)$ and $\theta = (\mu_1, \sigma_1^2, \beta_0, \beta_1, \beta_2, \sigma_{2\cdot 1}^2)$. The data are MCAR, the first m cases are complete, the next r_1 cases record Y_1 only, and the last r_2 cases record Y_2 only. Consider the properties of Buck's method, applied to (a) Y_1 and Y_2 and (b) Y_1, Y_2, and $Y_3 \equiv Y_1^2$, for deriving estimates of (i) the unconditional means $E(Y_1|\theta)$ and $E(Y_2|\theta)$, and (ii) the conditional means $E(Y_1|Y_2, \theta)$, $E(Y_1^2|Y_2, \theta)$ and $E(Y_2|Y_1, \theta)$.
13. Show that Buck's (1960) method yields consistent estimates of the means when the data are MCAR and the distribution of the variables has finite fourth moments.
14. Derive the expressions for the biases of Buck's (1960) estimators of σ_{jj} and σ_{jk} stated in Section 3.4.3.

CHAPTER 4

Nonresponse in Sample Surveys

4.1. INTRODUCTION

The problem of incomplete data in sample surveys differs from the problems we have considered in Chapters 2 and 3 in two basic respects. First, the population of interest is explicitly finite, and thus the entities being estimated are often finite population quantities, such as population means or totals. Second, the method of data collection has traditionally had a prominent role in guiding the analysis. Indeed, most of the literature on nonresponse in sample surveys adopts the *randomization* mode of inference, where the population values are treated as fixed and inferences are based on the distribution that determines sample selection. In contrast, Chapters 5–12 of this book derive inferences based on a stochastic model for values of the variables, and the method of sample selection enters the analysis only indirectly through its influence on the choice of models. We begin this chapter by outlining the differences in the two approaches to inference, both for completely observed data and for data subject to missing values. We then discuss common methods for handling nonresponse in the randomization literature. For a comprehensive discussion of survey nonresponse from both randomization and modeling perspectives, see Madow et al. (1983), in particular, Volume II.

4.2. RANDOMIZATION INFERENCE WITH COMPLETE DATA

Suppose inferences are required for a population with N cases or units, and let $Y = (y_{ij})$, where $y_i = (y_{i1}, \ldots, y_{ik})$ represents a vector of k items for unit i, $i = 1, \ldots, N$. For unit i, define the sample indicator function

$$I_i = \begin{cases} 1, & \text{unit } i \text{ included in the sample,} \\ 0, & \text{unit } i \text{ not included in the sample,} \end{cases}$$

and

$$I = (I_1, \ldots, I_N)^{\mathrm{T}}.$$

Let $inc = \{i | I_i = 1\}$. Sample selection processes can be characterized by a distribution for I given Y. For example, simple random sampling with sample size n is defined by the distribution

$$f(I | Y) = f(I) = \begin{cases} \binom{N}{n}^{-1}, & \text{if } \sum_{i=1}^{N} I_i = n \\ 0, & \text{otherwise} \end{cases},$$

where $\binom{N}{n}$ is the number of ways n units can be chosen from the population.

Simple random sampling does not make use of prior knowledge of the population in determining which units are selected. More efficient sampling designs make use of such information when it is available. For example, it may be possible to divide the population into strata within which units are relatively homogeneous. The units may then be selected by *stratified random sampling*. In stratum j with N_j units, n_j units are selected by simple random sampling, yielding a total sample size of $n = \sum_{j=1}^{J} n_j$. More generally, let $Z = (z_{ij})$, whose ith row, z_i, represents information about unit i known before the survey and used in the survey design. For example, in stratified sampling, z_i indicates the stratum to which the ith unit belongs. The sampling distribution in such designs is defined by the conditional distribution of I given Y and Z, written $f(I | Y, Z)$.

Randomization inference generally requires that units be selected by *probability sampling*, which is characterized by the following two properties:

1. The sampling distribution is determined by the sampler before any y values are known. In particular, $f(I | Y, Z) = f(I | Z)$, since the distribution cannot depend on the unknown values of items Y to be sampled in the survey.
2. Every unit has a positive (known) probability of selection. Writing $\pi_i = E(I_i | Y, Z) = \Pr(I_i = 1 | Y, Z)$, we require $\pi_i > 0$ for all i. In *equal probability* sample designs, such as simple random sampling, this probability is the same for all units.

The objective is to estimate finite population quantities, such as the population mean $\bar{Y}$ of a variable Y, by sample quantities, such as the sample mean. Inference is based on the distribution of the sample quantities in repeated sampling from the distribution of I, $f(I | Z)$. Detailed treatments of randomization inference can be found in sampling theory texts such as Cochran (1977) or Hansen, Hurwitz and Madow (1953). In outline, the following steps are involved in deriving inferences for a population quantity T:

(a) The choice of a statistic $t(Y_{\text{inc}})$, a function of the sampled values Y_{inc} of Y that is (approximately) unbiased for T in repeated samples. For example, if $T = \bar{Y}$, the population mean, then t might be the sample mean $\bar{y}$, which is unbiased for $\bar{Y}$ under equal probability sampling.

(b) The choice of a statistic $v(Y_{\text{inc}})$ that is (approximately) unbiased for the variance of $t(Y_{\text{inc}})$ in repeated sampling. For example, under simple random sampling it can be shown that the variance of the sample mean $\bar{y}$ is

$$\text{Var}(\bar{y}) = (n^{-1} - N^{-1})S_y^2,$$

where S_y^2 is the variance of the values $(y_1, \ldots, y_N)$ in the population. The statistic

$$v(Y_{\text{inc}}) = (n^{-1} - N^{-1})s_y^2, \tag{4.1}$$

where s_y^2 is the variance of the values of Y in the sample, is an unbiased estimate of $\text{Var}(\bar{y})$ for simple random sampling.

(c) The calculation of interval estimates of T, assuming t is approximately normally distributed with mean T and variance $v(Y_{\text{inc}})$. For example, a 95 percent confidence interval for $\bar{Y}$ assuming simple random sampling is given by

$$C_{95}(\bar{Y}) = \bar{y} \pm 1.96\sqrt{(n^{-1} - N^{-1})s_y^2}, \tag{4.2}$$

where 1.96 is the 97.5 percentile of the normal distribution. The normal approximation is justified by appealing to a finite population version of the central limit theorem (Hajek, 1960).

Note that throughout this process, the values $y_1, \ldots, y_N$ are treated as *fixed*. An attractive aspect of the randomization approach is the avoidance of a model specification for the population values, although (4.2) requires that the distribution of Y values in the population is sufficiently well behaved for the normality of t to apply for the observed sample size n and population size N.

An alternative approach to inference about finite population quantities is to specify, in addition to the sampling distribution, a model for Y often in the form of a density $f(Y|Z, \theta)$ indexed by unknown parameters θ. This model is then used to predict the nonsampled values of y excluded from the sample. For example, the population mean $\bar{Y}$ can be estimated by

$$\left(\sum_{I_i=1} y_i + \sum_{I_i=0} \hat{y}_i\right) \bigg/ N$$

where $\hat{y}_i$ is the expected value of y_i under the model, with θ replaced by an estimate $\hat{\theta}$, such as the mean of the posterior distribution of y_i when a full Bayesian model is specified. We discuss this approach to survey inference in Chapter 12.

4.3. QUASI-RANDOMIZATION INFERENCE FOR DATA WITH MISSING VALUES

The key ingredient of the randomization approach, a known probability distribution governing which values are observed and which are unobserved, is lost when some of the data are missing. As an illustration, suppose that n units are selected by simple random sampling and let

$$R_i = \begin{cases} 1, & y_i \text{ responds if sampled,} \\ 0, & \text{otherwise,} \end{cases}$$

and

$$R = (R_1, \ldots, R_N)^T.$$

Values of Y are recorded if and only if $R_i = I_i = 1$. It is now impossible to define a statistic that is a function of the recorded values of Y and is an unbiased estimate of the population mean $\bar{Y}$ with respect to the distribution of I. For example, Cochran (1963) shows that the mean of the responding units,

$$\bar{y}_R = \sum_{i=1}^{N} I_i R_i y_i \Big/ \sum_{i=1}^{N} I_i R_i, \tag{4.3}$$

is a biased estimate of $\bar{Y}$ with approximate bias

$$b(\bar{y}_R | Y, R) = \bar{Y}_R - \bar{Y} = (1 - \lambda_R)(\bar{Y}_R - \bar{Y}_{NR}), \tag{4.4}$$

where λ_R is the proportion of responding units in the population, and $\bar{Y}_{NR}$ is the mean of the nonresponding units in the population. In the absence of information about $\bar{Y}_{NR}$, we cannot correct $\bar{y}_R$ for this unknown bias.

Two ways around this difficulty are available. Either some modeling assumptions are made about the nonresponding portion of the population (e.g., that the means of Y in the responding and nonresponding parts of the population are equal), or a distribution is assumed for R and no assumptions are made about the distribution of Y in the population. Assumptions of the former kind relate to models for the y values and hence to the modeling approach to surveys with complete response. The latter approach, which formulates a distribution for R, is a more direct extension of randomization inference to the case of nonresponse; we shall use the term *quasi-randomization* inference to describe it, adopting the term suggested by Oh and Scheuren (1983).

The elements of the quasi-randomization approach are (1) a known distribution $f(I|Z)$ of sample selection, as for complete survey data; and (2) an assumed distribution for the response indicators R given I, Y, and Z. This distribution typically involves the assumption that, within identifiable subgroups of the population, the restriction to responding units corresponds to another stage of equal probability random sampling; (3) a function t of the recorded observations that is approximately unbiased for the parameter of interest T in repeated sampling from the joint distribution of I and R; and (4) a function v of the recorded data that is approximately unbiased for the variance of t in repeated sampling from the distribution of I and R. Confidence intervals for T can be constructed from t and v as before, under normality assumptions.

The perspective for the duration of this chapter will be the quasi-randomization approach; the modelling approach is discussed in chapter 12.

EXAMPLE 4.1. *Nonresponse as a Form of Random Subsampling.* Consider a population of size N with $\sum_{i=1}^{N} R_i = M$ respondents. A simple random sample of size n is selected, and $\sum_{i=1}^{N} R_i I_i = m$ of the sampled units respond. Suppose that the distribution of R given I and surveyed items Y is specified to be

$$f(R|I, Y) = \begin{cases} \binom{N}{M}^{-1}, & \sum R_i = M, \\ 0, & \sum R_i \neq M. \end{cases} \tag{4.5}$$

Note that this distribution does not depend on the values of I and Y, so that MCAR holds. The probability of response is M/N and does not depend on the units sampled or values of the items. Let $D_i = R_i I_i$ and $D = (D_1, \ldots, D_N)^{\mathrm{T}}$. The distribution of D given $\sum D_i = m$ is

$$f(D|I, Y) = \begin{cases} \binom{N}{m}^{-1}, & \sum D_i = m, \\ 0, & \text{otherwise}, \end{cases}$$

which is the distribution of a simple random sample of size m. Hence a 95% confidence interval for the population mean $\bar{Y}$ is $\bar{y}_R \pm 1.96\sqrt{(m^{-1} - N^{-1})s_{\mathrm{YR}}^2}$, where $\bar{y}_R$ and s_{YR}^2 are the mean and variance of the responding units.

The strong assumption underlying (4.5) is that R is independent of I and Y, that is

$$R \perp\!\!\!\perp (I, Y),$$

using Dawid's (1979) notation $\perp\!\!\!\perp$ for independence. This assumption is often unrealistic in practice. The weighting cell estimators discussed in the next section weaken this MCAR assumption by restricting it to hold only within subclasses of the population, so that MAR holds but MCAR does not.

4.4. WEIGHTING METHODS

4.4.1. Weighting Cell Estimators

One way of viewing probability sampling is that a unit selected with probability π_i is "representing" π_i^{-1} units in the population, and hence should be given the weight π_i^{-1} in estimates of population quantities. This view is certainly justified for stratified random sampling with n_j units chosen from the N_j units in stratum j, since then $\pi_i = n_j/N_j$ for units i in stratum j, and each selected unit represents N_j/n_j population units. In particular, in the absence of nonresponse the population total T of a variable Y can be estimated by

$$t = \sum_{i=1}^{N} y_i I_i \pi_i^{-1}, \tag{4.6}$$

which is called the *Horvitz–Thompson estimator* (Horvitz and Thompson, 1952). The population mean $\bar{Y}$ may be estimated by

$$\bar{y}_w = \sum_{i=1}^{N} w_i y_i, \tag{4.7}$$

where

$$w_i = I_i \pi_i^{-1} \Big/ \sum_k I_k \pi_k^{-1}$$

is the weight attached to the ith unit. Note that since $E(I_i|Y) = \pi_i$,

$$E(t|Y) = \sum_{i=1}^{N} y_i \pi_i \pi_i^{-1} = T,$$

so that the Horvitz–Thompson estimator of the total is unbiased in repeated sampling. The estimator $\bar{y}_w$ is unbiased for the mean $\bar{Y}$ in many sampling designs and is approximately unbiased in others. Of course, t and $\bar{y}_w$ can only be calculated in the absence of nonresponse so that y_i is observed whenever $I_i = 1$. Weighting cell estimators extend this approach to handle nonresponse by weighting responding units by the inverse of the probability of selection *and* response.

EXAMPLE 4.2. *Weighting Cell Estimators with Ignorable Nonresponse.* Suppose we can divide the population into J adjustment cells, within which response is independent of (Y, I). Define an adjustment cell variable C that takes value j for all units in cell j. Suppose that the distribution of response is

$$f(R|I, Y, C) = \begin{cases} \prod_{j=1}^{J} \binom{N_j}{M_j}^{-1}, & \sum_{i:c_i=j} R_i = M_j \text{ for all } j, \\ 0, & \sum_{i:c_i=j} R_i \neq M_j \text{ for any } j, \end{cases} \tag{4.8}$$

where N_j is the number of units in cell j, M_j is the number of units that respond if sampled, and $\phi_j = M_j/N_j$ is the response rate in cell j. If values of ϕ_j were known, Horvitz–Thompson estimators of means and totals would be obtained by weighting responding unit i in cell j by $(\pi_i \phi_j)^{-1}$. In practice the ϕ_j are not known but can be replaced by estimates $\hat{\phi}_j$ based on sample response rates in the cells. The resulting estimate of $\bar{Y}$ is

$$\sum_{j=1}^{J} \sum_{i \in R(j)} y_i \pi_i^{-1} \hat{\phi}_j^{-1} \Big/ \sum_{j=1}^{J} \sum_{i \in R(j)} \pi_i^{-1} \hat{\phi}_j^{-1}, \tag{4.9}$$

where $R(j)$ denotes the set of sampled units in adjustment cell j that respond. A natural estimate of ϕ_j is the sample response rate in cell j, that is, $\hat{\phi}_j = m_j/n_j$ where n_j is the number of sampled units and m_j is the number of sampled units that respond. For equal probability sample designs and this estimate of ϕ_j, (4.9) reduces to the *weighting cell* estimator of $\bar{Y}$:

$$\bar{y}_{\text{wc}} = n^{-1} \sum_{j=1}^{J} n_j \bar{y}_{jR}, \tag{4.10}$$

where $\bar{y}_{jR}$ is the respondent mean in cell j, and $n = \sum_{j=1}^{J} n_j$ is the sample size.

Suppose units are selected by simple random sampling and assumption (4.8) holds. Oh and Scheuren (1983) derive the mean and variance of $\bar{y}_{\text{wc}}$ over the distribution of R and I, given $\mathbf{m} = (m_1, \ldots, m_J)$ and $\mathbf{n} = (n_1, \ldots, n_J)$:

$$E(\bar{y}_{\text{wc}} \mid Y, C, \mathbf{m}, \mathbf{n}) = \sum_{j=1}^{J} \frac{n_j \bar{Y}_j}{n} = \bar{Y} + \sum_{j=1}^{J} \frac{n_j(\bar{Y}_j - \bar{Y})}{n}. \tag{4.11}$$

$$V(\bar{y}_{\text{wc}} \mid Y, C, \mathbf{m}, \mathbf{n}) = \sum_{j=1}^{J} \left(\frac{n_j}{n}\right)^2 \left(\frac{1}{m_j} - \frac{1}{N_j}\right) S_j^2, \tag{4.12}$$

where $\bar{Y}_j$ and S_j^2 denote the population mean and variance of Y in cell j. Note that $\bar{y}_{\text{wc}}$ is generally biased for $\bar{Y}$, conditional on $\mathbf{m}$ and $\mathbf{n}$. Oh and Scheuren (1983) propose the following estimate of the mean squared error of $\bar{y}_{\text{wc}}$:

$$\text{mse}(\bar{y}_{\text{wc}}) = \sum_{j=1}^{J} \left(\frac{n_j}{n}\right)^2 \left(1 - \frac{m_j n}{n_j N}\right) \frac{s_{jR}^2}{m_j} + \frac{N - n}{N - 1} \sum_{j=1}^{J} n_j \frac{(\bar{y}_{jR} - \bar{y}_{\text{wc}})^2}{n^2},$$

where s_{jR}^2 is the variance of sampled and responding units in cell j; $100(1 - \alpha)\%$ confidence intervals for $\bar{Y}$ may be constructed of the form $\bar{y}_{\text{wc}} \pm z_{1-\alpha/2}\{\text{mse}(\bar{y}_{\text{wc}})\}^{1/2}$, where $z_{1-\alpha/2}$ is the $100(1 - \alpha/2)$ percentile of the standard normal distribution, but their performance in applications is largely a matter of conjecture.

4.4.2. Choice of Adjustment Cells

Adjustment cells can be formed from survey design variables Z or from sampled items Y recorded for both respondents and nonrespondents. Weighting cell adjustments are used primarily to handle *unit* nonresponse, where none of the sampled items are recorded for nonrespondents. In these applications only survey design variables Z are available for forming adjustment cells.

A precise theory for the formation of adjustment cells is not available at present, but some general comments can be offered. Adjustment cells should be chosen so that (1) assumption (4.8) relating to the response distribution is satisfied, and (2) the mean squared error of estimates such as $\bar{y}_{\text{wc}}$ under assumption (4.8) is minimized. In samples selected with equal probability designs, the bias component in (4.11) of the mean squared error is likely to be dominated by the variance (4.12). In qualitative terms the variance (4.12) is minimized by choosing adjustment cells first that are homogeneous with respect to Y, so that S^2_{jR} is small, and second that avoid small respondent sample sizes, m_j. The key aspect of assumption (4.8) is that R is independent of Y and I within adjustment cells; that is,

$$R \perp\!\!\!\perp (Y, I) | C. \tag{4.13}$$

The theory of propensity scores (Rosenbaum and Rubin, 1983, 1985), discussed in the context of survey nonresponse in Little (1986), provides a prescription for choosing C so that (4.13) is approximately satisfied. Let X denote the set of variables observed for both respondents and nonrespondents, and suppose that

$$R \perp\!\!\!\perp (Y, I) | X, \tag{4.14}$$

so that (4.13) is satisfied when C is chosen to be X. However, adjustment cells cannot generally be formed for each distinct value of X, since then some cells with nonrespondents may have no respondents or a small number of respondents that results in estimates with inflated variance. Define the response propensity for unit i

$$p(x_i) = \Pr(R_i = 1 | x_i).$$

and let

$$p(X) = (p(x_i), \ldots, p(x_N))^{\text{T}}.$$

It can be shown that if $p(x_i)$ is positive for all i and (4.14) is satisfied, then $R \perp\!\!\!\perp (Y, I) | p(X)$, so that stratification on the propensity score $p(x_i)$ ensures that (4.13) holds.

In practice $p(x_i)$ requires estimation from sample data. An obvious procedure is to (1) estimate $p(x_i)$ by logistic or probit regression of the response indicator R_i on x_i; (2) form a grouped variable by coarsening the estimated $p(x_i)$ into five or six values; and (3) let C equal that grouped variable so that within adjustment cell j, all respondents and nonrespondents have the same value of the grouped variable.

4.4.3. Other Weighting Adjustments

Some variations on the weighting cell estimator (4.10) deserve mention. Cassel, Särndal and Wretman (1983) define nonresponse weights directly as the inverses of the estimated propensity scores $\hat{p}(x_i)$ of responding units. Under the modeling assumptions, this method removes nonresponse bias, but it may yield estimators with extremely high variance because respondents with very low estimated response propensity receive large nonresponse weights and may be unduly influential in estimates of means and totals. Also, weighting directly by $\hat{p}(x_i)$ may place more reliance on correct model specification of the regression of R_i on x_i than stratification, which uses $\hat{p}(x_i)$ only to form adjustment cells.

The population proportion N_j/N in each adjustment cell j is sometimes known, either from external sources or because adjustment cells are formed from stratifying variables Z in the survey design. In that case an alternative to $\bar{y}_{\text{wc}}$ is the poststratified mean

$$\bar{y}_{\text{ps}} = N^{-1} \sum_{j=1}^{J} N_j \bar{y}_{jR}. \tag{4.15}$$

Under the MAR assumption of Eq. (4.13), $\bar{y}_{\text{ps}}$ is unbiased for $\bar{Y}$ with variance

$$\text{Var}(\bar{y}_{\text{ps}} | Y, C, N_1, \ldots, N_J) = N^{-2} \sum_{j=1}^{J} N_j^2 \left(1 - \frac{m_j}{N_j}\right) \frac{S_{jR}^2}{m_j}. \tag{4.16}$$

An estimate of (4.16) is obtained by substituting the sample variance among respondents in cell j, s_{jR}^2, for S_{jR}^2. In most circumstances $\bar{y}_{\text{ps}}$ dominates $\bar{y}_{\text{wc}}$, except when the respondent sample sizes m_j and the between cell variance of Y are small. See Little (1986) for details.

An interesting variation on $\bar{y}_{\text{ps}}$ occurs when the adjustment cells are defined by the joint levels of two cross-classifying factors X_1 and X_2, with J and K levels, respectively. Suppose that n_{jk} units of N_{jk} in the population are sampled in the cell with $X_1 = j$, $X_2 = k$, for $j = 1, \ldots, J$, $k = 1, \ldots, K$. The value of a variable Y is recorded for m_{jk} out of the n_{jk} sampled units in each cell. The poststratified and weighting cell estimators take the form

$$\bar{y}_{\text{ps}} = \sum_{j=1}^{J} \sum_{k=1}^{K} N_{jk} \bar{y}_{jkR}/N$$

and

$$\bar{y}_{\text{wc}} = \sum_{j=1}^{J} \sum_{k=1}^{K} n_{jk} \bar{y}_{jkR}/n,$$

respectively, where $\bar{y}_{jkR}$ is the mean of responding units in subclass ($X_1 = j$, $X_2 = k$). An intermediate estimator is obtained if we assume that the population counts N_{jk} in the individual cells are not known, but the marginal counts for X_1 and X_2, $N_{j+} = \sum_{k=1}^{K} N_{jk}$ and $N_{+k} = \sum_{j=1}^{J} N_{jk}$, are known for all j and k, as from published census data (e.g., X_1 = sex, X_2 = race, where the marginal distributions of sex and race are available but the distribution in the sex by race table is not).

The method of *raking* applied to the cell counts n_{jk} consists in calculating estimates N_{jk}^* of N_{jk} that satisfy the marginal constraints

$$N_{j+}^* = \sum_{k=1}^{K} N_{jk}^* = N_{j+}, \qquad j = 1, \ldots, J,$$

$$N_{+k}^* = \sum_{j=1}^{J} N_{jk}^* = N_{+k}, \qquad k = 1, \ldots, K$$

and that differ from the observed counts n_{jk} by row and column factors, that is, can be expressed in the form

$$N_{jk}^* = a_j b_k n_{jk}, \qquad j = 1, \ldots, J; \quad k = 1, \ldots, K,$$

for certain row constants $\{a_j, j = 1, \ldots, J\}$ and column constants $\{b_k, k = 1, \ldots, K\}$. The N_{jk}^* table has margins equal to the known N_{j+} and N_{+k} margins but interactions equal to those in the n_{jk} table. The raked cell counts N_{jk}^* can be calculated by an iterative proportional fitting procedure, where current estimates are scaled by row or column factors to match the marginal totals N_{+k} or N_{j+}, respectively. That is, at the first step the estimators

$$N_{jk}^{(1)} = n_{jk}(N_{j+}/n_{j+}),$$

which match the row marginals N_{j+}, are calculated. Then the estimates

$$N_{jk}^{(2)} = N_{jk}^{(1)}(N_{+k}/N_{+k}^{(1)})$$

which match the column marginals N_{+k}, are constructed. Then

$$N_{jk}^{(3)} = N_{jk}^{(2)}(N_{j+}/N_{j+}^{(2)}),$$

and so on, until convergence. Convergence and statistical properties of this procedure are discussed by Ireland and Kullback (1968), who show, in particular, that the raked estimates N_{jk}^*/N of the cell proportions are optimal asymptotically normal estimates under a multinomial assumption for the cell counts n_{jk}, and as such are asymptotically equivalent to the (harder to calculate) maximum likelihood estimates under the multinomial model.

Raking the sample sizes n_{jk} yields the raked estimator of $\bar{Y}$:

$$\bar{y}_{RK} = \sum_{j=1}^{J} \sum_{k=1}^{K} N_{jk}^* \bar{y}_{jkR}/N,$$

which might be expected to have variance properties somewhere between $\bar{y}_{ps}$

and $\bar{y}_{wc}$. Note that this estimator is not defined when $m_{jk} = 0$, $n_{jk} \neq 0$, and in this situation some other estimator of the mean for that cell is required. Such estimators are discussed within the modeling approach to survey nonresponse, in Chapter 12.

4.5. IMPUTATION PROCEDURES

4.5.1. Introduction

We now discuss in general terms methods that impute (i.e., fill in) the values of items that are missing. Such methods were also discussed in Section 3.4 in the context of quick methods for multivariate data. Principal imputation methods in survey practice include:

(a) *Mean imputation*, where means from the responding units in the sample are substituted; also discussed in Sections 3.4.2 and 3.4.3. The means may be formed within cells analogous to the weighting cells formed for weighting procedures. Mean imputation then leads to estimates similar to those found by weighting, provided the sampling weights are constant within weighting classes.

(b) *Hot deck imputation* can be broadly defined as a method where an imputed value is selected from an estimated distribution for each missing value, in contrast with mean imputation, where the mean of the distribution is substituted. In most applications the empirical distribution consists of values from responding units, so that hot deck imputation involves substituting individual values drawn from similar responding units. Hot deck imputation is very common in practice and can involve very elaborate schemes for selecting units for imputation. Despite its popularity in practice, the literature on the theoretical properties of the method is very sparse; see Ernst (1980), Kalton and Kish (1981), Ford (1983), and David et al. (1986) for recent references.

(c) *Substitution* is a method for dealing with unit nonresponse at the field-work stage of the survey, by replacing nonresponding units by alternate units not selected into the sample. For example, if a household cannot be contacted, then a previously nonselected household in the same housing block may be substituted. The tendency to treat the resulting sample as complete should be resisted, since the substituted units are respondents and hence may differ systematically from nonrespondents. Hence at the analysis stage, substituted values should be regarded as imputed values of a particular type.

(d) *Cold deck imputation* replaces a missing value of an item by a constant value from an external source, such as a value from a previous realization of the same survey. As with substitution, current practice usually treats the resulting data as a complete sample, that is, ignores the consequences of imputation. Satisfactory theory for the analysis of data obtained by cold deck imputation is lacking.

(e) *Regression imputation* (also discussed in Section 3.4.3), replaces missing values by predicted values from a regression of the missing item on items observed for the unit, usually calculated from units with both observed and missing variables present. Mean imputation can be regarded as a special case of regression imputation where the predictor variables are dummy indicator variables for the cells within which the means are imputed. More generally, regression imputation is basically a modeling technique and as such will be discussed in more detail in Chapter 12.

(f) *Stochastic regression imputation* replaces missing values by a value predicted by regression imputation plus a residual, drawn to reflect uncertainty in the predicted value. With normal linear regression models, the residual will naturally be normal with zero mean and variance equal to the residual variance in the regression. With a binary outcome, as in logistic regression, the predicted value is a probability of 1 versus 0, and the imputed valued is a 1 or 0 drawn with that probability. Herzog and Rubin (1983) describe a two-stage procedure that uses stochastic regression for both normal and binary outcomes. Stochastic regression imputation is basically a modeling technique and so is discussed in Chapter 12.

(g) *Composite methods* can also be defined that combine ideas from different methods. For example, hot deck and regression imputation can be combined by calculating predicted means from a regression but then adding a residual randomly chosen from the empirical residuals to the predicted value when forming values for imputation. See, for example, the hybrid two-stage procedure attributed to Scheuren in Schieber (1978). Comparisons of composite methods with the Current Population Survey hot deck for earnings imputation are given in David et al. (1986).

(h) *Multiple imputation* methods (Rubin, 1978, 1987) impute more than one value for the missing items. An important limitation of single imputation methods is that standard variance formulas applied to the filled-in data systematically underestimate the variance of estimates, even if the model used to generate the imputations is correct. Multiple imputation methods allow valid estimates of the variance of estimates to be calculated using standard complete data procedures. They are discussed in Chapter 12 in the context of model-based methods.

4.5.2. Mean Imputation

Let y_{ij} be the value of Y for unit i in adjustment cell j, $i = 1, \ldots, N_j, j = 1, \ldots, J$. Mean imputation substitutes the mean $\bar{y}_{jR}$ of the m_j responding units in cell j for units that are sampled but that do not respond. For equally weighted designs, the population mean $\bar{Y}$ may be estimated by the mean of the observed and imputed units, namely,

$$\sum_{j=1}^{J} n_j \hat{\bar{y}}_j \Big/ \sum_{j=1}^{J} n_j,$$

where $\hat{\bar{y}}_j$ is the mean of the observed and imputed units in cell j. Now

$$\hat{\bar{y}}_j = [m_j \bar{y}_{jR} + (n_j - m_j)\bar{y}_{jR}]/n_j = \bar{y}_{jR},$$

so the resulting estimate of $\bar{Y}$ is simply the weighting cell estimator $\bar{y}_{\text{wc}}$, Eq. (4.10). If the proportions of the population in each cell are known, then the poststratified estimator $\bar{y}_{\text{ps}}$ can also be derived as an estimator based on mean imputation.

We have shown that for equally weighted designs, weighting responding units by the proportion responding in each adjustment cell yields the same estimates of the population mean and total as substituting respondent cell means for nonresponding units. The same remark applies for unequal probability designs, provided the sample weights are reflected in estimates of the proportion responding and in the imputed means. For more on the relationship between imputation and weighting, see Oh and Scheuren (1983); David et al. (1983); and Little (1986).

Mean imputation is a simple method to implement, but it has some undesirable properties, as discussed in Section 3.4.2. First, it is clear that valid estimates of the variance of $\bar{y}_{\text{wc}}$ (or $\bar{y}_{\text{ps}}$) cannot be derived from standard variance formulas applied to the filled-in data. Since the sample size is effectively reduced by nonresponse, standard variance formulas will underestimate the true variance. Second, estimates of quantities that are not linear in the data, such as the variance of Y or the correlation between a pair of variables, cannot be estimated consistently using standard complete-data methods on the completed data. Third, imputing means distorts the empirical distribution of the sampled Y values, which is important when studying the shape of the distribution of Y using histograms or other plots of the data. A related problem occurs if the values of Y are grouped into subclasses for cross-tabulation, since missing values in an adjustment cell are all replaced by a common mean value and hence are classified in the same subclass of Y. This problem suggests distributing the missing values using an imputation technique such as hot deck imputation. We now turn to this alternative method.

4.5.3. Hot Deck Imputation

With most hot deck procedures (and here we cannot be precise since the term does not have a well-defined common usage), missing values are replaced by values from similar responding units in the sample. The hot deck literally refers to the deck of matching computer cards for the donors available for a nonrespondent. Suppose as before that a sample of n out of N units is selected, and m out of the n sampled values of a variable Y are recorded, where n, N, and m are treated throughout this section as fixed. For simplicity let us label the first n units, $i = 1, \ldots, n$, as sampled, and the first $m < n$ units as respondents. Given an equal probability sampling scheme, the mean $\bar{Y}$ may be estimated as the mean of the responding and the imputed units. This may be written in the form

$$\bar{y}_{\text{HD}} = \{m\bar{y}_R + (n - m)\bar{y}^*_{\text{NR}}\}/n, \tag{4.17}$$

where $\bar{y}_R$ is the mean of the responding units, and

$$\bar{y}^*_{\text{NR}} = \sum_{i=1}^{m} \frac{H_i y_i}{n - m},$$

where H_i is the number of times y_i is used as a substitute for a missing value of Y. Note that $\sum_{i=1}^{m} H_i = n - m$, the number of missing units.

The properties of $\bar{y}_{\text{HD}}$ depend on the procedure used to generate the numbers $(H_1, \ldots, H_m)$. The simplest theory is obtained when imputed values can be regarded as selected from the values for the responding units by a probability sampling design, so that the distribution of $(H_1, \ldots, H_m)$ in repeated applications of the hot deck method is known.

For example, suppose that the H_i are defined by random sampling with replacement from the recorded values of Y. Conditioning on the sampled and the recorded values, the distribution of $(H_1, \ldots, H_m)$ in repetitions of the hot deck is multinomial with sample size $(n - m)$ and probabilities $(1/m, \ldots, 1/m)$. (See Cochran, 1977, sec. 2.8.). Hence

$$E(H_i | Y, R, I) = (n - m)/m,$$

$$\text{Var}(H_i | Y, R, I) = (n - m)(1 - 1/m)/m,$$

$$\text{Cov}(H_i, H_{i'} | Y, R, I) = -(n - m)/m^2, \qquad \text{for } i \neq i'.$$

Let $\bar{y}_{\text{HD1}}$ be the estimate (4.17) of $\bar{Y}$ for this distribution of $(H_1, \ldots, H_m)$. Then

$$E(\bar{y}_{\text{HD1}} | Y, R, I) = \bar{y}_R$$

and

$$\text{Var}(\bar{y}_{\text{HD1}} | Y, R, I) = (1 - m^{-1})(1 - m/n)s^2_{yR}/n. \tag{4.18}$$

The moments of $\bar{y}_{\text{HD1}}$ in repeated sampling from H, R, and I are

$$E(\bar{y}_{\text{HD1}} | Y) = E(E(\bar{y}_{\text{HD1}} | Y, R, I) | Y) \tag{4.19}$$

$$\text{Var}(\bar{y}_{\text{HD1}} | Y) = \text{Var}(E(\bar{y}_{\text{HD1}} | Y, R, I) | Y) + E(\text{Var}(\bar{y}_{\text{HD1}} | Y, R, I) | Y). \tag{4.20}$$

In particular, assuming simple random sampling and the MCAR response distribution of Eq. (4.5), we obtain

$$E(\bar{y}_{\text{HD}1} \mid Y) = \bar{Y}$$

$$\text{Var}(\bar{y}_{\text{HD}1} \mid Y) = (m^{-1} - N^{-1})S_y^2 + (1 - m^{-1})(1 - m/n)S_y^2/n.$$

Note that the hot deck procedure leads to an estimate with a variance larger than that of the estimate $\bar{y}_R$ obtained by mean imputation. From (4.20), the variance of any hot deck estimator $\bar{y}_{\text{HD}}$ for which $E(\bar{y}_{\text{HD}} \mid Y, R, I) = \bar{y}_R$ is greater than the variance of $\bar{y}_R$. The advantage of the hot deck method is that the imputed values do not distort the distribution of the sampled values of Y the way mean imputation does.

The added variance from sampling imputations with replacement, given by Eq. (4.18), is a nonnegligible quantity. It can be reduced by a more efficient choice of sampling scheme. In particular, suppose that imputed values are selected by sampling *without replacement*. If $(n - m) < m$, then we can select $(n - m)$ of the m recorded values of y without replacement, and $H_i = 1$ or 0 according to whether i is selected. To define the procedure in general, write

$$n - m = km + t,$$

where k is a nonnegative integer and $0 \leqslant t < m$. The hot deck without replacement selects all the recorded units k times, and then selects t additional units to yield the $(n - m)$ values required for the missing data. Thus

$$\bar{y}^*_{\text{NR}} = (km\bar{y}_R + t\bar{y}_t)/(n - m),$$

where $\bar{y}_t$ is the mean of the t supplementary values of Y. By the theory of simple random sampling,

$$E(\bar{y}_t \mid Y, R, I) = \bar{y}_R, \qquad \text{Var}(\bar{y}_t \mid Y, R, I) = (1 - t/m)s_{yR}^2/t.$$

If $\bar{y}_{\text{HD}2}$ is the estimate of $\bar{Y}$ from this procedure, then

$$\bar{y}_{\text{HD}2} = (k + 1)mn^{-1}\bar{y}_R + tn^{-1}\bar{y}_t,$$

and

$$E(\bar{y}_{\text{HD}2} \mid Y, R, I) = \bar{y}_R,$$

$$\text{Var}(\bar{y}_{\text{HD}2} \mid Y, R, I) = (t/n)(1 - t/m)s_{yR}^2/n, \tag{4.21}$$

which is always smaller than the corresponding additional variance component of $\bar{y}_{\text{HD}1}$ given by (4.18). Specifically, if we assume simple random sampling and a Bernoulli response mechanism, and ignore the finite population correction, then (1) the proportionate variance increase of $\bar{y}_{\text{HD}1}$ over $\bar{y}_R$ is at most 0.25, and this maximum is attained when $m/n = 0.5$; (2) the proportionate variance increase of $\bar{y}_{\text{HD}2}$ over $\bar{y}_R$ is at most 0.125, and this maximum is attained when $k = 0$, $t = n/4$, and $m = 3n/4$; that is, a quarter of the values of y are missing (Kalton and Kish, 1981).

Another method for generating imputations is the *sequential* hot deck, where responding and nonresponding units are treated in a sequence, and a missing value of Y is replaced by the nearest responding value preceding it in the sequence. For example, if $n = 6$, $m = 3$, y_1, y_4, and y_5 are present and y_2, y_3, and y_6 are missing, then y_2 and y_3 are replaced by y_1, and y_6 is replaced by y_5. If y_1 is missing, then some starting value is necessary, perhaps chosen by a cold deck procedure. The main advantage of the sequential hot deck procedure is computational simplicity. It formed the basis for early imputation schemes for the Census Bureau's Current Population Survey.

Suppose sampled units are regarded as randomly ordered, units are selected by simple random sampling, and a Bernoulli nonresponse mechanism is operating. Then Bailar, Bailey, and Corby (1978) show that the sequential hot deck estimate of $\bar{Y}$, say $\bar{y}_{\mathrm{HD3}}$, is unbiased for $\bar{Y}$ with variance (for large m and n and ignoring finite population corrections) given by

$$\mathrm{Var}(\bar{y}_{\mathrm{HD3}} | y) = (S_y^2/m)(1 + (n - m)/n).$$

Hence the proportionate increase in variance over $\bar{y}_R$ is $(n - m)/n$, the fraction of missing data.

Reductions in the additional variance from hot deck imputation can be achieved by selecting substitutes for missing units by using the y values themselves to form sampling strata (Bailar and Bailar, 1983; Kalton and Kish, 1981). The extreme form of stratification would be to order the recorded values of Y and then sample the t units by systematic sampling from this list.

The hot deck estimators we have discussed so far are unbiased only under the generally unrealistic assumption that the probability of response is not related to the values of Y. If covariate information is available for responding and nonresponding units, then this information may be used to reduce nonresponse bias. Two approaches are worthy of mention:

(a) *Hot Deck within Adjustment Cells.* Adjustment cells may be formed, and missing values within each cell replaced by recorded values in the same cell. Considerations relating to the choice of cells are similar to those in the choice of weighting cells for weighting estimates. The mean and variance of resulting hot deck estimates of $\bar{Y}$ can be found by applying previous formulas separately in each cell and then combining over cells. Since adjustment cells are formed from the joint levels of categorical variables, they are not ideal for interval scaled variables.

(b) *Nearest Neighbor Hot Deck.* Another approach is to define a metric to measure distance between units, based on the values of covariates, and then to choose imputed values that come from responding units close to the unit with the missing value. For example, let $x_{i1}, \ldots, x_{iJ}$ be the values of J appropriately scaled covariates for a unit i for which y_i is missing. Define the distance

$$d(i, i') = \max_j |x_{ij} - x_{i'j}|$$

between units i and i'. We might choose an imputed value for y_i from those units i' that are such that (1) $y_{i'}, x_{i'1}, \ldots, x_{i'J}$ are observed, and (2) $d(i, i')$ is less than some value d_0. The number of candidates i' can be controlled by varying the value of d_0. In Sande (1983), candidates are further restricted to satisfy certain logical constraints of the data (e.g., no negative ages). Such nearest neighbor schemes require considerable computing power, and their use is fairly recent in development. Substantial statistical literature on matching methods exists in the context of observational studies where treated units are matched to control units (Rubin, 1973a, b; Cochran and Rubin, 1973; Rubin, 1976a, b). Since imputed values are relatively complex functions of the responding items, quasi-randomization properties of estimates derived from such matching procedures remain largely unexplored.

EXAMPLE 4.3. *A Sequential Hot Deck Ordered by a Covariate* Colledge et al. (1978) present a case study where the hot deck method is used extensively in a Canadian survey of construction firms. The survey involved 50,538 firms, of which 41,432 were retained for the analysis. The survey items are divided into four groups: (a) Fully observed *key fields* from the tax return, including region, standard industrial classification (SIC), gross business income (GBI), net business income (NBI), and salary and wages indicator (SWI); (b) *basic financial* variables from the tax return, which were sometimes missing; (c) secondary financial variables; and (d) *survey* variables collected for different but overlapping subsamples, and sometimes missing. Only 908 of the 41,432 records provided information in all four variable groups; most of the records, 34,181, had only key fields observed; 2,316 records had key fields and basic financial variables observed, and 4,027 had key fields and survey variables observed. Hot deck imputation was conducted in several phases, where each phase replaced missing items in a particular group by the values of donor records for which all the items in that group were observed. In order to match donors with candidates, the field of all records was poststratified by province (or region), SIC, and SWI. The collection of donors (i.e., the hot deck) and the collection of candidates were identified for the particular phase. Within each poststratum, the records were ordered by GBI. In order to impute values for a particular candidate record, only the nearest 5 potential donors on either side were considered, yielding 10 possible donors with approximately the same value of GBI. From these 10 donors, one was chosen to minimize a distance function of the basic form

$$\mathrm{DIST}(c, d) = |\ln \mathrm{TEXP}_c - \ln \mathrm{TEXP}_d|,$$

where TEXP = GBI − NBI = total expenses, and the subscript c denotes candidate and d denotes donor. TEXP was used because many of the fields to be imputed were detailed expense breakdowns or were highly correlated with expenses. Note that the matching of donor to candidate was based entirely on key fields, which are all observed. More generally, DIST was expanded to depend on fields other than TEXP, and modified to spread donor usage by making it an increasing function of the number of times the potential donor d has already been used as an actual donor in the phase.

After a donor had been identified, the candidate missing fields for the group of variables in this phase were replaced by the corresponding fields from the donor record. Some adjustment or transformation was sometimes necessary to ensure that certain edit constraints were satisfied. For example, suppose three fields x, y, and z have to satisfy $x + y \leqslant z$, with x, y, and z nonnegative. The donor's values for these fields are x_d, y_d, and z_d, whereas the candidate has only z_c present. If the values x_d and y_d are simply written into the corresponding candidate's fields, we may find that $x_d + y_d > z_c$, which violates the edit constraint. In this situation x_d and y_d were prorated to satisfy the edit constraint, by substituting

$$x_c = (x_d/z_d)z_c; \qquad y_c = (y_d/z_d)z_c.$$

In other words, the proportions x_d/z_d and y_d/z_d rather than the values (x_d, y_d) were transferred to the candidate.

4.6 ESTIMATION OF SAMPLING VARIANCE IN THE PRESENCE OF NONRESPONSE

Most of our discussion of survey methods so far has focused on deriving estimates of population quantities in the presence of nonresponse. In this section we focus on the question of deriving estimates of sampling variance that incorporate the added variance due to nonresponse from the quasi-randomization perspective.

It is important to emphasize that in many applications the issue of nonresponse bias is often more crucial than that of variance. In fact, it has been argued that providing a valid estimate of sampling variance is worse than providing no estimate if the estimator has a large bias, which dominates the mean squared error. The variance estimates presented here all effectively assume that the method of adjustment for nonresponse has succeeded in eliminating nonresponse bias.

Explicit variance formulas that allow for nonresponse have generally been confined to simple random sampling, or simple random sampling within strata. Examples for weighting estimators have been given in Section 4.4. The added variance of hot deck procedures where the selection of imputed values is by a simple probability sampling scheme has been discussed in Section 4.5. There seems to be scope for further development in this area, although it is doubtful whether explicit estimators can be found for complicated sequential hot deck methods such as the one described in Example 4.3, except under overly simplified assumptions.

Even with complete response, the calculation of consistent estimates of variance is not a simple task for the complex sample designs often used in practice. As a result, approximate methods have been developed that can be applied to broad classes of sample designs. The simplicity of these methods results from restricting calculations to quantities calculated for collections of the sampled units known as *ultimate clusters* (UC's). These are the largest sampling units that are subsampled from the population. For example, the first stage of a design to sample households may involve the selection of census enumeration areas (EA's). The sample may include some self-representing EA's, which are included in the sample with probability 1, and some non-self-representing EA's, which are sampled from the population of EA's. The ultimate clusters then consist of the non-self-representing EA's and the sampling units that form the first stage of subsampling of self-representing EA's.

Estimates of variance calculated from estimates for UC's are based on the following lemma:

Lemma. Let $\hat{\theta}_1, \ldots, \hat{\theta}_k$ be random variables that are (1) uncorrelated and (2) have common mean μ. Let

$$\bar{\theta} = \sum_{j=1}^{k} \frac{\hat{\theta}_j}{k}$$

$$\hat{v}(\bar{\theta}) = \sum_{j=1}^{k} \frac{(\hat{\theta}_j - \bar{\theta})^2}{k(k-1)}.$$

Then (i) is an unbiased estimate of μ, and (ii) $\hat{v}(\bar{\theta})$ is an unbiased estimate of the variance of $\bar{\theta}$.

PROOF. $E(\bar{\theta}) = \sum_{j=1}^{k} E(\hat{\theta}_j)/k = \mu$, proving (i). To show (ii), note that

$$\sum_{j=1}^{k} (\hat{\theta}_j - \bar{\theta})^2 = \sum_{j=1}^{k} (\hat{\theta}_j - \mu)^2 - k(\bar{\theta} - \mu)^2.$$

Hence

$$E\left(\sum_{j=1}^{k}(\hat{\theta}_j-\bar{\theta})^2\right)-k(k-1)\,\mathrm{Var}(\bar{\theta})=\sum_{j=1}^{k}\mathrm{Var}(\hat{\theta}_j)-k\,\mathrm{Var}(\bar{\theta})-k(k-1)\,\mathrm{Var}(\bar{\theta})$$

$$=\sum_{j=1}^{k}\mathrm{Var}(\hat{\theta}_j)-k^2\,\mathrm{Var}(\bar{\theta}). \tag{4.22}$$

But

$$k^2\,\mathrm{Var}(\bar{\theta})=\mathrm{Var}\left(\sum_{j=1}^{k}\hat{\theta}_j\right)=\sum_{j=1}^{k}\mathrm{Var}(\hat{\theta}_j),$$

since the estimates $\hat{\theta}_j$ are uncorrelated. Hence the expression (4.22) equals zero, proving (ii). This lemma can be applied directly to linear estimators for sample designs that involve random sampling *with replacement* (rswr) of UC's. The most important case in practice is given by the next example.

EXAMPLE 4.4. *Standard Errors From Cluster Samples.* Suppose the population consists of K UC's, and the sample design includes k UC's by simple random sampling with replacement. Let t_j denote the total for a variable Y for UC j, and suppose we estimate the population total

$$t=\sum_{j=1}^{K}t_j$$

by the Horvitz–Thompson estimate

$$\hat{t}=\sum_{j=1}^{k}\hat{t}_j/\pi_j,$$

where the sum is over selected UC's (say, $1,\ldots,k$), $\hat{t}_j$ is an unbiased estimate of t_j, and π_j is the probability that UC j is selected. Then (1) $\hat{t}$ and $\{k\hat{t}_j/\pi_j, j=1,\ldots,k\}$ are all unbiased estimates of t, and (2) the estimates $\{k\hat{t}_j/\pi_j, j=1,\ldots,k\}$ are uncorrelated, by the method of sampling. Hence by the lemma,

$$\hat{v}(\hat{t}\mid Y)=\sum_{j=1}^{k}\frac{(k\hat{t}_j/\pi_j-\hat{t})^2}{k(k-1)} \tag{4.23}$$

is an unbiased estimator of the variance of $\hat{t}$.

Suppose in this example we have missing data and we derive estimates $\hat{t}_j$ of the UC totals by one of the missing-data techniques discussed earlier. Then we can still use (4.23) to estimate the variance, provided (1) the estimates $\hat{t}_j$ are unbiased over the distribution of I and R, that is, the imputation or weighting procedure does not lead to nonresponse bias; and (2) the imputations or weighting adjustments are carried out *independently within each UC*, so that estimates $\hat{t}_j$ remain uncorrelated. Thus in order to apply the lemma, adjustment cells must not cut across the ultimate clusters. This principle may lead to unacceptably small adjustment cells, particularly if the number of UC's is large. Thus the requirement of a valid estimate of variance may conflict with the need for an estimator with an acceptably small bias, at least using the techniques discussed thus far. The conflict is in a sense analogous to that which occurs in the choice of sample design, where systematic sampling may be the most efficient form of stratification but does not allow valid estimates of variance to be calculated without modeling assumptions.

In practice UC's are rarely sampled with replacement. If they are selected by simple random sampling without replacement (srswor), then UC estimates are negatively correlated, and estimates such as (4.23) based on the lemma overestimate the variance. We might hope that multiplication by the finite population correction $(1 - k/K)$ would correct the overestimate, but in fact this leads to an underestimate. An unbiased estimate requires information from the second and higher stages of sampling. Thus simple variance estimates based on UC's require that the proportion of UC's sampled is small, so that the overestimation introduced by sampling without replacement can be ignored. This is often the case in practical sample designs.

Most sampling designs also involve stratification in the selection of UC's. Again assuming that the proportion of UC's sampled in each stratum is small, valid estimates of the variance of linear statistics can be derived from UC estimates. Suppose that there are H strata, and let $\hat{t}_{hj}$ be an unbiased estimate of the total t_{hj} for UC j in stratum h, for $h = 1, \ldots, H, j = 1, \ldots, K_h$.

We can estimate t by

$$\hat{t} = \sum_{h=1}^{H} \sum_{j=1}^{k_h} \frac{\hat{t}_{hj}}{\pi_{hj}} = \sum_{h=1}^{H} \hat{t}_h,$$

say, where the summations are over the H strata and the k_h units sampled in stratum h, π_{hj} is the probability of selection of UC hj in stratum h, and $\hat{t}_h$ is the estimate of the total for stratum h. The variance of $\hat{t}$ is estimated by

$$\hat{v}(\hat{t} \mid y) = \sum_{h=1}^{H} \sum_{j=1}^{k_h} \frac{(k_h \hat{t}_{hj}/\pi_{hj} - \hat{t}_h)^2}{k_h(k_h - 1).}$$

In particular, with two UC's selected in each stratum, an especially popular design, the estimate of variance is

$$\hat{v}(\hat{t} \mid y) = \sum_{h=1}^{H} \frac{(\hat{t}_{h1}/\pi_{h1} - \hat{t}_{h2}/\pi_{h2})^2}{4}.$$

Conditions for using these estimates with imputed data are the same as those for random sampling. That is, imputations must be carried out independently in each UC.

Nonlinear estimators have been treated in the preceding framework by initial linearization using Taylor series methods, or other approximate techniques such as the jackknife, the bootstrap, or balanced repeated replication. For more information on these methods and references, see Cochran (1977) and Wolter (1985).

REFERENCES

Bailar, B. A., and Bailar, J. C. (1983). Comparison of the biases of the "hot deck" imputation procedure with an "equal weights" imputation procedure, in *Incomplete Data in Sample Surveys, Vol. III: Symposium on Incomplete Data, Proceedings* (W. G. Madow and I. Olkin, Eds.). New York: Academic Press.

Bailar, B. A., Bailey, L., and Corby, C. (1978). A comparison of some adjustment and weighting procedures for survey data, *American Statistical Association* 1978, *Proceedings of the Survey Research Methods Section*, pp. 175–200.

Cassel, C. M., Särndal, C. E., and Wretman, J. H. (1983). Some uses of statistical models in connection with the nonresponse problem, in *Incomplete Data in Sample Surveys, Vol. III: Symposium on Incomplete Data, Proceedings* (W. G. Madow and I. Olkin, Eds.) New York: Academic Press.

Cochran, W. G. (1963). *Sampling Techniques*, 2nd ed. New York: Wiley.

Cochran, W. G. (1977). *Sampling Techniques*, 3rd ed. New York: Wiley.

Cochran, W. G., and Rubin, D. B. (1973). Controlling bias in observational studies: A review, *Sankhya* **A35**, 417–446.

Colledge, M. J., Johnson, J. H., Paré, R., and Sande, I. G. (1978). Large scale imputation of survey data, *American Statistical Association* 1978, *Proceedings of the Survey Research Methods Section*, pp. 431–436.

David, M. H., Little, R. J. A., Samuhel, M. E., and Triest, R. K. (1983). Imputation methods based on the propensity to respond, *American Statistical Association* 1983, *Proceedings of the Business and Economics Section*.

David, M. H., Little, R. J. A., Samuhel, M. E., and Triest, R. K. (1986), Alternative methods for CPS income imputation, *J. Am. Statist. Associ.*, **81**, 29–41.

Dawid, A. P. (1979). Conditional independence in statistical theory (with discussion), *J. Roy. Statist. Soc.* **B41**, 1–31.

Ernst, L. R. (1980). Variance of the estimated mean for several imputation procedures, *American Statistical Association* 1980, *Proceedings of the Survey Research Methods Section*, pp. 716–720.

Ford, B. N. (1983). An overview of hot deck procedures, in *Incomplete Data in Sample Surveys, Vol. II: Theory and Annotated Bibliography* (W. G. Madow, I. Olkin, and D. B. Rubin, Eds.). New York: Academic Press.

Hajek, J. (1960). Limiting distributions in simple random sampling from a finite population, *Pub. Math. Inst. Hung. Acad. Sci.* **4**, 49–57.

Hansen, M. H., Hurwitz, W. N., and Madow, W. G. (1953). *Sample Survey Methods and Theory*. Two volumes. New York: Wiley.

Herzog, T. N., and Rubin, D. B. (1983). Using multiple imputations to handle nonresponse in sample surveys, in *Incomplete Data in Sample Surveys, Vol. II: Theory and Annotated Bibliography* (W. G. Madow, I. Olkin, and D. B. Rubin, Eds.). New York: Academic Press.

Horvitz, D. G., and Thompson, D. J. (1952). A generalization of sampling without replacement from a finite population, *J. Am. Statist. Associ.* **47**, 663–685.

Ireland, C. T., and Kullback, S. (1968). Contingency tables with given marginals, *Biometrika* **55**, 179–188.

Kalton, G., and Kish, L. (1981). Two efficient random imputation procedures. *American Statistical Association* 1981, *Proceedings of the Survey Research Methods Section*, pp. 146–151.

Little, R. J. A. (1986). Survey nonresponse adjustments, *Int. Statist. Rev.* **54**, 139–157.

Madow W. G., Olkin, I., Nisselson, H., and Rubin, D. B. (eds.) (1983). *Incomplete Data in Sample Surveys*, three volumes. New York: Academic Press.

Oh, H. L., and Scheuren, F. S. (1983). Weighting adjustments for unit nonresponse, in *Incomplete Data in Sample Surveys, Vol. II: Theory and Annotated Bibliography* (W. G. Madow, I. Olkin, and D. B. Rubin, eds.). New York: Academic Press.

Rosenbaum, P. R., and Rubin, D. B. (1983). The central role of the propensity score in observational studies for causal effects, *Biometrika* **70**, 41–55.

Rosenbaum, P. R., and Rubin, D. B. (1985), Constructing a control group using multivariate matched sampling incorporating the propensity score. *Am. Statist.* **39**, 33–38.

Rubin, D. B. (1973a). Matching to remove bias in observational studies, *Biometrics* **29**, 159–183.

Rubin, D. B. (1973b). The use of matched sampling and regression adjustment to remove bias in observational studies, *Biometrics* **29**, 185–203.

Rubin, D. B. (1976a). Multivariate matching methods that are equal percent bias reducing, I: Some examples, *Biometrics* **32**, 109–120. Printer's correction note p. 955.

Rubin, D. B. (1976b). Multivariate matching methods that are equal percent bias reducing, II: Maximums on bias reduction for fixed sample sizes," *Biometrics* **32**, 121–132. Printer's correction note p. 955.

Rubin, D. B. (1978). Multiple imputations in sample surveys, *American Statistical Association* 1978, *Proceedings of the Survey Research Methods Section*, pp. 20–34.

Rubin, D. B. (1987). *Multiple Imputation for Nonresponse in Surveys*. New York: Wiley.

Sande, I. G. (1983). Hot deck imputation procedures, in *Incomplete Data in Sample Surveys, Vol. III: Symposium on Incomplete Data, Proceedings* (W. G. Madow and I. Olkin, Eds.). New York: Academic Press.

Schieber, S. J. (1978). A comparison of three alternative techniques for allocating unreported social security income on the Survey of the Low-Income Aged and Disabled, *American Statistical Association* 1978, *Proceedings of the Survey Research Methods Section*, pp. 212–218.

Wolter, K. M. (1984). *Introduction to Variance Estimation*. New York: Springer-Verlag.

PROBLEMS

Data for Problems 1–3: A simple random sample of 100 individuals in a county are interviewed for a health survey, yielding the following data:

Age Group	Sample Size	Number of Respondents	Cholesterol Mean	Cholesterol S.D.
20–30	25	22	220	30
30–40	35	27	225	35
40–50	28	16	250	44
50–60	12	5	270	41

1. Compute the mean cholesterol for the respondent sample and its standard error. Assuming normality, compute a 95% confidence interval for the mean cholesterol for respondents in the county. Can this interval be applied to all individuals in the county?

2. Compute the weighting cell estimate of the mean cholesterol level in the population and its estimated mean squared error from the formula below (4.12). Hence construct an approximate 95% confidence interval for the population mean and compare it with the result of Problem 1. What assumptions are made about the nonresponse mechanism?

3. Suppose census data yield the following age distribution for the county of interest in (1): 20–30: 20%; 30–40: 40%; 40–50: 30%; 50–60: 10%. Calculate the poststratified estimate of mean cholesterol, its associated standard error, and a 95% confidence interval for the population mean.

4. Calculate Horvitz–Thompson and weighting cell estimators in the following artificial example of a stratified random sample, where the x_i and y_i values displayed are observed, the selection probabilities π_i are known, and the response probabilities ϕ_i are known for the Horvitz–Thompson estimator but unknown for the weighting class estimators. Note that various weighting cell estimators could be created.

Sample Values

x_i	1	2	3	4	5	6	7	8	9	10
y_i	1	4	3	2	6	10	14	?	?	?
π_i	.1	.1	.1	.1	.1	.5	.5	.5	.5	.5
ϕ_i	1	1	1	.9	.9	.8	.7	.6	.5	.1

5. Apply the Cassell, Särndal, and Wretman estimator of Section 4.4.3 to the data of Problem 4. Comment on the resulting weights as compared with those of the weighting cell estimator.

6. The following table shows respondent means of an incomplete variable Y (say, income in \$1000), and response rates (respondent sample size/sample size), classified by three fully observed covariates: Age (<30, >30), marital status (single, married), and gender (male, female). Note that adjustment cells cannot be based on age, marital status, and gender, since there is one cell with four units sampled, none of which respond. Calculate the following estimates of the mean of Y, both for the whole population and for the subpopulation of males:

(a) The unadjusted mean based on complete cases.
(b) The weighted mean from response propensity stratification, with three strata defined by combining cells in the table with response rates less than 0.4, between 0.4 and 0.8, and greater than 0.8.
(c) The mean from mean imputation within adjustment cells defined as for (b). Explain why adjusted estimates are higher than the unadjusted estimates.

Respondent Means and Response Rates, Classified by Age, Marital Status, and Gender

	Marital Status		Marital Status	
Age	Single	Married	Single	Married
<30	20.0	21.0	16.0	16.0
	24/25	5/16	11/12	2/4
>30	30.0	36.0	18.0	—
	15/20	2/10	8/12	0/4
	Gender-Male		Gender-Female	

7. Oh and Scheuren (1983) propose an alternative to the raked estimate $\bar{y}_{RK}$ in Section 4.4.3, where the estimated counts N_{jk}^* are found by raking the *respondent* sample sizes $\{m_{jk}\}$ instead of $\{n_{jk}\}$. Show that (i) unlike $\bar{y}_{RK}$, this estimator exists when $m_{jk} = 0$, $n_{jk} \neq 0$, and (ii) the estimator is biased unless the expectation of m_{jk}/n_{jk} can be written as a product of row and column effects.

8. Show that raking the cell sample sizes and raking the cell respondent sample sizes yields the same answer if and only if

$$p_{ij}p_{kl}/(p_{il}p_{jk}) = 1 \qquad \text{for all } i, j, k, \text{ and } l,$$

where p_{ij} is the response rate in cell (i, j) of the table.

9. Compute raked estimates of the cell counts from the sample counts and respondent counts in (a) and (b) below, using population marginal counts in (c):

(a) sample $\{n_{jk}\}$			(b) respondent $\{m_{jk}\}$			(c) population $\{N_{jk}\}$		
8	10	18	5	9	14	?	?	300
15	17	32	5	8	13	?	?	700
23	27	50	10	17	27	500	500	1000

10. For the data in Problem 9, compute the odds ratio of response rates discussed in Problem 8. Repeat the computation with the respondent counts 5 and 8 in the second row of (b) in Problem 8 interchanged. By comparing these odds ratios, predict which set of the raked respondent counts will be closer to the raked sample counts. Then compute the raked counts for (b) with the modified second row and check your prediction.

11. From the variance formulas for $\bar{y}_{\mathrm{HD1}}$ and $\bar{y}_{\mathrm{HD2}}$, prove the results in Section 4.5.3 about the maximum proportional increase in variance of these hot deck procedures over the variance of the respondent mean.

PART II

Likelihood-Based Approaches to the Analysis of Missing Data

CHAPTER 5

Theory of Inference Based on the Likelihood Function

5.1. THE COMPLETE-DATA CASE

Many methods of estimation for incomplete data can be viewed as maximizing the likelihood function under certain modeling assumptions. In this section we review some basic theory supporting the use of the maximum likelihood estimate and describe how it is implemented in the incomplete data setting. We begin by considering maximum likelihood estimation for complete data sets. Only basic results are given and mathematical details are omitted. For more material see, for example, Cox and Hinkley (1974).

Suppose that the data are denoted by Y, where Y may be scalar, vector valued, or matrix valued according to context. The data are assumed to be generated by a model described by a probability or density function $f(Y|\theta)$, indexed by a scalar or vector parameter θ. Given the model and parameter θ, $f(Y|\theta)$ is a function of Y that gives the probabilities or densities of various Y values.

Definition 5.1. Given the data value Y, the *likelihood function* $L(\theta|Y)$ is any function of θ proportional to $f(Y|\theta)$.

Note that the likelihood function, or more briefly, the likelihood, is a function of the parameter θ for fixed Y, whereas the probability or density is a function of Y for fixed θ. In both cases, the argument of the function is written first. It is slightly inaccurate to speak of "the" likelihood function, since it really consists of a set of functions that differ by any factor (that does not depend on θ).

Definition 5.2. The *loglikelihood function* $l(\theta | Y)$ is the natural logarithm (ln) of the likelihood function $L(\theta | Y)$.

It is somewhat more convenient to work with the loglikelihood than with the likelihood in many problems.

EXAMPLE 5.1. *Univariate Normal Sample.* The joint density of n independent and identically distributed observations, $Y = (y_1, \ldots, y_n)^T$, from a normal population with mean μ and variance σ^2 is

$$f(Y|\mu, \sigma^2) = (2\pi\sigma^2)^{-n/2} \exp\left\{-\frac{1}{2}\sum_{i=1}^{n}\frac{(y_i - \mu)^2}{\sigma^2}\right\}.$$

For given Y, the loglikelihood function is

$$l(\mu, \sigma^2 | Y) = \ln[f(Y|\mu, \sigma^2)]$$

or ignoring the additive constant,

$$l(\mu, \sigma^2 | Y) = -\frac{n}{2}\ln\sigma^2 - \frac{1}{2}\sum_{i=1}^{n}\frac{(y_i - \mu)^2}{\sigma^2} \tag{5.1}$$

considered as a function of $\theta = (\mu, \sigma^2)^T$ for fixed observed data Y.

EXAMPLE 5.2. *Exponential Sample.* The joint density of n independent and identically distributed observations from the exponential distribution with mean θ is

$$f(Y|\theta) = \theta^{-n}\exp\left(-\sum\frac{y_i}{\theta}\right).$$

Hence the loglikelihood of θ is

$$\begin{aligned} l(\theta | Y) &= \ln\left\{(\theta^{-n})\exp\left(-\sum\frac{y_i}{\theta}\right)\right\} \\ &= -n\ln\theta - \sum\frac{y_i}{\theta} \end{aligned} \tag{5.2}$$

considered as a function of θ for fixed observed data Y.

EXAMPLE 5.3. *Poisson Sample.* The probability of n independent observations $Y = (y_1, \ldots, y_n)^T$ from the Poisson distribution with mean θ is

$$f(Y|\theta) = \exp(-n\theta)\theta^{\Sigma y_i}\Big/\left(\prod_{i=1}^{n} y_i!\right),$$

where $y_i! = (y_i) \times (y_i - 1) \cdots \times (1)$. Hence the loglikelihood of θ is

$$l(\theta|Y) = -n\theta + \sum y_i \ln \theta \tag{5.3}$$

EXAMPLE 5.4. *Multivariate Normal Sample.* Let $Y = (y_{ij})$, where $i = 1, \ldots, n, j = 1, \ldots, K$, be a matrix representing an independent and identically distributed sample of n observations from the multivariate normal distribution with mean vector $\mu = (\mu_1, \ldots, \mu_K)$ and covariance matrix $\Sigma = \{\sigma_{jk}, j = 1, \ldots, K; k = 1, \ldots, K\}$. Thus y_{ij} represents the value of the jth variable for the ith observation in the sample. The density of Y is

$$f(Y|\mu, \Sigma) = (2\pi)^{-(1/2)nK}|\Sigma|^{-n/2} \exp\left\{-\sum_{i=1}^{n}(y_i - \mu)\Sigma^{-1}(y_i - \mu)^{\mathrm{T}}/2\right\}, \tag{5.4}$$

where $|\Sigma|$ denotes the determinant of Σ, T denotes the transpose of a matrix or vector, and y_i denotes the row vector of values for observation i. The likelihood of $\theta = (\mu, \Sigma)$ is this expression considered as a function of μ and Σ for fixed observed Y.

Suppose that for fixed data Y, two possible values of θ are being considered, θ' and θ''. Suppose further that $L(\theta'|Y) = 2L(\theta''|Y)$; it is reasonable to say that the observed outcome Y is twice as likely under θ' as under θ''. More generally, consider a value of θ, say $\hat{\theta}$, such that $L(\hat{\theta}|Y) \geqslant L(\theta|Y)$ for all other possible θ; the observed outcome Y is then at least as likely under $\hat{\theta}$ as under any other value of θ being considered. In some sense, such a value of θ is the one that is best supported by the data. This naturally leads to interest in the value of θ that maximizes the likelihood function. Further and more formal motivation is given in Section 5.2.

Definition 5.3. A *maximum likelihood (ML) estimate* of θ is a value of θ that maximizes the likelihood $L(\theta|Y)$, or equivalently, the loglikelihood $l(\theta|Y)$.

This definition is phrased to allow the possibility of more than one ML estimate. For many important models, however, the ML estimate is unique and furthermore, the likelihood function itself is differentiable and bounded above. In such cases, the ML estimate can be found by differentiating the likelihood (or the loglikelihood) with respect to θ, setting the result equal to zero, and solving for θ. The equation

$$S(\theta|Y) \equiv \frac{\partial l(\theta|Y)}{\partial\theta} = 0$$

involved in this process is called the *likelihood equation* and the derivative of the loglikelihood, $S(\theta|Y)$, is called the *score function.* Letting d be the number of components in θ, the likelihood equation is in fact a set of d simultaneous equations, defined by differentiating $l(\theta|Y)$ with respect to all d components of θ.

EXAMPLE 5.5. *Exponential Sample (Example 5.2 continued).* The loglikelihood for a sample from the exponential distribution is given by (5.2). Differentiating with respect to θ gives the likelihood equation

$$-\frac{n}{\theta} + \sum \frac{y_i}{\theta^2} = 0.$$

Solving for θ gives the ML estimate $\hat{\theta} = \bar{y} = \sum y_i/n$, the mean of the sampled values of Y.

EXAMPLE 5.6. *Poisson Sample (Example 5.3 continued).* The loglikelihood for the Poisson sample is given by (5.3). Differentiating with respect to θ gives the likelihood equation

$$-n + \sum y_i/\theta = 0.$$

Solving for θ gives the ML estimate $\hat{\theta} = \bar{y}$, the sample mean.

EXAMPLE 5.7. *Univariate Normal Sample (Example 5.1 continued).* From (5.1), the loglikelihood for a normal sample of n observations is

$$\begin{aligned} l(\mu, \sigma^2|Y) &= -\frac{n}{2}\ln \sigma^2 - \frac{1}{2}\sum_{i=1}^{n} \frac{(y_i - \mu)^2}{\sigma^2} \\ &= -\frac{n}{2}\ln \sigma^2 - \frac{1}{2}\frac{n(\bar{y} - \mu)^2}{\sigma^2} - \frac{1}{2}\frac{ns^2}{\sigma^2}, \end{aligned}$$

where $s^2 = n^{-1}\sum_{i=1}^{n}(y_i - \bar{y})^2$, the sample variance (with divisor n rather than $n - 1$, that is, uncorrected for degrees of freedom). Differentiating with respect to μ and setting the result equal to zero at $\mu = \hat{\mu}$ and $\sigma^2 = \hat{\sigma}^2$ gives

$$(\bar{y} - \hat{\mu})/\hat{\sigma}^2 = 0,$$

which implies that $\hat{\mu} = \bar{y}$. Differentiating with respect to σ^2 and setting the result equal to zero at $\mu = \hat{\mu}$ and $\sigma^2 = \hat{\sigma}^2$ gives

$$-\frac{n}{2\hat{\sigma}^2} + \frac{n(\bar{y} - \hat{\mu})^2}{2\hat{\sigma}^4} + \frac{ns^2}{2\sigma^4} = 0,$$

which, since $\hat{\mu} = \bar{y}$, implies that $\hat{\sigma}^2 = s^2$. Thus we obtain the ML estimates

$$\hat{\mu} = \bar{y}, \qquad \hat{\sigma}^2 = s^2.$$

EXAMPLE 5.8. *Multivariate Normal Sample (Example 5.4 continued).* Standard calculations in multivariate analysis (cf. Wilks, 1963; Rao, 1972; Anderson, 1965) show that maximizing (5.4) with respect to μ and Σ yields

$$\hat{\mu} = \bar{y}, \qquad \hat{\Sigma} = S,$$

where $\bar{y} = (\bar{y}_1, \ldots, \bar{y}_k)$ is the row vector of sample means, and $S = (s_{jk})$, the $(k \times k)$ sample covariance matrix with (j, k)th element $s_{jk} = n^{-1} \sum_{i=1}^{n} (y_{ij} - \bar{y}_j)(y_{ik} - \bar{y}_k)$.

Property 5.1. Let $g(\theta)$ be a one–one function of the parameter θ. Then the ML estimate of $g(\theta)$ is $g(\hat{\theta})$, the function evaluated at the ML estimate $\hat{\theta}$ of θ.

Property 5.1 is important for many problems and follows trivially from noting that the likelihood function of $\phi = g(\theta)$ is $L(g^{-1}(\phi) | Y)$, which is maximized when $\phi = g(\hat{\theta})$.

EXAMPLE 5.9. *A Conditional Distribution Derived from a Bivariate Normal Sample.* The data consist of n independent and identically distributed observations (y_{i1}, y_{i2}), $i = 1, \ldots, n$ from the bivariate normal distribution with mean (μ_1, μ_2) and covariance matrix

$$\Sigma = \begin{bmatrix} \sigma_{11} & \sigma_{12} \\ \sigma_{12} & \sigma_{22} \end{bmatrix}.$$

As in Example 5.4, the ML estimates are

$$\hat{\mu}_j = \bar{y}_j, \qquad j = 1, 2.$$

$$\hat{\sigma}_{jk} = s_{jk}, \qquad j, k = 1, 2,$$

where $\bar{y}_1$ and $\bar{y}_2$ are the sample means and $S = (s_{jk})$ is the sample covariance matrix with divisor n. By properties of the bivariate normal distribution, the conditional distribution of y_{i2} given y_{i1} is normal with mean $\mu_2 + \beta_{21\cdot1}(y_{i1} - \mu_1)$ and variance $\sigma_{22\cdot1}$, where

$$\beta_{21\cdot1} = \sigma_{12}/\sigma_{11} \qquad \text{and} \qquad \sigma_{22\cdot1} = \sigma_{22} - \sigma_{12}^2/\sigma_{11}$$

are, respectively, the slope and residual variance from the regression of y_{i2} on y_{i1}. By Property 5.1, the ML estimates of these quantities are

$$\hat{\beta}_{21\cdot1} = \hat{\sigma}_{12}/\hat{\sigma}_{11} = s_{12}/s_{11},$$

the least squares estimate of the slope, and

$$\hat{\sigma}_{22\cdot1} = \hat{\sigma}_{22} - \hat{\sigma}_{12}^2/\hat{\sigma}_{11} = s_{22\cdot1},$$

where $s_{22\cdot1} = n^{-1} \sum_{i=1}^{n} \{y_{i2} - \bar{y}_2 - \hat{\beta}_{21\cdot1}(y_{i1} - \bar{y}_1)\}^2$, the residual sum of squares from the regression based on the n sampled observations divided by n.

The ML estimates of $\beta_{21\cdot1}$ and $\sigma_{22\cdot1}$ can also be derived directly from the likelihood based on the conditional distribution of the y_{i2} given the y_{i1}. The connection between ML estimation for the normal linear regression model and least squares applies more generally, as discussed in the next example.

EXAMPLE 5.10. *Multiple Linear Regression.* The data consist of n observations $(y_i, x_{i1}, \ldots, x_{ip})$; $i = 1, \ldots, n$ on an outcome variable y and p predictor variables. We assume that, given $x_i = (x_{i1}, \ldots, x_{ip})$, the values of y_i are independent normal random variables with mean $\beta_0 + \sum_{j=1}^{p} \beta_j x_{ij}$ and variance σ^2. The loglikelihood of $\theta = (\beta_0, \beta_1, \ldots, \beta_p, \sigma^2)$ given observed data y_i, x_i; $i = 1, \ldots, n$ is

$$-\frac{n}{2}\ln\sigma^2 - \frac{1}{2}\sigma^{-2}\sum_{i=1}^{n}(y_i - \beta_0 - \beta_1 x_{i1} - \cdots - \beta_p x_{ip})^2. \tag{5.5}$$

Maximizing this expression with respect to θ, the ML estimates of $(\beta_0, \ldots, \beta_p)$ are found to be the least squares estimates of the intercept and the regression coefficients. The ML estimate of σ^2 is s^2, where ns^2 is the residual sum of squares from the least squares regression. Thus, as before, the ML estimate of σ^2 does not correct for the loss of degrees of freedom in estimating the $p + 1$ location parameters.

5.2. INFERENCE BASED ON ML ESTIMATES

5.2.1. Interval Estimation

In this section we outline some basic properties of ML estimates. Details can be found in texts such as Rao (1972) and Cox and Hinkley (1974).

Let $\hat{\theta}$ denote an ML estimate of θ based on observed data Y. The most important practical property of $\hat{\theta}$ is that in many cases, especially with large samples, the following approximation can be applied.

Approximation 5.1.

$$(\theta - \hat{\theta}) \sim N(0, C), \tag{5.6}$$

where C is the $d \times d$ covariance matrix for $(\theta - \hat{\theta})$. When introducing (5.6), Bayesian statisticians treat $\hat{\theta}$, the observed ML estimate, as fixed and θ as the random variable, whereas frequentist statisticians treat θ as fixed and unknown and the estimate $\hat{\theta}$ as a random variable. Both perspectives effectively treat C as fixed and of course condition on the model specification for $f(\cdot|\cdot)$ underlying the likelihood.

The frequency interpretation of (5.6) is that, under $f(\cdot|\cdot)$ in repeated samples, $\hat{\theta}$ will be approximately normally distributed with mean equal to the true value of θ and covariance matrix C, which has lower order variability than $\hat{\theta}$. The Bayesian interpretation of (5.6) is that conditional on $f(\cdot|\cdot)$ and on the observed values of the data, the posterior distribution of θ is normal with mean $\hat{\theta}$ and covariance matrix C, where $\hat{\theta}$ and C are statistics fixed at their observed values.

The Bayesian justification of Approximation 5.1 is based on a Taylor series expansion of the loglikelihood about the ML estimate, namely,

$$l(\theta|Y) = l(\hat{\theta}|Y) + (\theta - \hat{\theta})^{\mathrm{T}} S(\hat{\theta}|Y) - \tfrac{1}{2}(\theta - \hat{\theta})^{\mathrm{T}} I(\hat{\theta}|Y)(\theta - \hat{\theta}) + r(\theta|Y),$$

where $S(\hat{\theta}|Y)$ is the score function and $I(\theta|Y)$ is the observed information, defined by

$$I(\theta|Y) = -\frac{\partial^2 l(\theta|Y)}{\partial\theta\,\partial\theta}.$$

By definition, $S(\hat{\theta}|Y) = 0$. Hence, provided the remainder term $r(\theta|Y)$ can be neglected, and the prior distribution of θ is flat in the range of θ supported by the data, the posterior distribution of θ has density

$$f(\theta|Y) \propto \exp[-\tfrac{1}{2}(\theta - \hat{\theta})^{\mathrm{T}} I(\hat{\theta}|Y)(\theta - \hat{\theta})],$$

which is the normal distribution of Approximation 5.1 with covariance matrix

$$C = I^{-1}(\hat{\theta}|Y),$$

the inverse of the observed information evaluated at $\hat{\theta}$. More details are provided in Lindley (1965).

The frequentist justification for Approximation 5.1 involves a more complex argument. We first approximate $S(\hat{\theta}|Y)$ about the true value of θ by the first term of a Taylor series:

$$0 = S(\hat{\theta}|Y) = S(\theta|Y) - I(\theta|Y)(\hat{\theta} - \theta) + r(\hat{\theta}|Y).$$

If the remainder term $r(\hat{\theta}|Y)$ is negligible, we have

$$S(\theta|Y) \simeq I(\theta|Y)(\hat{\theta} - \theta).$$

Under certain regularity conditions, it can be shown by a central limit theorem that in repeated sampling $S(\theta|Y)$ is asymptotically normal with mean 0 and covariance matrix

$$J(\theta) = E(I(\theta|y)|\theta) = \int I(\theta|y) f(y|\theta)\,dy,$$

which is called the expected information matrix. A version of the law of large numbers implies that

$$J(\theta) \simeq J(\hat{\theta}) \simeq I(\hat{\theta}|Y).$$

Combining these facts leads to Approximation 5.1, with covariance matrix

$$C = J^{-1}(\hat{\theta}),$$

the inverse of the expected information evaluated at $\theta = \hat{\theta}$.

In any case, Approximation 5.1, with C fixed at $I^{-1}(\hat{\theta}|Y)$ or $J^{-1}(\hat{\theta})$ or some other approximation, can be used to provide interval estimates for θ. For example, 95% intervals for scalar θ are given by

$$\hat{\theta} \pm 1.96C^{1/2}, \tag{5.7}$$

where 1.96 can often be replaced by 2 in practice. For vector θ, 95% ellipsoids are given by the inequality

$$(\theta - \hat{\theta})^{\mathrm{T}} C^{-1}(\theta - \hat{\theta}) \leqslant \chi^2_{0.95,d} \tag{5.8}$$

where $\chi^2_{0.95,d}$ is the 95th percentile of the chi-squared distribution with degrees of freedom d, the dimensionality of θ. More generally, 95% ellipsoids for $q < d$ components of θ, say $\theta_{(1)}$, can be calculated as

$$(\theta_{(1)} - \hat{\theta}_{(1)})^{\mathrm{T}} C_{(11)}^{-1}(\theta_{(1)} - \hat{\theta}_{(1)}) \leqslant \chi^2_{0.95,q}, \tag{5.9}$$

where $\hat{\theta}_{(1)}$ is the ML estimate of $\theta_{(1)}$ and $C_{(11)}$ is the submatrix of C corresponding to $\theta_{(1)}$.

Under $f(\cdot|\cdot)$ and assuming large enough samples to make Approximation 5.1 appropriate, inferences based on 5.1 are not only appropriate, but optimal. It is thus not surprising that the ML estimator for θ with Approximation 5.1 constitutes a popular applied approach, especially considering that maximizing functions is a highly developed enterprise in many branches of applied mathematics. Throughout Part II, we primarily focus on large sample cases, and tend to avoid small sample complexities and distinctions between Bayesian and frequentist interpretations.

EXAMPLE 5.11. *Exponential Sample* (*Example 5.2 continued*). Differentiating (5.2) twice with respect to θ gives

$$I(\theta|Y) = -n/\theta^2 + 2\Sigma y_i/\theta^3.$$

Taking expectations over Y under the exponential specification for y_i gives

$$\begin{aligned} J(\theta) &= -n/\theta^2 + 2E(\Sigma y_i|\theta)/\theta^3 \\ &= -n/\theta^2 + 2n\theta/\theta^3. \end{aligned}$$

Substituting the ML estimate $\hat{\theta} = \bar{y}$ for θ gives

$$I(\hat{\theta}|Y) = J(\hat{\theta}) = n/\bar{y}^2.$$

Hence the large sample variance of $\theta - \hat{\theta}$ is $\bar{y}^2/n$.

EXAMPLE 5.12. *Univariate Normal Sample* (*Example 5.1 continued*). Differentiating (5.1) twice with respect to μ and σ^2, and substituting ML estimates of the parameters yields

$$I(\hat{\mu}, \hat{\sigma}^2 | y) = J(\hat{\mu}, \hat{\sigma}^2) = \begin{bmatrix} n/\hat{\sigma}^2 & 0 \\ 0 & 0.5n/\hat{\sigma}^4 \end{bmatrix}.$$

Hence inverting this matrix yields the large sample second moments $\text{Var}(\mu - \hat{\mu}) = \hat{\sigma}^2/n$, $\text{Cov}(\mu - \hat{\mu}, \sigma^2 - \hat{\sigma}^2) = 0$, $\text{Var}(\sigma^2 - \hat{\sigma}^2) = 2\hat{\sigma}^4/n$, where from Example 5.7, $\hat{\mu} = \bar{y}$ and $\hat{\sigma}^2 = s^2$.

5.2.2. Significance Levels for Null Values of θ

Especially when the number of components in θ, d, is greater than two, it is common to summarize evidence about the likely values of θ by significance levels rather than by ellipsoids such as (5.8). Specifically, for a null value θ_0 of θ, the distance from $\hat{\theta}$ to θ_0 can be calculated as

$$D_C(\theta_0, \hat{\theta}) = (\theta_0 - \hat{\theta})^{\mathrm{T}} C^{-1} (\theta_0 - \hat{\theta}),$$

which is the left-hand side of (5.8) evaluated at $\theta = \theta_0$. The associated percentile of the chi-squared distribution on d degrees of freedom is the significance level or p-value of the null value θ_0:

$$p_C = \Pr\{\chi_d^2 > D_C(\theta_0, \hat{\theta})\}.$$

From the frequentist perspective, the significance level provides the prior probability given $\theta = \theta_0$ that the ML estimate will be as far or farther from θ_0 than the observed ML estimate $\hat{\theta}$. A size α (two-sided) test of the null hypothesis H_0: $\theta = \theta_0$ is obtained by rejecting H_0 when the p-value p_C is less than α; common values of α are 0.1, 0.05, and 0.01.

From the Bayesian perspective, p_C gives the large sample posterior probability of the set of θ values with lower posterior density than θ_0: $\Pr[\theta \in \{\theta | f(\theta | Y) < f(\theta_0 | Y)\}]$; see Box and Tiao (1973) for discussion and examples.

Under Assumption 5.1, an asymptotically equivalent procedure for calculating significance levels is to use the likelihood ratio statistic to measure the distance between $\hat{\theta}$ and θ_0; this yields

$$p_L = \Pr\{\chi_d^2 > D_L(\theta_0, \hat{\theta})\},$$

where

$$D_L(\theta_0, \hat{\theta}) = 2 \ln[L(\hat{\theta} | Y)/L(\theta_0 | Y)] = 2[l(\hat{\theta} | Y) - l(\theta_0 | Y)].$$

More generally, suppose $\theta = (\theta_{(1)}, \theta_{(2)})$ and we are interested in evaluating the propriety of the null value of $\theta_{(1)}$, $\theta_{(1)0}$, where the number of components in $\theta_{(1)}$ is q. This situation commonly arises when comparing the fit of two models A and B, which are termed *nested* because the parameter space for model B is obtained from that for model A by setting $\theta_{(1)}$ to zero. Two asymptotically equivalent approaches derive significance levels corresponding to p_C and p_L as follows:

$$p_C(\theta_{(1)0}) = \Pr\{\chi_q^2 > (\theta_{(1)0} - \hat{\theta}_{(1)})^{\mathrm{T}} C_{(11)}^{-1} (\theta_{(1)0} - \hat{\theta}_{(1)})\},$$

where $C_{(11)}$ is the variance–covariance matrix of $\theta_{(1)}$ as in (5.9), and

$$p_L(\theta_{(1)0}) = \Pr\{\chi_q^2 > X^2\},$$

where

$$X^2 = 2\{l(\hat{\theta}|Y) - l(\tilde{\theta}|Y)\}$$

and $\tilde{\theta}$ is the value of θ that maximizes $l(\theta|Y)$ subject to the constraint that $\theta_{(1)} = \theta_{(1)0}$. Level α hypothesis tests reject H_0: $\theta_{(1)} = \theta_{(1)0}$ if the p-value for $\theta_{(1)0}$ is less than α.

EXAMPLE 5.13. *Univariate Normal Sample (Example 5.1 continued).* Suppose $\theta = (\mu, \sigma^2)$, $\theta_{(1)} = \mu$, $\theta_{(2)} = \sigma^2$. To test H_0: $\mu = \mu_0$, the likelihood ratio test statistic is

$$\begin{aligned} X^2 &= 2.0(-n/2 \ln s^2 - n/2 + n/2 \ln s_0^2 + n/2) \\ &= n \ln(s_0^2/s^2), \end{aligned}$$

where $s_0^2 = n^{-1} \sum_{i=1}^{n} (y_i - \mu_0)^2 = s^2 + (\bar{y} - \mu_0)^2$. Hence $X^2 = n \ln(1 + t^2/n)$, where $t^2 = n(\bar{y} - \mu_0)^2/s^2$ is, from Example 5.12, the test statistic for H_0 based on the asymptotic variance of $(\mu - \mu_0)$. Asymptotically, $X^2 = t^2$ and is chi-squared distributed with $q = 1$ degrees of freedom under H_0. An exact test is obtained in this case by comparing t^2 directly with an F distribution on 1 and $n - 1$ degrees of freedom. Such exact small sample tests are rarely available when we apply the likelihood ratio method to data sets with missing values.

5.3. LIKELIHOOD-BASED ESTIMATION FOR INCOMPLETE DATA

In one formal sense there is no difference between ML estimation for incomplete data and ML estimation for complete data: The likelihood for the parameters based on the incomplete data is derived and ML estimates are found by solving the likelihood equation. Asymptotic standard errors obtained from the information matrix are somewhat more questionable, however, since the observed data do not generally constitute an iid (independent, identically distributed) sample, and simple results that imply the large sample normality of the likelihood function do not immediately apply. Other complications arise from dealing with the process that creates missing data. Here to keep the notation simple we will be somewhat imprecise in our treatment of these complications. Rubin (1976) gives a mathematically precise treatment, which also encompasses frequentist approaches that are not based on the likelihood.

As before, let Y denote the data that would occur in the absence of missing values. We write $Y = (Y_{\text{obs}}, Y_{\text{mis}})$, where Y_{obs} denotes the observed values and Y_{mis} denotes the missing values. Let $f(Y|\theta) \equiv f(Y_{\text{obs}}, Y_{\text{mis}}|\theta)$ denote the probability or density of the joint distribution of Y_{obs} and Y_{mis}. The marginal probability density of Y_{obs} is obtained by integrating out the missing data Y_{mis}:

$$f(Y_{\text{obs}}|\theta) = \int f(Y_{\text{obs}}, Y_{\text{mis}}|\theta)\, dY_{\text{mis}}.$$

We define the likelihood of θ based on data Y_{obs} *ignoring the missing-data mechanism* to be any function of θ proportional to $f(Y_{\text{obs}}|\theta)$:

$$L(\theta|Y_{\text{obs}}) \propto f(Y_{\text{obs}}|\theta). \tag{5.10}$$

Inferences about θ can be based on this likelihood, $L(\theta|Y_{\text{obs}})$, providing the mechanism leading to incomplete data can be ignored, in a sense discussed below.

More generally, we can include in the model the distribution of a variable indicating whether each component of Y is observed or missing. We define for each component of Y a *missing-data indicator*, taking value 1 if the component is observed and 0 if it is missing. For example, if $Y = (Y_{ij})$, an $(n \times K)$ matrix of n observations measured for K variables, define the response indicator $R = (R_{ij})$, such that

$$R_{ij} = \begin{cases} 1, & y_{ij} \text{ observed,} \\ 0, & y_{ij} \text{ missing.} \end{cases}$$

The model treats R as a random variable and specifies the joint distribution of R and Y. The density of this distribution can be specified as the product of the densities of the distribution of Y and the conditional distribution of R given Y, that is,

$$f(Y, R|\theta, \psi) = f(Y|\theta)f(R|Y, \psi).$$

We call the conditional distribution of R given Y indexed by an unknown parameter ψ, the distribution for the missing-data mechanism. In some situations the distribution is known, and ψ is unnecessary.

The actual observed data consist of the values of the variables (Y_{obs}, R). The distribution of the observed data is obtained by integrating Y_{mis} out of the joint density of $Y = (Y_{\text{obs}}, Y_{\text{mis}})$ and R. That is,

$$f(Y_{\text{obs}}, R|\theta, \psi) = \int f(Y_{\text{obs}}, Y_{\text{mis}}|\theta) f(R|Y_{\text{obs}}, Y_{\text{mis}}, \psi)\, dY_{\text{mis}}. \tag{5.11}$$

The likelihood of θ and ψ is any function of θ and ψ proportional to (5.11):

$$L(\theta, \psi | Y_{\text{obs}}, R) \propto f(Y_{\text{obs}}, R|\theta, \psi). \tag{5.12}$$

The question now arises as to when should inference for θ be based on the likelihood $L(\theta, \psi | Y_{\text{obs}}, R)$ in (5.12) and when can it be based on the simpler $L(\theta | Y_{\text{obs}})$ in (5.10), which ignores the missing-data mechanism. Observe that if the distribution of the missing-data mechanism does not depend on the missing values Y_{mis}, that is, if

$$f(R|Y_{\text{obs}}, Y_{\text{mis}}, \psi) = f(R|Y_{\text{obs}}, \psi), \tag{5.13}$$

then from (5.11),

$$\begin{aligned} f(Y_{\text{obs}}, R, \theta, \psi) &= f(R|Y_{\text{obs}}, \psi) \times \int f(Y_{\text{obs}}, Y_{\text{mis}}, \theta)\, dY_{\text{mis}} \\ &= f(R|Y_{\text{obs}}, \psi) f(Y_{\text{obs}}|\theta). \end{aligned}$$

In many important practical applications, the parameters θ and ψ are distinct, in the sense that the joint parameter space of (θ, ψ) is the product of the parameter space of θ and the parameter space of ψ. If θ and ψ are distinct, then likelihood-based inferences for θ from $L(\theta, \psi | Y_{\text{obs}}, R)$ will be the same as likelihood-based inferences for θ from $L(\theta | Y_{\text{obs}})$. That is, if Eq. (5.13) is satisfied, the missing-data mechanism is ignorable in that the resulting likelihoods are proportional.

Rubin (1976) defines the missing data to be missing at random (MAR) when Eq. (5.13) is satisfied. Note in particular that if (5.13) is true, then the probability that a particular component of Y is missing cannot depend on the value of that component when it is missing. Equation (5.13) is a more precise expression of the intuitive notion of MAR which was discussed in the first chapter. Of practical importance is the fact that likelihood-based inferences that ignore the missing-data mechanism do not require MCAR for their validity but only the weaker MAR.

EXAMPLE 5.14. *Incomplete Exponential Sample.* Suppose we have an incomplete univariate sample with $Y_{\text{obs}} = (y_1, \ldots, y_m)^{\text{T}}$ observed and $Y_{\text{mis}} = (y_{m+1}, \ldots, y_n)^{\text{T}}$ missing. To fix ideas, we assume that the y_i are exponential random variables. Hence, as in Example 5.2,

$$f(Y|\theta) = \theta^{-n} \exp\left(-\sum_1^n \frac{y_i}{\theta} \right).$$

The likelihood ignoring the missing-data mechanism is proportional to the distribution of Y_{obs} given θ given by

$$f(Y_{\text{obs}}|\theta) = \theta^{-m} \exp\left(-\sum_1^m \frac{y_i}{\theta} \right). \tag{5.14}$$

Now $R = (R_1, \ldots, R_n)^{\text{T}}$, where $R_i = 1$, $i = 1, \ldots, m$ and $R_i = 0$, $i = m + 1, \ldots, n$.

Suppose that each unit is observed with probability ψ so that (5.13) holds. Then

$$f(R|Y, \psi) = \psi^m (1 - \psi)^{n-m},$$

and

$$f(Y_{\text{obs}}, R|\theta, \psi) = \psi^m (1 - \psi)^{n-m} \theta^{-m} \exp\left(-\sum_1^m \frac{y_i}{\theta} \right).$$

If ψ and θ are distinct, inferences about θ can be based on $f(Y_{\text{obs}}|\theta)$, ignoring the missing-data mechanism. In particular, the ML estimate of θ is simply $\sum_1^m y_i/m$, the mean of the responding values of Y.

Suppose, instead, that the incomplete data are created by censoring at some known censoring point c, so that only values less than c are recorded. Then

$$f(R|Y, \psi) = \prod_{i=1}^n f(R_i|y_i, \psi),$$

where

$$f(R_i|y_i, \psi) = \begin{cases} 1, & R_i = 1 \text{ and } y_i < c, \quad \text{or } R_i = 0 \text{ and } y_i > c; \\ 0, & \text{otherwise.} \end{cases}$$

Hence

$$\begin{aligned} f(Y_{\text{obs}}, R|\theta) &= \prod_{i=1}^m f(y_i, R_i|\theta) \prod_{i=m+1}^n f(R_i|\theta) \\ &= \prod_{i=1}^m f(y_i|\theta) f(R_i|y_i) \prod_{i=m+1}^n \Pr(y_i > c|\theta) \\ &= \theta^{-m} \exp\left(-\sum_1^m \frac{y_i}{\theta} \right) \exp\left[-\frac{(n-m)c}{\theta} \right], \end{aligned} \tag{5.15}$$

since $\Pr(y_i > c|\theta) = \exp(-c/\theta)$, using the properties of the exponential distribution. In this case the missing-data mechanism is not ignorable, and the correct likelihood (5.15) differs from (5.14). Maximizing (5.15) with respect to θ gives the ML estimate $\hat{\theta} = (\sum_1^m y_i + (n - m)c)/m$, which can be compared with the previous estimate $\sum_1^m y_i/m$. The inflation of the sample mean reflects the censoring of the missing values.

EXAMPLE 5.15. *Bivariate Normal Sample with One Variable Subject to Nonresponse.* Suppose we have a bivariate normal sample as in Example 5.9, but the values y_{i2}, $i = m + 1, \ldots, n$ of the second variable are missing. We thus have the monotone pattern of Figure 1.3. The loglikelihood ignoring the missing data mechanism is

$$l(\mu, \Sigma \mid Y_{\text{obs}}) = -\frac{1}{2} m \ln|\Sigma| - \frac{1}{2}\sum_{i=1}^{m} (y_i - \mu)\Sigma^{-1}(y_i - \mu)^{\mathrm{T}}$$
$$-\frac{1}{2}(n - m)\ln \sigma_{11} - \frac{1}{2}\sum_{i=m+1}^{n} \frac{(y_{i1} - \mu_1)^2}{\sigma_{11}}. \qquad (5.16)$$

This loglikelihood is appropriate for inferences provided the distribution of R (and in particular, the probability that y_{i2} is observed) does not depend on the values of y_{i2}, although it may depend on the values of y_{i1}, and $\theta = (\mu, \Sigma)$ is distinct from parameters of the response mechanism. Under these conditions, ML estimates of μ and Σ can be found by maximizing (5.16). A simpler approach to this maximization based on factoring the likelihood is described in the next chapter.

5.4. MAXIMIZING OVER THE PARAMETERS AND THE MISSING DATA

5.4.1. The Method

A different approach to handling incomplete data occasionally encountered in the literature is to treat the missing data as parameters and to maximize the complete-data likelihood over the missing data and parameters. That is, let

$$L_{\text{mis}}(\theta, Y_{\text{mis}} \mid Y_{\text{obs}}) = L(\theta \mid Y_{\text{obs}}, Y_{\text{mis}}) = f(Y_{\text{obs}}, Y_{\text{mis}} \mid \theta) \qquad (5.17)$$

be regarded as a function of (θ, Y_{mis}) for fixed Y_{obs}, and estimate θ by maximizing $L_{\text{mis}}(\theta, Y_{\text{mis}} \mid Y_{\text{obs}})$ over both θ and Y_{mis}. When the missing data are not MAR, or θ is not distinct from ψ, θ would be estimated by maximizing

$$L_{\text{mis}}(\theta, \psi, Y_{\text{mis}} \mid Y_{\text{obs}}, R) = L(\theta, \psi \mid Y_{\text{obs}}, Y_{\text{mis}}, R)$$
$$= f(Y_{\text{obs}}, Y_{\text{mis}} \mid \theta) f(R \mid Y_{\text{obs}}, Y_{\text{mis}}, \psi) \qquad (5.18)$$

over $(\theta, \psi, Y_{\text{mis}})$. Although useful in particular problems, this is not a generally reliable approach to the analysis of incomplete data. In particular, Little and

Rubin (1983) show that it does not share the optimal properties of ML estimation, except under the trivial asymptotics in which the proportion of missing data goes to zero as the sample size increases.

5.4.2. Background

The classic example of this approach is the treatment of missing plots in analysis of variance where missing outcomes Y_{mis} are treated as parameters and estimated along with the model parameters to allow computationally efficient methods to be used for analysis (see Chapter 2). More recently, DeGroot and Goel (1980) propose this approach as one possibility for the analysis of a mixed-up bivariate normal sample, where the missing data are the indices that allow the values of the two variables to be paired, and a priori all pairings are assumed equally likely. Press and Scott (1976) present a Bayesian analysis of an incomplete multivariate normal sample, which is equivalent to maximizing L_{mis} in (5.17) over (θ, Y_{mis}). Box, Draper, and Hunter (1970) and Bard (1974) apply the same approach in a more general setting where the multivariate normal mean vector has a nonlinear regression on covariates.

Formally, $L(\theta | Y_{obs})$ defined in Eq. (5.10) is the true likelihood of θ based on the observed data Y_{obs} if the data are MAR; the function L_{mis} is not a likelihood since the argument includes random variables Y_{mis}, which have a distribution under the model and hence should not be treated as fixed parameters. From this perspective, maximization of L_{mis} with respect to θ and Y_{mis} is *not* a maximum likelihood procedure.

A serious problem with treating both θ and Y_{mis} as parameters is that the number of parameters increases with the number of observations. The maximization of L_{mis} only has the optimal properties of maximum likelihood when the fraction of missing values tends to zero as the sample size increases. In contrast, the parameter θ does not depend on the amount of data, and hence, loosely speaking, standard likelihood asymptotics based on the maximization of $L(\theta | Y_{obs})$ apply, provided the information increases with the sample size. This deficiency when treating Y_{mis} as a parameter is easily illustrated in simple examples.

5.4.3. Examples

EXAMPLE 5.16. *Univariate Normal Sample with Missing Data.* Suppose that $Y = (Y_{obs}, Y_{mis})$ consists of n realizations from a normal distribution with mean μ and variance σ^2, where Y_{obs} consists of m observed values and Y_{mis} represents $n - m$ missing values, which are assumed MAR; θ is (μ, σ^2), which we assume is distinct from the parameters of the missing-data mechanism. Since

$$f(Y|\theta) = \prod_{i=1}^{n} f(y_i|\theta) = \prod_{i=1}^{m} f(y_i|\theta) \times \prod_{i=m+1}^{n} f(y_i|\theta), \tag{5.19}$$

it follows that $f(Y_{\text{obs}}|\theta) = \prod_{i=1}^{m} f(y_i|\theta)$ and $f(Y_{\text{mis}}|\theta) = \prod_{i=m+1}^{n} f(y_i|\theta)$. Thus $L(\theta|Y_{\text{obs}})$ is identical in form to the likelihood for a sample of size m from a normal distribution. By Example 5.7, maximizing $L(\theta|Y_{\text{obs}})$ over θ thus leads to ML estimates

$$\hat{\mu} = \sum_{i}^{m} \frac{y_i}{m} \qquad \text{and} \qquad \hat{\sigma}^2 = \sum_{1}^{m} \frac{(y_i - \hat{\mu})^2}{m}. \tag{5.20}$$

In contrast,

$$L_{\text{mis}}(\theta, Y_{\text{mis}}|Y_{\text{obs}}) = f(Y_{\text{obs}}|\theta) f(Y_{\text{mis}}|\theta), \tag{5.21}$$

which is to be maximized over θ and Y_{mis}. The maximization over Y_{mis} in the second factor in (5.21) gives maximizing values

$$\tilde{y}_i = \tilde{\mu} \qquad \text{for } i = m+1, \ldots, n \tag{5.22}$$

where $\tilde{\mu}$ is the maximizing value of μ. From Example 5.1, the maximizing values for μ and σ^2 are

$$\tilde{\mu} = \left(\sum_{1}^{m} y_i + \sum_{m+1}^{n} \tilde{y}_i \right) \bigg/ n \quad \text{and} \quad \tilde{\sigma}^2 = \left[\sum_{1}^{m} (y_i - \tilde{\mu})^2 + \sum_{m+1}^{n} (\tilde{y}_i - \tilde{\mu})^2 \right] \bigg/ n. \tag{5.23}$$

Substituting (5.22) into (5.23) and comparing with (5.20) gives

$$\tilde{\mu} = \hat{\mu} \qquad \text{and} \qquad \tilde{\sigma}^2 = \hat{\sigma}^2 m/n.$$

Thus the ML estimate of the mean is obtained, but the ML estimate of the variance is multiplied by the fraction of data observed. When the fraction of missing data is substantial (e.g., $m/n = .5$), the estimated variance $\tilde{\sigma}^2$ is badly biased, and this bias does not vanish as $n \to \infty$ unless $m/n \to 1$; more relevant asymptotics would fix m/n as the sample size increases.

EXAMPLE 5.17. *Missing-Plot Analysis of Variance.* Suppose we add to the previous example a set of covariates X that is observed for all n observations. We assume that the value of Y for observation i with covariate values x_i is normal with mean $\beta_0 + x_i\beta$ and variance σ^2, and write $\theta = (\beta_0, \beta, \sigma^2)$. The maximum likelihood estimates of β_0, β, and σ^2 obtained by maximizing $L(\theta|Y_{\text{obs}})$ are obtained by applying least squares regression to the m observed data points. The estimates of β_0 and β obtained by maximizing L_{mis} are the same as the ML estimates. As in Example 5.16, however the estimate of variance is the ML estimate multiplied by the proportion of values observed.

EXAMPLE 5.18. *An Exponential Sample with Censored Values.* In Examples 5.16 and 5.17 estimation based on maximizing L_{mis} at least yields reasonable estimates of location, even though estimates of the scale parameter need adjustment. In other examples, however estimates of location also can be badly biased. For example, consider, as in Example 5.14, a censored sample from an exponential distribution with mean θ, where Y_{obs} represents the m observed values, which lie below a known censoring point c, and Y_{mis} represents the $n - m$ values beyond c, which are censored. The ML estimate of θ is $\hat{\theta} = \bar{y} + (n - m)c/m$. Maximization of L_{mis} in (5.18) over θ and Y_{mis} (ψ is null) leads to estimating censored values of Y at the censoring point c and estimating θ by $(m/n)\hat{\theta}$. Thus in this case the estimate of the mean is inconsistent unless the proportion of missing values tends to zero as the sample size increases.

Biased estimates of location parameters from L_{mis} can also occur in problems involving the normal distribution, as Press and Scott (1976) and Little and Rubin (1983) show.

REFERENCES

Anderson, T. W. (1965). *An Introduction to Multivariate Statistical Analysis*. New York: Wiley.

Bard, Y. (1974). *Nonlinear Parameter Estimation*. New York: Academic Press.

Box, M. J., Draper, N. R., and Hunter, W. G. (1970), Missing values in multi-response nonlinear data fitting, *Technometrics* **12**, 613–620.

Box, G. E. P. and Tiao, G. C. (1973). *Bayesian Inference in Statistical Analysis*. Reading, MA: Addison-Wesley.

Cox, D. R., and Hinkley, D. V. (1974). *Theoretical Statistics*. New York: Wiley.

DeGroot, M. H., and Goel, K. (1980). Estimation of the correlation coefficient from a broken random sample, *Ann. Statist*. **8**, 264–278.

Little, R. J. A., and Rubin, D. B. (1983). On jointly estimating parameters and missing data by maximizing the complete-data likelihood, *Am. Statist*. **37**, 218–220.

Lindley, D. V. (1965). *Introduction to Probability and Statistics from a Bayesian Viewpoint, Part 2, Inference*. Cambridge, Cambridge University Press.

Press, S. J., and Scott, A. J. (1976). Missing variables in Bayesian regression, II, *J. Am. Statist. Associ*. **71**, 366–369.

Rao, C. R. (1972). *Linear Statistical Inference*. New York: Wiley.

Rubin, D. B. (1976). Inference and missing data, *Biometrika* **63**, 581–592.

Wilks, S. S. (1963). *Mathematical Statistics*. New York: Wiley.

PROBLEMS

1. Write down the likelihood function for an iid sample from the (a) beta distribution; (b) multinomial distribution; (c) Cauchy distribution.
2. Find the score function for the distributions in Problem 1, and the ML estimates for those distributions that have closed form estimates.
3. For a univariate normal sample, find the ML estimate of the coefficient of variation, σ/μ.

4. (a) Relate ML and least squares estimates for the model of Example 5.10.
(b) Show that if the data $\{y_1, \ldots, y_n\}$ are iid with the Laplace (double exponential) distribution

$$f(y_i|\theta) = 0.5\exp(-|y_i - \mu(x_i)|),$$

where $\mu(x_i) = \beta_0 + \beta_1 x_{1i} + \cdots + \beta_k x_{ki}$, then ML estimates of $\beta_0, \ldots, \beta_k$ are obtained by minimizing the sum of absolute deviations of the y values from their expected values.

5. Summarize the theoretical and practical differences between the frequentist and Bayesian interpretation of Approximation 5.1.

6. For the distributions of Problem 1, calculate the observed information and the expected information.

7. Show that for random samples from regular distributions (differentials can be passed through the integral) the expected squared score function equals the expected information.

8. In Example 5.13 show that for large n, $X^2 = t^2$.

9. Find variance estimates for the two ML estimates in Example 5.14.

10. For a bivariate normal sample on (Y_1, Y_2) with parameters $\theta = (\mu_1, \mu_2, \sigma_{11}, \sigma_{22}, \sigma_{12})$ and values of Y_2 missing, state for the following missing-data mechanisms whether (i) the data are MAR, and (ii) the missing-data mechanism is ignorable for likelihood-based inference.

(a) $\Pr(y_2 \text{ missing}|y_1, y_2, \theta, \psi) = \exp(\psi_0 + \psi_1 y_1)/\{1 + \exp(\psi_0 + \psi_1 y_1)\}$, $\psi = (\psi_0, \psi_1)$ distinct from θ.
(b) $\Pr(y_2 \text{ missing}|y_1, y_2, \theta, \psi) = \exp(\psi_0 + \psi_1 y_1)/\{1 + \exp(\psi_0 + \psi_1 y_2)\}$, $\psi = (\psi_0, \psi_1)$ distinct from θ.
(c) $\Pr(y_2 \text{ missing}|y_1, y_2, \theta, \psi) = 0.5\exp(\mu_1 + \psi y_1)/\{1 + \exp(\mu_1 + \psi y_1)\}$, scalar ψ distinct from θ.

11. Suppose that given sets of (possibly overlapping) covariates X_1 and X_2, y_{i1} and y_{i2} are bivariate normal with means $x_{i1}\beta_1$ and $x_{i2}\beta_2$, variances σ_1^2 and $\sigma_2^2 = 1$, and correlation ρ. The data consist of a random sample of observations i with x_{i1} and x_{i2} always present, y_{i2} always missing and y_{i1} present if and only if $y_{i2} > 0$. Show that, given the parameters

$$\Pr(y_{i1} \text{ observed}|y_{i1}, x_{i1}, x_{i2}) = 1 - \Phi\left[\frac{-x_{i2}\beta_2 - \rho(y_{i1} - x_{i1}\beta_1)}{\sqrt{(1 - \rho^2)}}\right],$$

where Φ is the standard normal cumulative distribution function. Hence give conditions on the parameters under which the data are MAR and for when the missing-data mechanism is ignorable for likelihood-based inference. (This model is considered in detail in Example 11.4.)

CHAPTER 6

Methods Based on Factoring the Likelihood, Ignoring the Missing-Data Mechanism

6.1. INTRODUCTION

The loglikelihood $l(\theta | Y_{obs})$ based on incomplete data Y_{obs} can be a complicated function with no obvious maximum and an apparently complicated form for the information matrix. For certain models and incomplete-data patterns, however, analyses of $l(\theta | Y_{obs})$ can employ standard complete-data techniques. The general idea will be described here in Section 6.1, and specific examples will be given in the remainder of this chapter for normal data and in Chapter 9 for multinomial (i.e., cross-classified) data.

For a variety of models and incomplete data patterns, an alternative parameterization $\phi = \phi(\theta)$, where ϕ is a one–one monotone function of θ, can be found such that the loglikelihood decomposes into components

$$l(\phi | Y_{obs}) = l_1(\phi_1 | Y_{obs}) + l_2(\phi_2 | Y_{obs}) + \cdots + l_J(\phi_J | Y_{obs}), \tag{6.1}$$

where

1. $\phi_1, \phi_2, \ldots, \phi_J$ are distinct parameters, in the sense that the joint parameter space of $\phi = (\phi_1, \phi_2, \ldots, \phi_J)$ is the product of the individual parameter spaces for ϕ_j, $j = 1, \ldots, J$.
2. The components $l_j(\phi_j | Y_{obs})$ correspond to loglikelihoods for complete data problems, or more generally, for easier incomplete-data problems.

If a decomposition with these properties can be found, then since $\phi_1, \ldots, \phi_J$ are distinct, $l(\phi | Y_{obs})$ can be maximized by maximizing $l_j(\phi_j | Y_{obs})$ separately for each j. If $\hat{\phi}$ is the resulting ML estimate of ϕ, then the ML estimate of any function $\theta(\phi)$ of ϕ is obtained by applying Property 5.1, that is, substituting $\hat{\theta} = \theta(\hat{\phi})$.

The decomposition (6.1) can also be used to calculate the approximate covariance matrix for ML estimates, C in (5.6). Differentiating (6.1) twice with respect to $\phi_1, \ldots, \phi_J$ yields a block diagonal information matrix of the form

$$I(\phi \mid Y_{\text{obs}}) = \begin{bmatrix} I(\phi_1 \mid Y_{\text{obs}}) & & 0 \\ & I(\phi_2 \mid Y_{\text{obs}}) \ddots & \\ 0 & & I(\phi_J \mid Y_{\text{obs}}) \end{bmatrix}.$$

Hence the covariance matrix C is also block diagonal, with the form

$$C(\hat{\phi} \mid Y_{\text{obs}}) = \begin{bmatrix} I^{-1}(\hat{\phi}_1 \mid Y_{\text{obs}}) & & 0 \\ & I^{-1}(\hat{\phi}_2 \mid Y_{\text{obs}}) \ddots & \\ 0 & & I^{-1}(\hat{\phi}_J \mid Y_{\text{obs}}) \end{bmatrix}. \tag{6.2}$$

Since the components of this matrix correspond to complete-data problems, they are relatively easy to calculate. The approximate covariance matrix of the ML estimate of a function $\theta = \theta(\phi)$ of ϕ can be found using the formula

$$C(\hat{\theta} \mid Y_{\text{obs}}) = D(\hat{\theta}) C(\hat{\phi} \mid Y_{\text{obs}}) D^{\text{T}}(\hat{\theta}), \tag{6.3}$$

where D is the matrix of partial derivatives of θ with respect to ϕ:

$$D(\theta) = \{d_{jk}(\theta)\}, \qquad \text{where } d_{jk}(\theta) = \frac{\partial \theta_j}{\partial \phi_k},$$

and θ is expressed as a column vector.

6.2. BIVARIATE NORMAL DATA WITH ONE VARIABLE SUBJECT TO NONRESPONSE: ML ESTIMATION

Anderson (1957) first introduced factored likelihoods for the normal data of Example 5.15.

EXAMPLE 6.1. *Bivariate Normal Sample with One Variable Subject to Nonresponse (Example 5.15 continued).* The loglikelihood for a bivariate normal sample with m complete bivariate observations $\{(y_{i1}, y_{i2}); i = 1, \ldots, m\}$ and $n - m$ univariate observations $\{y_{i1}; i = m+1, \ldots, n\}$ is given by (5.16). ML estimates of μ and Σ can be found by maximizing this function with respect to μ and Σ. The likelihood equations, however, do not have an obvious solution. Anderson (1957) factors the joint distribution of y_{i1} and y_{i2} into the marginal distribution of y_{i1} and the conditional distribution of y_{i2} given y_{i1}:

$$f(y_{i1}, y_{i2}|\mu, \Sigma) = f(y_{i1}|\mu_1, \sigma_{11})f(y_{i2}|y_{i1}, \beta_{20\cdot 1}, \beta_{21\cdot 1}, \sigma_{22\cdot 1}),$$

where, by properties of the bivariate normal distribution discussed in Example 5.9, $f(y_{i1}|\mu_1, \sigma_{11})$ is the normal distribution with mean μ_1 and variance σ_{11}, and $f(y_{i2}|y_{i1}, \beta_{20\cdot 1}, \beta_{21\cdot 1}, \sigma_{22\cdot 1})$ is the normal distribution with mean

$$\beta_{20\cdot 1} + \beta_{21\cdot 1}y_{i1}$$

and variance $\sigma_{22\cdot 1}$. The parameter

$$\phi = (\mu_1, \sigma_{11}, \beta_{20\cdot 1}, \beta_{21\cdot 1}, \sigma_{22\cdot 1})^{\mathrm{T}}$$

is a one–one monotone function of the original parameter

$$\theta = (\mu_1, \mu_2, \sigma_{11}, \sigma_{12}, \sigma_{22})^{\mathrm{T}}$$

of the joint distribution of y_{i1} and y_{i2}. In particular, μ_1 and σ_{11} are common to both parameterizations, and the other components of ϕ are given by the following functions of components of θ.

$$\begin{aligned} \beta_{21\cdot 1} &= \sigma_{12}/\sigma_{11}, \\ \beta_{20\cdot 1} &= \mu_2 - \beta_{21\cdot 1}\mu_1, \\ \sigma_{22\cdot 1} &= \sigma_{22} - \sigma_{12}^2/\sigma_{11}. \end{aligned} \tag{6.4}$$

Similarly, the components of θ other than μ_1 and σ_{11} can be expressed as the following functions of the components of ϕ:

$$\begin{aligned} \mu_2 &= \beta_{20\cdot 1} + \beta_{21\cdot 1}\mu_1, \\ \sigma_{12} &= \beta_{21\cdot 1}\sigma_{11}, \\ \sigma_{22} &= \sigma_{22\cdot 1} + \beta_{21\cdot 1}^2\sigma_{11}. \end{aligned} \tag{6.5}$$

The density of the data Y_{obs} factors in the following way:

$$\begin{aligned} f(Y_{\text{obs}}|\theta) &= \prod_{i=1}^{m} f(y_{i1}, y_{i2}|\theta) \prod_{i=m+1}^{n} f(y_{i1}|\theta) \\ &= \left[\prod_{i=1}^{m} f(y_{i1}|\theta)f(y_{i2}|y_{i1}, \theta)\right]\left[\prod_{i=m+1}^{n} f(y_{i1}|\theta)\right] \\ &= \left[\prod_{i=1}^{n} f(y_{i1}|\mu_1, \sigma_{11})\right]\left[\prod_{i=1}^{m} f(y_{i2}|y_{i1}, \beta_{20\cdot 1}, \beta_{21\cdot 1}, \sigma_{22\cdot 1})\right]. \end{aligned} \tag{6.6}$$

The first bracketed factor in (6.6) is the density of an independent sample of size n from the normal distribution with mean μ_1 and variance σ_{11}. The second factor is the density for m observations from the conditional normal distribution with mean $\beta_{20\cdot 1} + \beta_{21\cdot 1}y_{i1}$ and variance $\sigma_{22\cdot 1}$. Furthermore, if the parameter space for θ is the standard parameter space with no prior restrictions, then (μ_1, σ_{11}) and $(\beta_{20\cdot 1}, \beta_{21\cdot 1}, \sigma_{22\cdot 1})$ are distinct, since knowledge of (μ_1, σ_{11}) does not imply any information about $(\beta_{20\cdot 1}, \beta_{21\cdot 1}, \sigma_{22\cdot 1})$. Hence ML estimates of ϕ can be obtained by independently maximizing the likelihoods corresponding to these two components.

Maximizing the first factor yields

$$\hat{\mu}_1 = n^{-1} \sum_{i=1}^{n} y_{i1}$$

and (6.7)

$$\hat{\sigma}_{11} = n^{-1} \sum_{i=1}^{n} (y_{i1} - \hat{\mu}_1)^2,$$

that is, the sample mean and sample variance of the n observations $y_{11}, \ldots, y_{n1}$.

Maximizing the second factor uses standard regression results and yields (cf. Example 5.9):

$$\begin{aligned} \hat{\beta}_{21\cdot 1} &= s_{12}/s_{11}, \\ \hat{\beta}_{20\cdot 1} &= \bar{y}_2 - \hat{\beta}_{21\cdot 1}\bar{y}_1, \\ \hat{\sigma}_{22\cdot 1} &= s_{22\cdot 1}, \end{aligned} \tag{6.8}$$

where $\bar{y}_j = m^{-1} \sum_{i=1}^{m} y_{ij}$, $s_{jk} = m^{-1} \sum_{i=1}^{m} (y_{ij} - \bar{y}_j)(y_{ik} - \bar{y}_k)$ for j, $k = 1, 2$, and $s_{22\cdot 1} = s_{22} - s_{12}^2/s_{11}$.

The ML estimates of other parameters can now be obtained using Property 5.1. In particular, from (6.5),

$$\hat{\mu}_2 = \hat{\beta}_{20\cdot 1} + \hat{\beta}_{21\cdot 1}\hat{\mu}_1,$$

or from (6.7) and (6.8),

$$\hat{\mu}_2 = \bar{y}_2 + \hat{\beta}_{21\cdot 1}(\hat{\mu}_1 - \bar{y}_1); \tag{6.9}$$

from (6.5),

$$\hat{\sigma}_{22} = \hat{\sigma}_{22\cdot 1} + \hat{\beta}_{21\cdot 1}^2 \hat{\sigma}_{11},$$

or from (6.7) and (6.8),

$$\hat{\sigma}_{22} = s_{22} + \hat{\beta}_{21\cdot 1}^2 (\hat{\sigma}_{11} - s_{11}). \tag{6.10}$$

Finally, from (6.5), we have for the correlation

$$\rho \equiv \sigma_{12}(\sigma_{11}\sigma_{22})^{-1/2} = \beta_{21\cdot 1}\sigma_{11}^{1/2}(\sigma_{22\cdot 1} + \beta_{21\cdot 1}^2 \sigma_{11})^{-1/2},$$

so from (6.7) and (6.8)

$$\hat{\rho} = [s_{12}(s_{11}s_{22})^{-1/2}](\hat{\sigma}_{11}/s_{11})^{1/2}(s_{22}/\hat{\sigma}_{22})^{1/2}. \tag{6.11}$$

Table 6.1 Data on Size of Apple Crop (y_{i1}) and 100 × Percentage of Wormy Fruit (y_{i2})

Tree Number	Size of Crop (100s of Fruits) (y_{i1})	100 × Percentage Wormy Fruits (y_{i2})	Regression Prediction (y_{i2})
1	8	59	56.1
2	6	58	58.2
3	11	56	53.1
4	22	53	42.0
5	14	50	50.1
6	17	45	47.0
7	18	43	46.0
8	24	42	39.9
9	19	39	45.0
10	23	38	41.0
11	26	30	37.9
12	40	27	23.7
13	4	—	60.2
14	4	—	60.2
15	5	—	59.2
16	6	—	58.2
17	8	—	56.1
18	10	—	54.1

$\bar{y}_1 = 19$; $\bar{y}_2 = 45$; $\hat{\mu}_2 = 49.3333$; $\hat{\mu}_1 = 14.7222$
$s_{11} = 77.0$; $s_{12} = -78.0$; $s_{22} = 101.8333$; $\hat{\sigma}_{11} = 89.5340$

Source: Adapted from Snedecor and Cochran (1967, Table 6.9.1).

The first terms on the right side of (6.9), (6.10) and (6.11) are the ML estimates of μ_2, σ_{22} and ρ with the $(n - m)$ incomplete observations discarded. Thus the remaining terms represent adjustments based on the additional information from the $n - m$ extra values of y_{i1}.

EXAMPLE 6.2. *Bivariate Normal Numerical Illustration.* The first $m = 12$ observations in Table 6.1, taken from Snedecor and Cochran (1967, Table 6.9.1), give measurements on the size of crop on apple trees, in hundreds of fruits (y_{i1}), and 100 times the percentage of wormy fruits (y_{i2}). These observations suggest a negative association between the size of crop and the percentage of wormy fruits. Suppose that the objective is to estimate the mean of y_{i2}, but for some of the trees with smaller crops, numbered 13 to 18 in the table, the value of y_{i2} was not determined. The sample mean, $\bar{y}_2 = 45$, underestimates the percentage of wormy fruits, since the percentage for the six omitted trees is expected to be larger than the percentage for the measured trees because the omitted trees tended to be smaller (i.e., the data may be MAR but do not appear to be MCAR). The ML estimate is $\hat{\mu}_2 = 49.33$, which can be compared with the estimate $\bar{y}_2 = 45.0$ from the complete observations. This analysis should be taken only as a numerical illustration; a serious analysis of these data would consider issues such as whether transformations of y_{i1} and y_{i2} (e.g., log, square root) would better meet the underlying normality assumptions.

The ML estimate (6.9) of the mean of y_{i2} is of particular interest. It can be written in the form

$$\hat{\mu}_2 = n^{-1}\left\{\sum_{i=1}^{m} y_{i2} + \sum_{i=m+1}^{n} \hat{y}_{i2}\right\}, \tag{6.12}$$

where

$$\hat{y}_{i2} = \bar{y}_2 + \hat{\beta}_{21\cdot 1}(y_{i1} - \bar{y}_1).$$

Hence $\hat{\mu}_2$ is a type of regression estimator commonly used in sample surveys (e.g., Cochran, 1977), which effectively imputes the predicted values $\hat{y}_{i2}$ for missing y_{i2} from linear regression of observed y_{i2} on observed y_{i1}.

6.3. BIVARIATE NORMAL DATA WITH ONE VARIABLE SUBJECT TO NONRESPONSE: PRECISION OF ESTIMATION

An important development of the examples in Section 6.2 concerns the precision of the resulting ML estimates.

6.3.1. Large-Sample Covariance Matrix

The large-sample covariance matrix C of $(\phi - \hat{\phi})$ is found by calculating and inverting the information matrix. The loglikelihood of ϕ is from Eq. (6.6)

$$l(\phi \mid Y_{\text{obs}}) = -(2\sigma_{22\cdot 1})^{-1} \sum_{i=1}^{m} (y_{i2} - \beta_{20\cdot 1} - \beta_{21\cdot 1} y_{i1})^2 - \tfrac{1}{2} m \ln \sigma_{22\cdot 1}$$

$$- (2\sigma_{11})^{-1} \sum_{i=1}^{n} (y_{i1} - \mu_1)^2 - \tfrac{1}{2} n \ln \sigma_{11}.$$

Differentiating twice with respect to ϕ gives

$$I(\hat{\phi} \mid Y_{\text{obs}}) = \begin{bmatrix} I(\hat{\mu}_1, \hat{\sigma}_{11} \mid Y_{\text{obs}}) & 0 \\ 0 & I(\hat{\beta}_{20\cdot 1}, \hat{\beta}_{21\cdot 1}, \hat{\sigma}_{22\cdot 1} \mid Y_{\text{obs}}) \end{bmatrix},$$

where

$$I(\hat{\mu}_1, \hat{\sigma}_{11} | Y_{\text{obs}}) = \begin{bmatrix} n/\hat{\sigma}_{11} & 0 \\ 0 & n/(2\hat{\sigma}_{11}^2) \end{bmatrix},$$

and

$$I(\hat{\beta}_{20\cdot 1}, \hat{\beta}_{21\cdot 1}, \hat{\sigma}_{22\cdot 1} | Y_{\text{obs}}) = \begin{bmatrix} m/\hat{\sigma}_{22\cdot 1} & m\bar{y}_1/\hat{\sigma}_{22\cdot 1} & 0 \\ m\bar{y}_1/\hat{\sigma}_{22\cdot 1} & \sum_{i=1}^m y_{i1}^2/\hat{\sigma}_{22\cdot 1} & 0 \\ 0 & 0 & m/(2\hat{\sigma}_{22\cdot 1}^2) \end{bmatrix}.$$

Inverting these matrices yields

$$C = \begin{bmatrix} I^{-1}(\hat{\mu}_1, \hat{\sigma}_{11} | Y_{\text{obs}}) & 0 \\ 0 & I^{-1}(\hat{\beta}_{20\cdot 1}, \hat{\beta}_{21\cdot 1}, \hat{\sigma}_{22\cdot 1} | Y_{\text{obs}}) \end{bmatrix},$$

where

$$I^{-1}(\hat{\mu}_1, \hat{\sigma}_{11} | Y_{\text{obs}}) = \begin{bmatrix} \hat{\sigma}_{11}/n & 0 \\ 0 & 2\hat{\sigma}_{11}^2/n \end{bmatrix},$$

and

$$I^{-1}(\hat{\beta}_{20\cdot 1}, \hat{\beta}_{21\cdot 1}, \hat{\sigma}_{22\cdot 1} | Y_{\text{obs}})$$
$$= \begin{bmatrix} \hat{\sigma}_{22\cdot 1}(1 + \bar{y}_1^2/s_{11})m^{-1} & -\bar{y}_1\hat{\sigma}_{22\cdot 1}/ms_{11} & 0 \\ -\bar{y}_1\hat{\sigma}_{22\cdot 1}/ms_{11} & \hat{\sigma}_{22\cdot 1}/ms_{11} & 0 \\ 0 & 0 & 2\hat{\sigma}_{22\cdot 1}^2/m \end{bmatrix}.$$

The large-sample covariance matrix of $(\theta - \hat{\theta})$ can be found using Eq. (6.3). To illustrate the calculations we consider the parameter μ_2, the mean of the incompletely observed variable. Since $\mu_2 = \beta_{20\cdot 1} + \beta_{21\cdot 1}\mu_1$, we have

$$D(\mu_2) = \left(\frac{\partial \mu_2}{\partial \mu_1}, \frac{\partial \mu_2}{\partial \sigma_{11}}, \frac{\partial \mu_2}{\partial \beta_{20\cdot 1}}, \frac{\partial \mu_2}{\partial \beta_{21\cdot 1}}, \frac{\partial \mu_2}{\partial \sigma_{22\cdot 1}}\right)$$
$$= (\hat{\beta}_{21\cdot 1}, 0, 1, \hat{\mu}_1, 0),$$

substituting ML estimates of μ_1 and $\beta_{21\cdot 1}$. Hence with some calculation, the large sample variance of $(\mu_2 - \hat{\mu}_2)$ is

$$\text{DCD}^{\text{T}} = \hat{\sigma}_{22\cdot 1}\left[\frac{1}{m} + \frac{\hat{\rho}^2}{n(1 - \hat{\rho}^2)} + \frac{(\bar{y}_1 - \hat{\mu}_1)^2}{ms_{11}}\right]. \tag{6.13}$$

The third term in the brackets is $0(m^{-2})$ if the data are MAR *and* OAR (i.e., MCAR), because $(\bar{y}_1 - \hat{\mu}_1)^2$ is $0(m^{-1})$ in this case. Ignoring this term yields

$$\hat{\sigma}_{22\cdot 1}\left[\frac{1}{m} + \frac{\hat{\rho}^2}{n(1 - \hat{\rho}^2)}\right] = \frac{\hat{\sigma}_{22}}{m}\left(1 - \hat{\rho}^2\frac{n - m}{n}\right). \tag{6.14}$$

This expression can be compared with the variance of $\bar{y}_2$, σ_{22}/m. Thus in large samples, the percent reduction in variance obtained by also using the $n - m$ observations on y_1 alone is proportional to ρ^2 times the proportion of incomplete observations, $(n - m)/m$.

6.3.2. Small-Sample Inference for the Parameters

Given large samples, interval estimates for the parameters can be obtained by applying Approximation 5.1 [Eq. (5.6)], as discussed in Section 5.2. In particular, a 95% interval for μ_2 takes the form

$$\hat{\mu}_2 \pm 1.96\sqrt{\mathrm{Var}(\hat{\mu}_2 - \mu_2)}, \tag{6.15}$$

where $\mathrm{Var}(\hat{\mu}_2 - \mu_2)$ is given by Eq. (6.13).

Small-sample inference is more problematic. Consider, for example, inference about μ_2 from a frequentist perspective. The quantity $(\hat{\mu}_2 - \mu_2)/\sqrt{\mathrm{Var}(\hat{\mu}_2 - \mu_2)}$ obtained from (6.13) is standard normally distributed in large samples, but its distribution in small samples is complex and depends on the parameters. The t distribution with $m - 1$ degrees of freedom has been suggested as a useful approximate reference distribution for this quantity, and performs reasonably well in simulations (Little, 1976). The same reference t distribution has also been proposed for inference about the difference in means $\mu_2 - \mu_1$, based on $(\hat{\mu}_2 - \hat{\mu}_1 - \mu_2 + \mu_1)/\sqrt{\mathrm{Var}(\hat{\mu}_2 - \hat{\mu}_1 - \mu_2 + \mu_1)}$. Approximate small-sample confidence intervals for other parameters, such as ρ, do not appear to have been developed.

An alternative approach to small-sample interval estimation is to specify a prior distribution for the parameters and then derive the posterior distribution given the data. Specifically, suppose we assume that a priori μ_1, σ_{11}, $\beta_{20\cdot 1}$, $\beta_{21\cdot 1}$, and $\sigma_{22\cdot 1}$ are a priori independent with the reference prior distribution with density

$$f(\mu_1, \sigma_{11}, \beta_{20\cdot 1}, \beta_{21\cdot 1}, \sigma_{22\cdot 1}) \propto \sigma_{11}^{-a}\sigma_{22\cdot 1}^{-c}. \tag{6.16}$$

The choice $a = c = 1$ yields the Jeffreys prior for the factored density (Box and Tiao, 1973).

Applying standard Bayesian theory to the random sample $\{y_{i1}: i = 1, \ldots, n\}$, the posterior distribution of (μ_1, σ_{11}) is such that (1) $n\hat{\sigma}_{11}/\sigma_{11}$ has a chi-squared distribution with $n + 2a - 3$ degrees of freedom, and (2) the posterior distribution of μ_1 given σ_{11} is normal with mean $\hat{\mu}_1$ and variance σ_{11}/n. Applying standard Bayesian regression theory to the random sample $\{(y_{i1}, y_{i2}): i = 1, \ldots, m\}$, the posterior distribution of $(\beta_{20\cdot 1}, \beta_{21\cdot 1}, \sigma_{22\cdot 1})$ is such that (3) $m\hat{\sigma}_{22\cdot 1}/\sigma_{22\cdot 1}$ has a chi-squared distribution with $m + 2c - 4$ degrees of freedom; (4) the posterior distribution of $\beta_{21\cdot 1}$ given $\sigma_{22\cdot 1}$ is normal with mean $\hat{\beta}_{21\cdot 1}$ and variance $\sigma_{22\cdot 1}/(ms_{11})$, and (5) the posterior distribution of $\beta_{20\cdot 1}$ given $\beta_{21\cdot 1}$ and $\sigma_{22\cdot 1}$ is normal with mean $\bar{y}_2 - \beta_{21\cdot 1}\bar{y}_1$ and variance $\sigma_{22\cdot 1}/m$. For derivations of these results, see Lindley (1965). Furthermore, (6) (μ_1, σ_{11}) and $(\beta_{20\cdot1}, \beta_{21\cdot1}, \sigma_{22\cdot1})$ are a posteriori independent.

Results 1–6 imply that the posterior distribution of any function $\psi(\phi)$ of the parameters ϕ can be simulated by creating values ψ_i as follows:

1. Draw independently χ^2_{1t}, χ^2_{2t} from chi-squared distributions with, respectively, $n + 2a - 3$ and $m + 2c - 4$ degrees of freedom. Draw three independent standard normal deviates z_{1t}, z_{2t}, and z_{3t}.
2. Compute $\phi^{(t)} = (\sigma_{11}^{(t)}, \mu_1^{(t)}, \sigma_{22\cdot1}^{(t)}, \beta_{20\cdot1}^{(t)}, \beta_{21\cdot1}^{(t)})^{\mathrm{T}}$, where

$$\begin{aligned}
\sigma_{11}^{(t)} &= n\hat{\sigma}_{11}/\chi^2_{1t} \\
\mu_1^{(t)} &= \hat{\mu}_1 + z_{1t}(\sigma_{11}^{(t)}/n)^{1/2} \\
\sigma_{22\cdot1}^{(t)} &= m\hat{\sigma}_{22\cdot1}/\chi^2_{2t} \\
\beta_{21\cdot1}^{(t)} &= \hat{\beta}_{21\cdot1} + z_{2t}(\sigma_{22\cdot1}^{(t)}/(ms_{11}))^{1/2} \\
\beta_{20\cdot1}^{(t)} &= \bar{y}_2 - \beta_{21\cdot1}^{(t)}\bar{y}_1 + z_{3t}(\sigma_{22\cdot1}^{(t)}/m)^{1/2}.
\end{aligned}$$

3. Compute $\psi_t = \psi(\phi^{(t)})$. For example, if $\psi(\phi) \equiv \mu_2 = \beta_{20\cdot1} + \beta_{21\cdot1}\mu_1$, then $\mu_{2t} = \beta_{20\cdot1}^{(t)} + \beta_{21\cdot1}^{(t)}\mu_1^{(t)}$.

6.3.3. Numerical Illustration

The methods of the previous two sections are now applied to the data in Table 6.1.

EXAMPLE 6.3. *Interval Estimation for the Bivariate Normal (Example 6.2 continued).* Table 6.2 shows 95% intervals for μ_2, σ_{22}, and ρ for the data in Table 6.1. Intervals based on four methods are presented: (1) asymptotic theory based on the inverse of the observed information matrix, as in (6.15) for μ_2; (2) the t-approximation for intervals for μ_2 obtained by replacing the normal percentile 1.96 in (6.15) by the 97.5th percentile of the t distribution on $m - 1 = 11$ degrees of freedom, namely, 2.201; (3) the 2.5th to 97.5th percentile of the Bayesian posterior distribution for the prior (6.16) with $a = c = 1$, simulated using the method of Section 6.3.2 with 9999 simulated values; and (4) normal intervals obtained by fitting normal distributions to the simulated posterior distributions from method 3, using the simulated posterior mean and variance. Methods 3 and 4 are repeated for two independent sets of random numbers (A and B) to give some idea of the simulation variance.

As one would expect, the narrowest intervals are obtained from asymptotic theory (method 1), and thus these intervals presumably have lower than the stated 95% coverage. The other intervals for μ_2 are fairly similar. The simulated

Table 6.2 95% Intervals for Parameters of Bivariate Normal Distribution, Based on Data in Table 6.1

Method	Parameters		
	μ_2	σ_{22}	ρ
Asymptotic theory	(43.98, 54.69)	(30.68, 198.71)	(−1.0018, −0.7882)
t-approximation	(43.38, 55.28)	—	—
Bayes simulation A	(43.71, 54.42)	(60.06, 289.73)	(−0.964, −0.662)
Normal approximation to A	(43.50, 55.21)	(15.91, 256.68)	(−1.029, −0.717)
Bayes simulation B	(43.54, 55.68)	(59.18, 293.93)	(−0.964, −0.656)
Normal approximation to B	(43.42, 55.22)	(14.62, 257.19)	(−1.033, −0.710)
ML estimates	49.33	114.70	−0.895

Bayesian probability intervals for σ_{22} have 2.5 percentiles much closer to the ML estimate (114.7) than the 97.5th percentiles, reflecting marked right skewness in the posterior distribution of σ_{22}. The normal approximations cannot reflect this asymmetry, but they have one realistic feature: They are considerably wider than the interval from asymptotic theory, and hence should have closer to 95% coverage in repeated sampling. Similar remarks apply to the intervals for ρ. Note that the lower limits for ρ from all the normal-based intervals are less than the minimum possible values (-1.0) and should be replaced by -1.0 in practice. The simulated Bayesian intervals for ρ look more reasonable. Coverage properties of these intervals in repeated sampling require further study. Little (1986) considers coverage properties of t approximations to the posterior distribution of μ_2 for various choices of a and c.

6.4. MONOTONE DATA WITH MORE THAN TWO VARIABLES

The methods described in Sections 6.2 and 6.3 can be readily generalized to the monotone pattern of data in Figure 6.1, where for each observation i, y_{ij} is recorded if $y_{i,j+1}$ is recorded $(j = 1, \dots, J-1)$, so that Y_1 is more observed than Y_2, which is more observed than Y_3, and so on. We confine attention to ML estimation. Precision of estimation can be addressed using straightforward extensions of the methods of Section 6.3.

The appropriate factorization for this pattern is

$$\prod_{i=1}^{n} f(y_{i1}, \dots, y_{iJ} | \phi)$$
$$= \prod_{i=1}^{n} f(y_{i1} | \phi_1) \prod_{i=1}^{m_2} f(y_{i2} | y_{i1}, \phi_2) \cdots \prod_{i=1}^{m_J} f(y_{iJ} | y_{i1}, \dots, y_{i,J-1}, \phi_J),$$

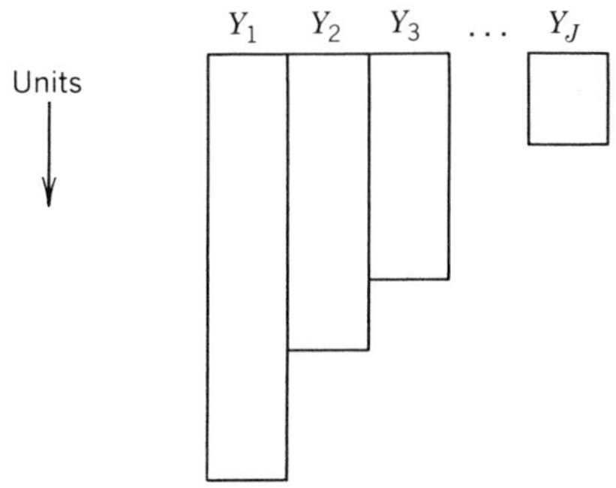

Figure 6.1. Schematic representation of a monotone data pattern.

where for $j = 1, \ldots, J$, $f(y_{ij}|y_{i1}, \ldots, y_{i,j-1}, \phi_j)$ is the conditional distribution of y_{ij} given $y_{i1}, \ldots, y_{i,j-1}$, indexed by the parameter ϕ_j. If $(y_{i1}, \ldots, y_{iJ})$ follows a multivariate normal distribution, then $f(y_{ij}|y_{i1}, \ldots, y_{i,j-1}, \phi_j)$ is a normal distribution with mean linear in $y_{i1}, \ldots, y_{i,j-1}$ and with constant variance. With the usual unrestricted parameter space of ϕ, the ϕ_j are distinct, and so ML estimates of ϕ_j are obtained by regressing the y_{ij} on the $y_{i1}, \ldots, y_{i,j-1}$, using the set of observations for which $y_{i1}, \ldots, y_{ij}$ are all observed.

EXAMPLE 6.4. *$K + 1$ Variables, One Subject to Nonresponse.* A simple but important extension of Example 6.1 replaces y_{i1} by a set of K completely observed variables, as for the data pattern in Figure 1.3. This results in a special case of monotone data with $J = 2$ and Y_1 representing K variables. The ML estimates of μ_2 and σ_{22} are

$$\begin{aligned} \hat{\mu}_2 &= \bar{y}_2 + (\hat{\mu}_1 - \bar{y}_1)\hat{\beta}_{21\cdot 1}, \\ \hat{\sigma}_{22} &= s_{22} + \hat{\beta}_{21\cdot 1}^{\mathrm{T}}(\hat{\sigma}_{11} - s_{11})\hat{\beta}_{21\cdot 1}, \end{aligned} \tag{6.17}$$

where $\hat{\mu}_1, \bar{y}_1$ are $(1 \times K)$ mean vectors, $\hat{\beta}_{21\cdot 1}$ is the $(K \times 1)$ vector of regression coefficients from the multiple regression of y_{i2} on y_{i1}, and $\hat{\sigma}_{11}$ and s_{11} are $(K \times K)$ covariance matrices, $\hat{\sigma}_{11}$ based on all n observations of y_{i1} and s_{11} based on the m observations of y_{i1} where y_{i2} is also observed. The ML estimate $\hat{\mu}_2$ corresponds to imputing the missing values of y_{i2} using the ML estimates of the multiple regression of y_{i2} on y_{i1}.

The estimate (6.17) is ML if (y_{i1}, y_{i2}) are iid $(K + 1)$-variate normally distributed, and the data are MAR. More generally, it is also ML if the data are MAR and

1. y_{i2} given y_{i1} is normal with mean $(\beta_{20\cdot 1} + y_{i1}\beta_{21\cdot 1})$ and variance $\sigma_{22\cdot 1}$.
2. y_{i1} has any distribution such that (i) $\hat{\mu}_1$ is the ML estimate of the mean of y_{i1}, and (ii) μ_1 and $\beta_{20\cdot 1}, \beta_{21\cdot 1}, \sigma_{22\cdot 1}$ are distinct from the parameters of this distribution.

An important special case arises with *dummy variable regression*, where y_{i1} represents K dummy variables indicating $K + 1$ groups. The kth component of y_{i1} is defined to be 1 if the ith observation belongs to group k and is zero otherwise: For an observation in group 1, $y_{i1} = (1, 0, 0, \ldots, 0)$; for an observation in group 2, $y_{i1} = (0, 1, 0, \ldots, 0)$; for an observation in group $K - 1$, $y_{i1} = (0, 0, \ldots, 0, 1)$; and for an observation in group K, $y_{i1} = (0, 0, \ldots, 0, 0)$; group K is often called the *reference* group.

With these definitions, $\hat{\mu}_1$ is a vector consisting of the proportions of the n sampled observations in each of the first K groups, μ_1 is the corresponding vector of expected proportions, and condition 2 above is satisfied. Condition

1 is equivalent to assuming that all values of y_{i2} in group k are normal with common mean μ_k and variance $\sigma_{22\cdot 1}$.

By the properties of dummy variable regression, the predicted value of y_{i2} for an observation in group k is the mean of the observed values of y_{i2} in group k. Thus the ML estimate corresponds to imputing the subclass means for missing values of y_{i2}, a form of mean imputation that we discussed when considering nonresponse in sample surveys in Chapter 4.

EXAMPLE 6.5. *Monotone Multivariate Normal Data.* A numerical illustration of ML estimation for monotone data with $J > 2$ is given in an analysis by Marini, Olsen and Rubin (1980) of panel study data with 4352 cases. The data pattern, given in Table 1.2, does not have a monotone pattern. But as noted in Chapter 1, a monotone pattern can be achieved by discarding some data, in particular, those superscripted by the letter b in the table. The resulting pattern is monotone as in Figure 6.1, with $J = 4$. Assuming normality, ML estimates of the mean and covariance matrix of the variables can be found by the following procedure:

1. Calculate the mean vector and covariance matrix for the fully observed block 1 variables, from all the observations.
2. Calculate the multivariate linear regression of the next most observed variables, block 2 on block 1, from observation with both block 1 and block 2 variables recorded.
3. Calculate the multivariate linear regression of block 3 on blocks 1 and 2, from observations with blocks 1–3 recorded.
4. Calculate the multivariate linear regression of block 4 on blocks 1–3 from observations with all variables recorded.

ML estimates of the means and covariance matrix of all the variables can be obtained as functions of the parameter estimates in 1 to 4. The computational details, involving the powerful sweep operator, are deferred until Section 6.5. Results are displayed in Table 6.3.

The first column gives a description of the variables. The next two columns give ML estimates of the means (μ_{ML}) and the standard deviations (σ_{ML}) of each variable. The rest of the table compares estimates of two alternative methods of estimation. Estimates (μ_A, σ_A) from the available-cases method are the sample means and standard deviations using all the observations available for each variable (Cf. Section 3.3). The two columns after the estimates indicate the magnitude of differences between the ML and available-case methods, measured in percent standard deviations. Estimates from available cases are quite close to the ML estimates, indicating some virtue for the former method. However, the method is not recommended for measures

Table 6.3 Maximum Likelihood Estimates of Means and Standard Deviations Obtained for the Total Original Sample and Comparisons with Two Alternative Sets of Estimates

Variable	Maximum Likelihood Estimates for Total Sample		Estimates Based on Available Cases				Estimates Based on Complete Cases			
	Mean	S.D.	Mean	S.D.	$\frac{(\mu_A - \mu_{ML})}{\sigma_{ML}} \times 100$	$\frac{(\sigma_A - \sigma_{ML})}{\sigma_{ML}} \times 100$	Mean	S.D.	$\frac{(\mu_C - \mu_{ML})}{\sigma_{ML}} \times 100$	$\frac{(\sigma_C - \sigma_{ML})}{\sigma_{ML}} \times 100$
Block 1 Variables: Measured during Adolescence										
Father's education	11.702	3.528	11.702	3.528	0.0	0.0	12.050	3.449	9.9	−1.6
Mother's education	11.508	2.947	11.508	2.947	0.0	0.0	11.864	2.865	12.1	−2.4
Father's occupation	6.115	2.904	6.115	2.904	0.0	0.0	6.407	2.868	10.1	−1.2
Intelligence	106.625	12.910	106.625	12.910	0.0	0.0	109.036	11.174	18.7	−13.4
College preparatory curriculum	0.411	0.492	0.411	0.492	0.0	0.0	0.528	0.499	13.4	1.4
Time spent on homework	1.589	0.814	1.589	0.814	0.0	0.0	1.633	0.795	5.4	−2.3
Grade point average	2.324	0.773	2.324	0.773	0.0	0.0	2.594	0.701	34.9	−9.3
College plans	0.488	0.500	0.488	0.500	0.0	0.0	0.595	0.491	21.4	−1.8
Friends' college plans	0.512	0.369	0.512	0.369	0.0	0.0	0.572	0.354	16.3	−4.1
Participation in extracurricular activities	0.413	0.492	0.413	0.492	0.0	0.0	0.492	0.500	15.8	1.4
Membership in top leading crowd	0.088	0.283	0.088	0.283	0.0	0.0	0.131	0.338	8.6	19.4
Membership in intermediate leading crowd	0.170	0.376	0.170	0.376	0.0	0.0	0.198	0.399	5.6	4.1
Cooking/drinking	0.570	1.032	0.570	1.032	0.0	0.0	0.483	0.835	−8.4	−9.4
Dating frequency at time of survey	4.030	4.802	4.030	4.802	0.0	0.0	3.701	4.523	−6.8	−5.8
Liking for self	2.366	0.525	2.366	0.525	0.0	0.0	2.364	0.515	−0.4	−1.9
Grade in school	2.432	1.048	2.432	1.048	0.0	0.0	2.496	1.064	6.1	1.5

Block 2 Variables: Measured for All Follow-up Respondents										
Educational attainment	13.625	2.295	13.274	2.262	6.5	−1.4	14.196	2.204	24.9	−4.0
Occupational prestige	44.405	13.008	45.085	12.893	5.2	−0.9	47.056	12.745	20.4	−2.0
Marital status	0.940	0.238	0.940	0.238	0.0	0.0	0.940	0.237	0.0	−0.4
Number of children	1.991	1.306	1.973	1.304	−1.4	−0.2	1.928	1.242	−4.8	−4.9
Age	30.629	1.221	30.655	1.225	2.1	0.3	30.726	1.152	7.9	−5.4
Father's occupational prestige	43.998	14.821	44.258	14.786	1.8	−0.2	44.782	14.333	5.3	−3.2
Block 3 Variables: Measured Only for Initial Questionnaire Respondents to the Follow-up										
Personal esteem	3.128	0.377	3.148	0.378	5.2	0.3	3.148	0.373	5.3	−1.1
Dating frequency during last two years of high school	4.374	3.408	4.202	3.261	−5.1	−1.4	4.213	3.352	−4.7	−1.6
Number of siblings	2.219	1.748	2.099	1.744	−6.9	−0.2	2.055	1.660	−9.4	−5.0
Block 4 Variables: Measured on Parents' Questionnaire										
Family income	4.092	1.530	4.075	1.538	−1.1	0.5	4.215	1.570	8.0	2.6
Parental encouragement to go to college	0.714	0.434	0.706	0.455	−1.6	4.8	0.754	0.431	8.0	−0.7
Number of children in family of origin	3.039	1.539	3.067	1.671	1.8	8.6	2.975	1.551	−4.2	0.8

of association, such as covariances or regression coefficients, as noted in Chapter 3.

The last four columns of the table present and compare estimates based only on the 1594 complete observations, the complete-case method discussed in Chapter 3. Estimates of means from this procedure can differ markedly from the ML estimates. For example, the estimate for grade point average is 0.35 of a standard deviation higher than the ML estimate, indicating the fact that students lost to follow-up appear to have lower scores than average.

6.5. COMPUTATION FOR MONOTONE NORMAL DATA VIA THE SWEEP OPERATOR

In this section we review the use of the *sweep operator* (Beaton, 1964) in linear regression with complete observations and show how this operator provides a simple and convenient way of performing the ML calculations for incomplete normal data. The version of sweep we describe is not exactly the one originally defined in Beaton (1964); rather, it is the one defined by Dempster (1969); another accessible reference is Goodnight (1979). The sweep operator will also be useful in Chapter 8 when we consider ML estimation for normal data with a general pattern of incompleteness.

The sweep operator is defined for symmetric matrices as follows. A $p \times p$ symmetric matrix G is said to be *swept on row and column* k if it is replaced by another symmetric $p \times p$ matrix H with elements defined as follows:

$$\begin{aligned} h_{kk} &= -1/g_{kk}, \\ h_{jk} &= h_{kj} = g_{jk}/g_{kk}, \qquad k \neq j, \\ h_{jl} &= g_{jl} - g_{jk}g_{kl}/g_{kk}, \qquad k \neq j, k \neq l. \end{aligned} \tag{6.18}$$

To illustrate (6.18) consider the 3×3 case:

$$G = \begin{bmatrix} g_{11} & g_{12} & g_{13} \\ g_{12} & g_{22} & g_{23} \\ g_{13} & g_{23} & g_{33} \end{bmatrix},$$

$$H = \mathrm{SWP}[1]G = \begin{bmatrix} -1/g_{11} & g_{12}/g_{11} & g_{13}/g_{11} \\ g_{12}/g_{11} & g_{22} - g_{12}^2/g_{11} & g_{23} - g_{13}g_{12}/g_{11} \\ g_{13}/g_{11} & g_{23} - g_{13}g_{12}/g_{11} & g_{33} - g_{13}^2/g_{11} \end{bmatrix},$$

We use the notation $\mathrm{SWP}[k]G$ to denote the matrix H defined by (6.18). Also the result of successively applying the operations $\mathrm{SWP}[k_1]$, $\mathrm{SWP}[k_2]$, ..., $\mathrm{SWP}[k_t]$ to the matrix G will be denoted by $\mathrm{SWP}[k_1, k_2, \ldots, k_t]G$. In actual computations, the sweep operation is most efficiently achieved by first replacing g_{kk} by $h_{kk} = -1/g_{kk}$, then replacing the remaining elements g_{jk} and g_{kj} in row and column k by $h_{jk} = h_{kj} = -g_{jk}h_{kk}$, and finally replacing elements g_{jl} that are neither in row k nor in column k by $h_{jl} = g_{jl} - h_{jk}g_{kl}$. Storage space can be saved by storing the distinct elements of the symmetric $p \times p$ matrices in vectors of length $p(p+1)/2$, so that for $k \leqslant j$, the (j,k)th element of the matrix is stored as the $(j(j-1)/2 + k)$th element of the vector.

Some algebra shows that the sweep operator is commutative, that is,

$$\text{SWP}[j,k]G = \text{SWP}[k,j]G.$$

It follows more generally that

$$\text{SWP}[j_1,\ldots,j_t]G = \text{SWP}[k_1,\ldots,k_t]G,$$

where $j_1, \ldots, j_t$ is any permutation of the set $k_1, \ldots, k_t$. That is, the order in which a set of sweeps is carried out does not affect the final answer algebraicly, although some orders may be computationally more accurate than others.

The sweep operator is closely related to linear regression. For example, suppose that G is a (2×2) covariance matrix of two variables Y_1 and Y_2. If $H = \text{SWP}[1]G$, then h_{12} is the regression coefficient of Y_1 from the regression of Y_2 on Y_1, and h_{22} is the residual variance of Y_2. Furthermore, if G is a sample covariance matrix from n independent observations, then $-h_{11}h_{22}/n$ is the estimated variance of the sample regression coefficient, h_{12}.

More generally, suppose we have a sample of n observations on p variables $Y_1, \ldots, Y_p$. Let G denote the $(p+1) \times (p+1)$ matrix

$$G = \begin{bmatrix} 1 & \bar{y}_1 & \cdots & \bar{y}_j & \cdots & \bar{y}_p \\ \bar{y}_1 & n^{-1}\sum y_1^2 & & & \cdots & n^{-1}\sum y_p y_1 \\ \vdots & \vdots & \ddots & & & \vdots \\ \bar{y}_k & & & n^{-1}\sum y_j y_k & & \\ \bar{y}_p & n^{-1}\sum y_1 y_p & & & \cdots \ddots & n^{-1}\sum y_p^2 \end{bmatrix},$$

where $\bar{y}_1, \ldots, \bar{y}_p$ are the sample means, and summations are over the n observations. For convenience we index the rows and columns from 0 to p, so that row and column j corresponds to variable Y_j.

Sweeping on row and column 0 yields

$$\text{SWP}[0]G = \begin{bmatrix} -1 & \bar{y}_1 & \cdots & \bar{y}_j & \cdots & \bar{y}_p \\ \bar{y}_1 & s_{11} & & & \cdots & s_{p1} \\ \vdots & \vdots & \ddots & & & \vdots \\ \bar{y}_k & & & s_{jk} & & \\ \vdots & & & & \ddots & \\ \bar{y}_p & s_{1p} & & & \cdots & s_{pp} \end{bmatrix}, \tag{6.19}$$

where s_{jk} is the sample covariance of Y_j and Y_k, with factor n^{-1} rather than $(n-1)^{-1}$. This operation corresponds to correcting the scaled cross products matrix of $Y_1, \ldots, Y_p$, G, for the means of $Y_1, \ldots, Y_p$ to create the covariance matrix. In terms of regression, the means in the first row and column of $\text{SWP}[0]G$ are coefficients from the regression of $Y_1, \ldots, Y_p$ on the constant term $Y_0 \equiv 1$, and the corrected, scaled cross-products matrix $\{s_{jk}\}$ is the residual covariance matrix from this regression. Thus we also call this process *sweeping on the constant term*. We shall call (6.19) the *augmented covariance matrix* of the variables $Y_1, \ldots, Y_p$.

Sweeping on row and column 1, corresponding to Y_1, yields the symmetric matrix

$$\text{SWP}[0,1]G$$

$$= \begin{bmatrix} -(1+\bar{y}_1^2/s_{11}) & \bar{y}_1/s_{11} & \bar{y}_2 - (s_{12}/s_{11})\bar{y}_1 & \cdots & \bar{y}_p - (s_{1p}/s_{11})\bar{y}_1 \\ & -1/s_{11} & s_{12}/s_{11} & \cdots & s_{1p}/s_{11} \\ & & s_{22} - s_{12}^2/s_{11} & \cdots & s_{2p} - s_{1p}s_{12}/s_{11} \\ & & & & \vdots \\ \bar{y}_p - (s_{1p}/s_{11})\bar{y}_1 & & & & s_{pp} - s_{1p}^2/s_{11} \end{bmatrix}$$

$$= \begin{bmatrix} -A & B \\ B^T & C \end{bmatrix},$$

where A is 2×2, B is $2 \times (p-1)$, and C is $(p-1) \times (p-1)$. This matrix yields results for the (multivariate) regression of $Y_2, \ldots, Y_p$ on Y_1. In particular, the jth column of B gives the intercept and slope for the regression of Y_{j+1} on Y_1, for $j = 1, \ldots, p-1$. The matrix C gives the residual covariance matrix of $Y_2, \ldots, Y_p$ given Y_1. Finally, the elements of A, when multiplied by the appropriate residual variance or covariance in C and divided by n, yield variances and covariances of the estimated regression coefficients in B.

Sweeping the constant term and the first q elements yields results for the multivariate regression of $Y_{q+1}, \ldots, Y_p$ on $Y_1, \ldots, Y_q$. Specifically, letting

$$\text{SWP}[0, 1, \ldots, q]G = \begin{bmatrix} -D & E \\ E^{\text{T}} & F \end{bmatrix},$$

where D is $(q+1) \times (q+1)$, E is $(q+1) \times (p-q)$, and F is $(p-q) \times (p-q)$, the jth column of E gives the least squares intercept and slopes of the regression of Y_{j+q} on $Y_1, \ldots, Y_q$, for $j = 1, 2, \ldots, p-q$; the matrix F is the residual covariance matrix of $Y_{q+1}, \ldots, Y_p$; and the elements of D can be used as above to give variances and covariances of the estimated regression coefficients in E.

In summary, ML estimates for the multivariate linear regression of $Y_{q+1}, \ldots, Y_p$ on $Y_1, \ldots, Y_q$ can be found by sweeping the rows and columns corresponding to the constant term and the predictor variables $Y_1, \ldots, Y_q$ out of the scaled cross products matrix G.

The operation of sweeping on a variable in effect turns that variable from an outcome (or dependent) variable into a predictor (or independent) variable. There is also an operator inverse to sweep that turns predictor variables into outcome variables. This operator is called *reverse sweep* (RSW) and is defined by

$$H = \text{RSW}[k]G,$$

where

$$\begin{aligned} h_{kk} &= -1/g_{kk}, \\ h_{jk} = h_{kj} &= -g_{jk}/g_{kk}, \qquad k \neq j, \\ h_{jl} &= g_{jl} - g_{jk}g_{kl}/g_{kk}, \qquad k \neq j, l \neq j. \end{aligned} \tag{6.20}$$

It is readily verified that reverse sweep is also commutative and is the inverse operator to sweep; that is

$$(\text{RSW}[k])(\text{SWP}[k])G = (\text{SWP}[k])(\text{RSW}[k])G = G.$$

EXAMPLE 6.6. *Bivariate Normal Monotone Data (Example 6.1 continued).* Various parameterizations of the bivariate normal distribution are easily related using the sweep and reverse sweep operators. Thus the parameters θ and ϕ of Example 6.1 and the relationships in (6.4) and (6.5) can be compactly expressed using the SWP[ ] and RSW[ ] notation. Also important, numerical values of ML estimates can be simply computed using these operators.

Suppose we arrange $\theta = (\mu_1, \mu_2, \sigma_{11}, \sigma_{12}, \sigma_{22})^{\text{T}}$ in the following symmetric matrix, which is the population analog of (6.19):

$$\theta^* = \begin{bmatrix} -1 & \mu_1 & \mu_2 \\ \mu_1 & \sigma_{11} & \sigma_{12} \\ \mu_2 & \sigma_{21} & \sigma_{22} \end{bmatrix}.$$

The matrix θ^* represents the parameters of the bivariate normal with the constant swept. If θ^* is swept on row and column 1 we obtain from (6.18):

$$\text{SWP}[1]\theta^* = \begin{bmatrix} -(1 + \mu_1^2/\sigma_{11}) & \mu_1/\sigma_{11} & \mu_2 - \mu_1\sigma_{12}/\sigma_{11} \\ \mu_1/\sigma_{11} & -\sigma_{11}^{-1} & \sigma_{12}/\sigma_{11} \\ \mu_2 - \mu_1\sigma_{12}/\sigma_{11} & \sigma_{21}/\sigma_{11} & \sigma_{22} - \sigma_{12}^2/\sigma_{11} \end{bmatrix}.$$

An examination of (6.4) reveals that row and column 2 of $\text{SWP}[1]\theta^*$ provide the intercept $(\mu_2 - \mu_1\sigma_{12}/\sigma_{11})$, the slope of the regression of Y_2 on Y_1 $(\sigma_{12}/\sigma_{11})$ and the residual variance $\sigma_{22\cdot 1} = \sigma_{22} - \sigma_{21}^2/\sigma_{11}$. Also, although not in a particularly familiar form, the 2×2 submatrix formed by rows and columns 0 and 1 provides the parameters of the distribution of Y_1. To see this, write

$$\phi^* = \mathrm{SWP}[1]\theta^* = \begin{bmatrix} \mathrm{SWP}[1]\begin{bmatrix} -1 & \mu_1 \\ \mu_1 & \sigma_{11} \end{bmatrix} & \begin{matrix} \beta_{20\cdot 1} \\ \beta_{21\cdot 1} \end{matrix} \\ \begin{matrix} \beta_{20\cdot 1} & \beta_{21\cdot 1} \end{matrix} & \sigma_{22\cdot 1} \end{bmatrix}, \tag{6.21}$$

where ϕ^* is a slightly modified version of $\phi = (\mu_1, \sigma_{11}, \beta_{20\cdot 1}, \beta_{21\cdot 1}, \sigma_{22\cdot 1})^{\mathrm{T}}$, displayed as a matrix. By Property 5.1, a similar expression relates the ML estimates of θ to the ML estimates of ϕ:

$$\hat{\phi}^* = \mathrm{SWP}[1]\hat{\theta}^* = \begin{bmatrix} \mathrm{SWP}[1]\begin{bmatrix} -1 & \hat{\mu}_1 \\ \hat{\mu}_1 & \hat{\sigma}_{11} \end{bmatrix} & \begin{matrix} \hat{\beta}_{20\cdot 1} \\ \hat{\beta}_{21\cdot 1} \end{matrix} \\ \begin{matrix} \hat{\beta}_{20\cdot 1} & \hat{\beta}_{21\cdot 1} \end{matrix} & \hat{\sigma}_{22\cdot 1} \end{bmatrix}.$$

Applying the RSW[1] operator to both sides yields

$$\hat{\theta}^* = \mathrm{RSW}[1]\begin{bmatrix} \mathrm{SWP}[1]\begin{bmatrix} -1 & \hat{\mu}_1 \\ \hat{\mu}_1 & \hat{\sigma}_{11} \end{bmatrix} & \begin{matrix} \hat{\beta}_{20\cdot 1} \\ \hat{\beta}_{21\cdot 1} \end{matrix} \\ \begin{matrix} \hat{\beta}_{20\cdot 1} & \hat{\beta}_{21\cdot 1} \end{matrix} & \hat{\sigma}_{22\cdot 1} \end{bmatrix}. \tag{6.22}$$

Expression (6.22) defines the transformation from $\hat{\phi}$ to $\hat{\theta}$ in terms of the sweep and reverse sweep operators and thus shows how these operators can be used to compute $\hat{\theta}$ from $\hat{\phi}$.

EXAMPLE 6.7. *Multivariate Monotone Normal Data* (*Example 6.5 continued*). We now extend Example 6.6 to show how the sweep and reverse sweep operators can be applied to find ML estimates of the mean and covariance matrix of a multivariate normal distribution from data with a monotone pattern. We assume that the data have the pattern of Figure 6.1, after suitable arrangement of the variables. Also for simplicity we consider the case with $J = 3$ blocks of variables. The extension to more than three blocks of variables is immediate.

STEP 1. Find the ML estimates $\hat{\mu}_1$ and $\hat{\Sigma}_{11}$ of the mean μ_1 and covariance matrix Σ_{11} of the first block of variables, which are completely observed. These are simply the sample mean and covariance matrix of Y_1 based on all the observations.

STEP 2. Find the ML estimates $\hat{\beta}_{20\cdot 1}$, $\hat{\beta}_{21\cdot 1}$, and $\hat{\Sigma}_{22\cdot 1}$ of the intercepts, regression coefficients, and residual covariance matrix for the regression of Y_2 on Y_1. These can be found by sweeping the variables Y_1 out of the augmented covariance matrix of Y_1 and Y_2 based on the observations with Y_1 and Y_2 both observed.

STEP 3. Find the ML estimates $\hat{\beta}_{30\cdot 12}$, $\hat{\beta}_{31\cdot 12}$, $\hat{\beta}_{32\cdot 12}$, and $\hat{\Sigma}_{33\cdot 12}$ of the intercepts, regression coefficients, and residual covariance matrix for the regression of Y_3 on Y_1 and Y_2. These can be found by sweeping the variables Y_1 and Y_2 out of the augmented covariance matrix of Y_1, Y_2, and Y_3 based on the complete observations with Y_1, Y_2, and Y_3 observed.

STEP 4. Calculate the matrix

$$A = \mathrm{SWP}[\mathbf{1}]\begin{bmatrix} -1 & \hat{\mu}_1^{\mathrm{T}} \\ \hat{\mu}_1 & \hat{\Sigma}_{11} \end{bmatrix} = \begin{bmatrix} a_{11} & a_{12} \\ a_{21} & A_{22} \end{bmatrix},$$

where SWP[**1**] is shorthand for sweeping on the set of variables Y_1.

STEP 5. Calculate the matrix

$$B = \mathrm{SWP}[\mathbf{2}]\begin{bmatrix} a_{11} & a_{12} & \hat{\beta}_{20\cdot 1}^{\mathrm{T}} \\ a_{21} & A_{22} & \hat{\beta}_{21\cdot 1}^{\mathrm{T}} \\ \hat{\beta}_{20\cdot 1} & \hat{\beta}_{21\cdot 1} & \hat{\Sigma}_{22\cdot 1} \end{bmatrix} = \begin{bmatrix} c_{11} & c_{12} & c_{13} \\ c_{21} & c_{22} & c_{23} \\ c_{31} & c_{32} & c_{33} \end{bmatrix},$$

where SWP[**2**] is shorthand for sweeping on the set of variables Y_2.

STEP 6. Finally, the ML estimate of the augmented covariance matrix of Y_1, Y_2, and Y_3 is given by

$$\begin{bmatrix} -1 & \hat{\mu}^{\mathrm{T}} \\ \hat{\mu} & \hat{\Sigma} \end{bmatrix} = \mathrm{RSW}[\mathbf{1},\mathbf{2}]\begin{bmatrix} c_{11} & c_{12} & c_{13} & \hat{\beta}_{20\cdot 1}^{\mathrm{T}} \\ c_{21} & c_{22} & c_{23} & \hat{\beta}_{31\cdot 12}^{\mathrm{T}} \\ c_{31} & c_{32} & c_{33} & \hat{\beta}_{32\cdot 12}^{\mathrm{T}} \\ \hat{\beta}_{30\cdot 12} & \hat{\beta}_{31\cdot 12} & \hat{\beta}_{32\cdot 12} & \hat{\Sigma}_{33\cdot 12} \end{bmatrix}.$$

This matrix contains the ML estimates of the mean and covariance matrix of Y_1, Y_2, and Y_3, as indicated.

Steps 4–6 can be represented concisely by the equation

$$\begin{bmatrix} -1 & \hat{\mu}^{\mathrm{T}} \\ \hat{\mu} & \hat{\Sigma} \end{bmatrix}$$

$$= \mathrm{RSW}[\mathbf{1},\mathbf{2}]\left[\mathrm{SWP}[\mathbf{2}]\left[\mathrm{SWP}[\mathbf{1}]\begin{bmatrix} -1 & \hat{\mu}_1^{\mathrm{T}} \\ \hat{\mu}_1 & \hat{\Sigma}_{11} \end{bmatrix} \begin{matrix} \hat{\beta}_{20\cdot 1}^{\mathrm{T}} \\ \hat{\beta}_{21\cdot 1}^{\mathrm{T}} \end{matrix} \atop \begin{matrix} \hat{\beta}_{20\cdot 1} & \hat{\beta}_{21\cdot 1} & \hat{\Sigma}_{22\cdot 1} \end{matrix}\right] \begin{matrix} \hat{\beta}_{30\cdot 12}^{\mathrm{T}} \\ \hat{\beta}_{31\cdot 12}^{\mathrm{T}} \\ \hat{\beta}_{32\cdot 12}^{\mathrm{T}} \end{matrix} \atop \begin{matrix} \hat{\beta}_{30\cdot 12} & \hat{\beta}_{31\cdot 12} & \hat{\beta}_{32\cdot 12} & \hat{\Sigma}_{33\cdot 12} \end{matrix}\right],$$

with obvious generalizations to more than three blocks of variables. This equation defines the transformation from $\hat{\phi}$ to $\hat{\theta}$ for this problem.

EXAMPLE 6.8. *A Numerical Example.* Rubin (1976) presents the calculations described above for the data in Table 6.4, taken from Draper and Smith

Table 6.4 The Data for Example 6.8[a]

	Variables				
Units	X_1	X_2	X_3	X_4	$Y = X_5$
1	7	26	6	60	78.5
2	1	29	15	52	74.3
3	11	56	8	20	104.3
4	11	31	8	47	87.6
5	7	52	6	33	95.9
6	11	55	9	22	109.2
7	3	71	17	(6)	102.7
8	1	31	22	(44)	72.5
9	2	54	18	(22)	93.1
10	(21)	(47)	4	(26)	115.9
11	(1)	(40)	23	(34)	83.8
12	(11)	(66)	9	(12)	113.3
13	(10)	(68)	8	(12)	109.4

[a] Values in parentheses are considered missing in the example.

(1968). The original labeling of the variables is $X_1, \ldots, X_5$. The data have the pattern of Figure 6.1 with $J = 3$ and $Y_1 = (X_3, X_5)$, $Y_2 = (X_1, X_2)$ and $Y_3 = X_4$. Step 1 gives the ML estimates of the marginal distribution of (X_3, X_5) as

$$\hat{\mu}_3 = 11.769, \qquad \hat{\mu}_5 = 95.423, \quad \hat{\sigma}_{33} = 37.870,$$

$$\hat{\sigma}_{35} = -47.566, \quad \hat{\sigma}_{55} = 208.905.$$

Step 2 is based on observations 1–9 and yields regression coefficients from the regression of (Y_1, Y_2) on (Y_3, Y_5)

$$\hat{\beta}_{10\cdot35} = 2.802, \qquad \hat{\beta}_{13\cdot35} = -0.526, \quad \hat{\beta}_{15\cdot35} = 0.105,$$

$$\hat{\beta}_{20\cdot35} = -74.938, \quad \hat{\beta}_{23\cdot35} = 1.062, \qquad \hat{\beta}_{25\cdot35} = 1.178,$$

and estimated residual covariance matrix

$$\hat{\Sigma}_{12\cdot35} = \begin{matrix} & Y_1 & Y_2 \\ Y_1 & 3.804 & -8.011 \\ Y_2 & -8.011 & 24.382 \end{matrix}.$$

Step 3 is based on observations 1–6 and yields the following estimated coefficients and residual variance for the regression of Y_4 on the other variables:

$$\hat{\beta}_{40\cdot1235} = 85.753, \quad \hat{\beta}_{41\cdot1235} = -1.863, \quad \hat{\beta}_{42\cdot1235} = -1.324$$

$$\hat{\beta}_{43\cdot1235} = -1.533, \quad \hat{\beta}_{45\cdot1235} = 0.397, \qquad \hat{\sigma}_{44\cdot1235} = 0.046.$$

Steps 4–6 yield

$$\begin{bmatrix} -1 & \hat{\mu}^{\mathrm{T}} \\ \hat{\mu} & \hat{\Sigma} \end{bmatrix}$$

$$\begin{matrix} x_3 & x_5 & x_1 & x_2 & x_4 \end{matrix}$$

$$= \mathrm{RSW}[1235]\left[\begin{array}{cc} \mathrm{SWP}[12]\left[\begin{array}{cc} \mathrm{SWP}[35]\begin{bmatrix} -1 & 11.769 & 95.423 \\ 11.769 & 37.870 & -47.566 \\ 95.423 & -47.566 & 208.905 \end{bmatrix} & \begin{matrix} 2.802 & -74.938 \\ -0.526 & 1.062 \\ 0.105 & 1.178 \end{matrix} \\ \begin{matrix} 2.802 & -0.526 & 0.105 \\ -74.938 & 1.062 & 1.178 \end{matrix} & \begin{matrix} 3.804 & -8.011 \\ -8.011 & 24.382 \end{matrix} \end{array}\right] & \begin{matrix} 85.753 \\ -1.533 \\ 0.397 \\ -1.863 \\ -1.324 \end{matrix} \\ \begin{matrix} 85.753 & -1.533 & 0.397 & -1.863 & -1.324 \end{matrix} & 0.046 \end{array}\right]$$

Calculating the right side and reordering the variables gives ML estimates

$$\begin{matrix} & x_1 & x_2 & x_3 & x_4 & x_5 \end{matrix}$$

$$\hat{\mu}^{\mathrm{T}} = [\begin{matrix} 6.655 & 49.965 & 11.769 & 27.047 & 95.423 \end{matrix}]$$

$$\hat{\Sigma} = \begin{bmatrix} 21.826 & 20.864 & -24.900 & -11.473 & 46.953 \\ 20.864 & 238.012 & -15.817 & -252.072 & 195.604 \\ -24.900 & -15.817 & 37.870 & -9.599 & -47.556 \\ -11.473 & -252.072 & -9.599 & 294.183 & -190.599 \\ 46.953 & 195.604 & -47.556 & -190.599 & 208.905 \end{bmatrix}.$$

6.6. FACTORIZATIONS FOR SPECIAL NONMONOTONE PATTERNS

Nonmonotone patterns of incomplete data where factorizations of the likelihood are possible have been noted by Anderson (1957), where each factor is a complete-data likelihood and the data are normal, and Rubin (1974) more generally. The basic case is given by Figure 6.2, adapted from Rubin (1974). It has variables arranged into three blocks (Y_1, Y_2, Y_3) such that

1. Y_3 is *more observed* than Y_1, in the sense that for any unit for which Y_1 is at least partially observed, Y_3 is fully observed.

	Y_3	Y_2	Y_1
	$1 \cdots 1$	$0 \cdots 0$	$X \cdots X$
Units	$\vdots \quad \vdots$	$\vdots \quad \vdots$	$\vdots \quad \vdots$
↓	$1 \cdots 1$	$0 \cdots 0$	$X \cdots X$
	$X \cdots X$	$X \cdots X$	$0 \cdots 0$
	$\vdots \quad \vdots$	$\vdots \quad \vdots$	$\vdots \quad \vdots$
	$X \cdots X$	$X \cdots X$	$0 \cdots 0$

1 = Observed
0 = Missing
X = Possibly observed

Figure 6.2. Data pattern where Y_3 is more observed than Y_1, and Y_1 and Y_2 are never jointly observed.

2. Y_1 and Y_2 *are never jointly observed*, in the sense that for any unit for which Y_2 is at least partially observed, Y_1 is completely missing, and vice versa.
3. The rows of Y_1 are conditionally independent given Y_3 with the same set of parameters.

When Y_2 is ignored and Y_1 and Y_3 are scalar, Figure 6.2 reduces to bivariate monotone data. Under MAR, the loglikelihood of the data decomposes into two terms: the first term, for the marginal distribution of Y_2 and Y_3 with parameter ϕ_{23}, is based on all the units; the second term for the conditional distribution of Y_1 given Y_3 with parameter $\phi_{1\cdot 3}$ is based on units with Y_3 fully observed. The proof of this result, which encompasses a proof of factorizations for monotone data, is given in Rubin (1974, §2).

The parameters ϕ_{23} and $\phi_{1\cdot 3}$ are often distinct, as ϕ_{23} can be reparameterized (in the obvious notation) as $\phi_{2\cdot 3}$ and ϕ_3, and the parameters $\phi_{1\cdot 3}$, $\phi_{2\cdot 3}$, and ϕ_3 are often distinct. An important aspect of this example is that ϕ_{23} and $\phi_{1\cdot 3}$ do not provide a complete reparametrization of the parameters of the joint distribution of Y_1, Y_2, and Y_3, in that the parameters of conditional association (e.g., partial correlation) between Y_1 and Y_2 given Y_3 are not included. These parameters do not appear in the loglikelihood and cannot be estimated from the data.

Rubin (1974) shows how repeated reductions of the pattern of Figure 6.2 can be used to factorize the likelihood as fully as possible. Although in general not all of the resultant factors can be dealt with independently using complete-data methods, we illustrate the main ideas using two examples that do reduce to complete-data problems.

EXAMPLE 6.9. *A Normal Three-Variable Example.* Lord (1955) and Anderson (1957) consider a trivariate normal sample with the pattern of Figure 6.2, with Y_1, Y_2, and Y_3 univariate, no complete observations, m_1 observations on Y_1 and Y_3, and m_2 observations on Y_2 and Y_3, with $n = m_1 + m_2$. Assuming the data are MAR, the likelihood factorizes into three components: (1) $m_1 + m_2$ observations on the marginal normal distribution of Y_3, with parameters μ_3 and σ_{33}; (2) m_1 observations on the conditional distribution of Y_1 given Y_3, with intercept $\beta_{10\cdot 3}$, slope $\beta_{13\cdot 3}$ and variance $\sigma_{11\cdot 3}$; and (3) m_2 observations on the conditional distribution of Y_2 given Y_3, with intercept $\beta_{20\cdot 3}$, slope $\beta_{23\cdot 3}$ and variance $\sigma_{22\cdot 3}$. These three components involve eight distinct parameters, whereas the original joint distribution of Y_1, Y_2, and Y_3 involves nine parameters, namely, three means, three variances, and three covariances. The missing parameter in the reparameterization is the partial (conditional) correlation between Y_1 and Y_2 given Y_3, $\sigma_{12\cdot 3}/\sqrt{\sigma_{11\cdot 3}\sigma_{22\cdot 3}}$, about which there is no information in the data.

Data having such a pattern of incompleteness are not uncommon. One context, where each Y_i is multivariate, is the *file-matching* problem, which arises when combining large government data bases. For example, suppose we have one file that is a random sample of Internal Revenue Service (IRS) records (with unit identifiers removed) and another file that is a random sample of Social Security Administration (SSA) records (also with unit identifiers removed). The IRS file has detailed income information (Y_1) and background information (Y_3), whereas the SSA file has detailed work history information (Y_2) and the same background information (Y_3). The merged file can be viewed as a sample with Y_3 observed on all units but Y_1 and Y_2 never jointly observed. The term *file matching* is used to describe this situation because an attempt is often made to fill in the missing Y_1 and Y_2 values by matching units across files on the basis of Y_3 and imputing the values from matching units. Such problems are discussed in Rubin (1986).

EXAMPLE 6.10. *An Application to Educational Data.* In educational testing problems, such as given in Rubin and Thayer (1978), it is common that several new tests will be evaluated on different random samples from the same population. Specifically, let $X = (X_1, \ldots, X_p)$ represent p standard tests given to all sampled subjects, and suppose new test Y_1 is given to the first sample of m_1 subjects, new test Y_2 is given to the second sample of m_2 subjects, and so on up to Y_q, where the samples have no individual in common; because of the random sampling, the missing Y values are MCAR. Figure 6.3 displays the case with $q = 3$, which is a simple extension of the pattern in Example 6.9.

The partial correlations among the Y_j's given X are inestimable in the strict sense that there is no information about their values in the data. The simple correlations among the Y_j's are often of more interest in educational testing

		Standard Tests			New Tests		
Subsample	Subject	X_1,	...,	X_p	Y_1	Y_2	Y_3
1	1	1	⋯	1	1	0	0
	⋮	⋮		⋮	⋮	⋮	⋮
	m_1	1	⋯	1	1	0	0
2	$m_1 + 1$	1	⋯	1	0	1	0
	⋮	⋮		⋮	⋮	⋮	⋮
	$m_1 + m_2$	1	⋯	1	0	1	0
3	$m_1 + m_2 + 1$	1	⋯	1	0	0	1
	⋮	⋮		⋮	⋮	⋮	⋮
	$m_1 + m_2 + m_3$	1	⋯	1	0	0	1

Figure 6.3. Data structure with three new tests: 1 = score observed, 0 = score missing.

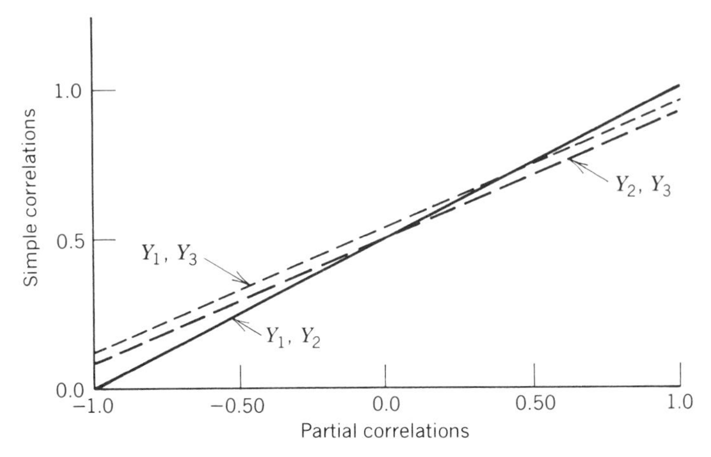

Figure 6.4. Simple correlations as a function of partial correlations (Rubin and Thayer, 1978).

problems. Although these correlations do not have unique ML estimates, there is information in the data about their values.

Straightforward algebra shows that the simple correlation between Y_j and Y_k depends on the partial correlation between Y_j and Y_k but not on the partial correlation between any other pair of variables. As the partial correlation between Y_j and Y_k increases, the simple correlation between Y_j and Y_k increases; furthermore, this relationship is linear. Hence, given estimates of the simple correlation for two different values of the partial correlation (e.g., 0 and 1), one can estimate the simple correlation for any other value of the partial correlation using linear interpolation (or extrapolation, depending on the chosen values). Figure 6.4 displays plots of the estimated simple correlations as a function of the partial correlations for Education Testing Service data with the structure of Figure 6.3, with $m_1 = 1325$, $m_2 = 1345$, $m_3 = 2000$ and bivariate X (Rubin and Thayer, 1978).

As with monotone normal data, the sweep operator is an extremely useful notational and computational device for creating this figure. Computations can be described as follows:

STEP 1. Find the ML estimates of the marginal distribution of X, μ_x, and Σ_{xx}. These are simply the sample mean and covariance of all n observations, $\hat{\mu}_x$ and $\hat{\Sigma}_{xx}$. This step yielded $\hat{\mu}_x^T = (43.27, 26.79)$ and

$$\hat{\Sigma}_{xx} = \begin{bmatrix} 330.33 & 118.92 \\ 118.92 & 138.13 \end{bmatrix}.$$

STEP 2. Find the ML estimates $\hat{\beta}_{10\cdot x}$, $\hat{\beta}_{1x\cdot x}$, and $\hat{\sigma}_{11\cdot x}$ of the regression coefficients and residual variance for the regression of Y_1 on X. These can be found by sweeping the variables X out of the augmented sample covariance matrix of Y_1 and X based on the m_1 observations with X and Y_1 both observed. This step yielded $(\hat{\beta}_{10\cdot x}, \hat{\beta}_{1x\cdot x}^{\mathrm{T}}) = (0.9925, 0.1010, 0.1718)$ and $\hat{\sigma}_{11\cdot x} = 11.0887$.

STEP 3. Find the ML estimates $\hat{\beta}_{20\cdot x}$, $\hat{\beta}_{2x\cdot x}$, and $\hat{\sigma}_{22\cdot x}$ of the regression coefficients and residual variance for the regression of Y_2 on X. These can be found by sweeping the variables X out of the augmented sample covariance matrix of Y_2 and X based on the m_2 observations with X and Y_2 both observed. This step yielded $(\hat{\beta}_{20\cdot x}, \hat{\beta}_{2x\cdot x}^{\mathrm{T}}) = (-0.4444, 0.1760, 0.2278)$ and $\hat{\sigma}_{22\cdot x} = 27.3818$.

STEP 4. Find the ML estimates $\hat{\beta}_{30\cdot x}$, $\hat{\beta}_{3x\cdot x}$, and $\hat{\sigma}_{33\cdot x}$ of the regression coefficients and residual variance for the regression of Y_3 on X. These can be found by sweeping the variables X out of the augmented sample covariance matrix of Y_3 and X based on the m_3 observations with X and Y_3 both observed. This step yielded $(\hat{\beta}_{30\cdot x}, \hat{\beta}_{3x\cdot x}^{\mathrm{T}}) = (0.3309, 0.2298, 0.5731)$ and $\hat{\sigma}_{33\cdot x} = 71.4943$.

STEP 5. Fix all inestimable partial correlations at zero; then find unique ML estimates of the mean vector $\mu = \begin{pmatrix} \mu_x \\ \mu_y \end{pmatrix}$ and covariance matrix

$$\Sigma = \begin{bmatrix} \Sigma_{xx} & \Sigma_{xy} \\ \Sigma_{yx} & \Sigma_{yy} \end{bmatrix}$$

of all variables, as follows:

$$\begin{bmatrix} -1 & \hat{\mu}_{(0)}^{\mathrm{T}} \\ \hat{\mu}_{(0)} & \hat{\Sigma}_{(0)} \end{bmatrix} = \mathrm{RSW}[x] \begin{bmatrix} \mathrm{SWP}\begin{bmatrix} -1 & \hat{\mu}_x^{\mathrm{T}} \\ \hat{\mu}_x & \hat{\Sigma}_{xx} \end{bmatrix} & \begin{matrix} \hat{\beta}_{10\cdot x} & \hat{\beta}_{20\cdot x} & \hat{\beta}_{30\cdot x} \\ \hat{\beta}_{1x\cdot x} & \hat{\beta}_{2x\cdot x} & \hat{\beta}_{3x\cdot x} \end{matrix} \\ \begin{matrix} \hat{\beta}_{10\cdot x} & \hat{\beta}_{1x\cdot x}^{\mathrm{T}} \\ \hat{\beta}_{20\cdot x} & \hat{\beta}_{2x\cdot x}^{\mathrm{T}} \\ \hat{\beta}_{30\cdot x} & \hat{\beta}_{3x\cdot x}^{\mathrm{T}} \end{matrix} & \begin{matrix} \hat{\sigma}_{11\cdot x} & 0 & 0 \\ 0 & \hat{\sigma}_{22\cdot x} & 0 \\ 0 & 0 & \hat{\sigma}_{11\cdot x} \end{matrix} \end{bmatrix}, \quad (6.23)$$

where the zeroes on the left side of (6.23) refer to the estimates being conditional on the zero partial correlations. Step 5 yielded $\hat{\mu}_y^{\mathrm{T}} = (9.96, 13.27, 25.63)$,

$$\hat{\Sigma}_{yy} = \begin{bmatrix} 22.66 & 17.61 & 32.84 \\ 17.61 & 54.31 & 49.61 \\ 32.84 & 49.61 & 165.64 \end{bmatrix},$$

$$\hat{\Sigma}_{xy} = \begin{bmatrix} 53.78 & 85.22 & 144.08 \\ 36.74 & 52.39 & 106.50 \end{bmatrix}.$$

STEP 6. Fix all inestimable correlations at 1 and find the corresponding ML estimates

$$\begin{bmatrix} -1 & \hat{\mu}_{(1)}^{\mathrm{T}} \\ \hat{\mu}_{(1)} & \hat{\Sigma}_{(1)} \end{bmatrix}.$$

These estimates are obtained by replacing the lower-right 3×3 submatrix on the right side of (6.23) with

$$\begin{bmatrix} \hat{\sigma}_{11 \cdot x} & \sqrt{\hat{\sigma}_{11 \cdot x}\hat{\sigma}_{22 \cdot x}} & \sqrt{\hat{\sigma}_{11 \cdot x}\hat{\sigma}_{33 \cdot x}} \\ \sqrt{\hat{\sigma}_{11 \cdot x}\hat{\sigma}_{22 \cdot x}} & \hat{\sigma}_{22 \cdot x} & \sqrt{\hat{\sigma}_{22 \cdot x}\hat{\sigma}_{33 \cdot x}} \\ \sqrt{\hat{\sigma}_{11 \cdot x}\hat{\sigma}_{33 \cdot x}} & \sqrt{\hat{\sigma}_{22 \cdot x}\hat{\sigma}_{33 \cdot x}} & \hat{\sigma}_{33 \cdot x} \end{bmatrix}.$$

This step yielded the same value of $\hat{\mu}_y$ and the diagonal of $\hat{\Sigma}_{yy}$ but different estimates of the other parameters. In particular, the estimated correlations among the Y variables were .999, .996, .990. The corresponding estimates from step 5 were .50, .54, .52.

Linear interpolation between the correlations reported in steps 5 and 6 produces Figure 6.4. Other parameters, such as multiple correlations, were also considered in Rubin and Thayer (1978). In general, these are not linear in the inestimable partial correlations, but they are still simple to compute.

REFERENCES

Anderson, T. W. (1957). Maximum likelihood estimates for the multivariate normal distribution when some observations are missing, *J. Am. Statist. Assoc.* **52**, 200–203.

Beaton, A. E. (1964). The use of special matrix operations in statistical calculus, Educational Testing Service Research Bulletin, RB-64-51.

Box, G. E. P., and Tiao, G. C. (1973). *Bayesian Inference in Statistical Analysis*, Reading MA: Addison-Wesley.

Cochran, W. G. (1977). *Sampling Techniques*, 3rd ed. New York: Wiley.

Dempster, A. P. (1969). *Elements of Continuous Multivariate Analysis*. Reading, MA: Addison-Wesley.

Dixon, W. J. (Ed.) (1983). *BMDP Statistical Software*, 1983 revised printing. Berkeley: University of California Press.

Draper, N. R., and Smith, H. (1968). *Applied Regression Analysis*, New York: Wiley.

Goodnight, J. H. (1979). A tutorial on the SWEEP operator. *American Statistician*, **33**, 149–158

Lindley, D. V. (1965). *Introduction to Probability and Statistics from a Bayesian Viewpoint*, Vol. 2. Cambridge: Cambridge University Press.

Little, R. J. A. (1976), Inference about means from incomplete multivariate data, *Biometrika* **63**, 593–604.

Little, R. J. A. (1986). Some methods for interval estimation with incomplete data.

Little, R. J. A. and Rubin, D. B. (1983). On jointly estimating parameters and missing data by maximizing the complete-data likelihood. *American Statistician* **37**, 218–220.

Lord, F. M. (1955). Estimation of parameters from incomplete data, *J. Am. Statist. Assoc.* **50**, 870–876.

Marini, M. M., Olsen, A. R., and Rubin, D. B. (1980). Maximum likelihood estimation in panel studies with missing data, in *Sociological Methodology 1980*. San Francisco: Jossey-Bass.

Morrison, D. F. (1971). Expectations and variances of maximum likelihood estimates of the multivariate normal distribution parameters with missing data, *J. Am. Statist. Assoc.* **66**, 602–4.

Rubin, D. B. (1974), Characterizing the estimation of parameters in incomplete data problems, *J. Am. Statist. Assoc.* **69**, 467–474.

Rubin, D. B. (1976). Comparing regressions when some predictor variables are missing, *Technometrics* **18**, 201–206.

Rubin, D. B., and Thayer, D. (1978). Relating tests given to different samples, *Psychometrika* **43**, 3–10.

Rubin, D. B. (1986). Statistical matching using file concatenation with adjusted weights and multiple imputation, *J. Business Econ. Stat.* **4**, 87–94.

Snedecor, G. W., and Cochran, W. G. (1967). *Statistical Methods*, 6th ed. Ames: Iowa State University Press.

PROBLEMS

1. Review the literature on the data structure of Example 6.1 prior to Anderson (1957).

2. Assume the data in Example 6.1 are MAR. Show that given $(y_{11}, \ldots, y_{n1})$, $\hat{\beta}_{20\cdot 1}$ and $\hat{\beta}_{21\cdot 1}$ are unbiased for $\beta_{20\cdot 1}$ and $\beta_{21\cdot 1}$. Hence show that $\hat{\mu}_2$ is unbiased for μ_2.

3. Assume the data in Example 6.1 are MCAR. By first conditioning on $(y_{11}, \ldots, y_{n1})$, find the exact small sample variance of $\hat{\mu}_2$. [*Hint:* If u is chi-squared on d degrees of freedom, then $E(1/u) = 1/(d-2)$.] (See Morrison, 1971). Hence show that $\hat{\mu}_2$ has a smaller variance than $\bar{y}_2$ if and only if $\rho^2 > 1/(m-2)$, where m is the number of complete cases.

4. Compare the asymptotic variance of $\hat{\mu}_2 - \mu_2$ given by (6.13) and (6.14) with the small sample variance computed in problem 3.

5. Prove results 1–6 in Section 6.3.2. [For help see Chapter 2 of Box and Tiao (1973).]

6. Derive the asymptotic variance of the ML estimate of $\ln \sigma_{22}$ for the data of Section 6.2 by expressing $\ln \sigma_{22}$ as a function of ϕ and applying the method of Section 6.3.1. Hence calculate an asymptotic 95% confidence interval for $\ln \sigma_{22}$ for the data in Table 6.1, and transform it into an interval for σ_{22}. Compare this interval with other intervals for σ_{22} in Table 6.2.

7. For the bivariate normal distribution, express the regression coefficient $\beta_{12\cdot 2}$ of X_1 on X_2 in terms of the parameters ϕ in Section 6.2, and hence derive its ML estimate for the data of Example 6.1.

8. Show that for the setup of Problem 7, the estimate of $\beta_{12\cdot 2}$ obtained by maximizing the complete data loglikelihood over parameters and missing data is $\tilde{\beta}_{12\cdot 2} = \hat{\beta}_{12\cdot 2}\hat{\sigma}_{22}/\hat{\sigma}^{*}_{22}$, where (in the notation of Section 6.2),

$$\hat{\sigma}^{*}_{22} = \hat{\beta}^{2}_{21\cdot 1}\hat{\sigma}_{11} + n^{-1}\sum_{i=1}^{m}[y_{i2} - \bar{y}_2 - \hat{\beta}_{21\cdot 1}(y_{i1} - \bar{y}_1)]^2.$$

Hence show that $\tilde{\beta}_{12\cdot 2}$ is not consistent for $\beta_{12\cdot 2}$ unless the fraction of missing data tends to zero as $n \to \infty$. (Cf. Section 5.4; for help see Little and Rubin, 1983.)

9. Show that the factorization of Example 6.1 does not yield distinct parameters $\{\phi_j\}$ for a bivariate normal sample with means (μ_1, μ_2), correlation ρ and *common* variance σ^2, with missing values on Y_2.

10. Using the computer or otherwise, generate a bivariate normal sample of 20 cases with parameters $\mu_1 = \mu_2 = 0$, $\sigma_{11} = 1$, $\sigma_{22} = 2$, $\sigma_{12} = 1$, and delete values of Y_2 so that $\Pr(y_2 \text{ missing}|y_1, y_2)$ equals 0.2 if $y_1 < 0$ and 0.8 if $y_1 \geqslant 0$.

(a) Construct a test for whether the data are MCAR and carry out the test on your data set. [For help see the program BMDP8D in the BMDP Statistical Software manual (Dixon, 1983).]

(b) Compute 95% confidence intervals for μ_2, using (i) the data before values were deleted; (ii) the complete cases; and (iii) the t-approximation in (b) of Table 6.2. Summarize the properties of these intervals for this missing-data mechanism.

11. Prove that SWP is commutative and conclude that the order in which a set of sweeps is taken is irrelevant algebraically. (However, it can be shown that the order can matter for computational ease and accuracy.)

12. Show that RSW is the inverse operation to SWP.

13. Show how to compute partial correlations and multiple correlations using SWP.

14. Estimate the parameters of the distribution of X_1, X_2, X_3, and X_5 in Example 6.8, pretending X_4 is never observed. Would the calculations be more or less work if X_3 rather than X_4 was never observed?

15. Create a factorization table (see Rubin, 1974) for the data in Example 6.10. State why the estimates produced in Example 6.10 are ML.

16. If data are MAR and the data analyst discards values to yield a data set with all complete-data factors, then are the resultant missing data necessarily MAR? Provide an example to illustrate important points.

CHAPTER 7

Maximum Likelihood for General Patterns of Missing Data: Introduction and Theory with Ignorable Nonresponse

7.1. ALTERNATIVE COMPUTATIONAL STRATEGIES

Patterns of incomplete data in practice often do not have the particular forms that allow explicit ML estimates to be calculated by exploiting factorizations of the likelihood. Furthermore, for some models a factorization exists, but the parameters ϕ_j in the factorization are not distinct, and thus maximizing the factors separately does not maximize the likelihood. In this chapter we consider iterative methods of computation for situations without explicit ML estimates. In some cases these methods can be applied to incomplete-data factors discussed in Section 6.6.

Suppose as before that we have a model for the complete data Y, with associated density $f(Y|\theta)$ indexed by unknown parameter θ. We write $Y = (Y_{obs}, Y_{mis})$ where Y_{obs} represents the observed part of Y and Y_{mis} denotes the missing values. In this chapter we assume for simplicity that the data are MAR and that the objective is to maximize the likelihood

$$L(\theta | Y_{obs}) = \int f(Y_{obs}, Y_{mis} | \theta)\, dY_{mis} \tag{7.1}$$

with respect to θ. Similar considerations apply to the more general case where the data are not MAR and consequently a term representing the missing-data mechanism is included in the model; such cases are considered in Chapter 11.

If the likelihood is differentiable and unimodal, ML estimates can be found by solving the likelihood equation

$$S(\theta | Y_{\text{obs}}) \equiv \frac{\partial \ln L(\theta | Y_{\text{obs}})}{\partial \theta} = 0. \tag{7.2}$$

When a closed-form solution of (7.2) cannot be found, iterative methods can be applied. Let $\theta^{(0)}$ be an initial estimate of θ, for example, an estimate based on the completely observed observations. Let $\theta^{(t)}$ be the estimate at the tth iteration. The *Newton–Raphson* algorithm is defined by the equation

$$\theta^{(t+1)} = \theta^{(t)} + I^{-1}(\theta^{(t)} | Y_{\text{obs}}) S(\theta^{(t)} | Y_{\text{obs}}), \tag{7.3}$$

where $I(\theta | Y_{\text{obs}})$ is the observed information,

$$I(\theta | Y_{\text{obs}}) = -\frac{\partial^2 l(\theta | Y_{\text{obs}})}{\partial \theta \, \partial \theta}.$$

If the loglikelihood function is concave and unimodal, then the sequence of iterates $\theta^{(t)}$ converges to the ML estimate $\hat{\theta}$ of θ, in one step if the loglikelihood is a quadratic function of θ. A variant of this procedure is the *Method of Scoring*, where the observed information in (7.3) is replaced by the expected information:

$$\theta^{(t+1)} = \theta^{(t)} + J^{-1}(\theta^{(t)}) S(\theta^{(t)} | Y_{\text{obs}}), \tag{7.4}$$

where

$$J(\theta) = E\{I(\theta | Y_{\text{obs}}) | \theta\} = -\int \frac{\partial^2 l(\theta | Y_{\text{obs}})}{\partial \theta \, \partial \theta} f(Y_{\text{obs}} | \theta) \, dY_{\text{obs}}.$$

Both these methods involve calculating the matrix of second derivatives of the loglikelihood. For complex patterns of incomplete data, the entries in this matrix tend to be complicated functions of θ. Also, the matrix is large when θ has high dimension. As a result, to be practicable the methods can require careful algebraic manipulations and efficient programming.

An alternative algorithm (Berndt et al., 1974) exploits the fact that the sampling covariance matrix of the score $S(\theta | Y_{\text{obs}})$ is a consistent estimate of the information in the neighborhood of $\hat{\theta}$. The resulting iterative equation is

$$\theta^{(t+1)} = \theta^{(t)} + \lambda_t Q^{-1}(\theta^{(t)}) S(\theta^{(t)} | Y_{\text{obs}})$$

where $Q(\theta) = \sum_{i=1}^{n} (\partial l_i / \partial \theta)(\partial l_i / \partial \theta)^{\text{T}}$; l_i is the loglikelihood of the ith observation; and λ_t is a positive step size designed to ensure convergence to a local maximum. Other variants of the Newton–Raphson algorithm approximate the derivatives of the loglikelihood numerically, by using first and second differences between successive iterates.

An alternative computing strategy for incomplete-data problems, which does not require second derivatives to be calculated or approximated, is the *Expectation–Maximization (EM)* algorithm, a method that relates ML estimation of θ from $l(\theta | Y_{\text{obs}})$ to ML estimation based on the complete-data loglikelihood $l(\theta | Y)$. In many important cases, the EM algorithm is remarkably simple, both conceptually and computationally. The remainder of this chapter is devoted to the EM algorithm.

7.2. THE EM ALGORITHM: BACKGROUND MATERIAL

The EM algorithm is a very general iterative algorithm for ML estimation in incomplete-data problems. In fact, the range of problems that can be attacked by EM is very broad and includes problems not usually considered to be ones arising from missing or incomplete data (e.g., variance components estimation, iteratively reweighted least squares).

The EM algorithm formalizes a relatively old ad hoc idea for handling missing data: (1) replace missing values by estimated values, (2) estimate parameters, (3) reestimate the missing values assuming the new parameter estimates are correct, (4) reestimate parameters, and so forth, iterating until convergence. Such methods are EM algorithms for models where the complete-data loglikelihood $l(\theta | Y_{\text{obs}}, Y_{\text{mis}}) = \ln L(\theta | Y_{\text{obs}}, Y_{\text{mis}})$ is linear in Y_{mis}; more generally, missing sufficient statistics rather than individual observations need to be estimated, and even more generally, the loglikelihood $l(\theta | Y)$ itself needs to be estimated at each iteration of the algorithm.

Because the EM algorithm is so closely tied to the intuitive idea of filling in missing values and iterating, it is not surprising that the algorithm has been proposed for many years in special contexts. The earliest reference seems to be McKendrick (1926), which considers it in a medical application. Hartley (1958) considers the general case of counted data and develops the theory quite extensively; many of the key ideas can be found in this article. Baum et al. (1970) use the algorithm in a Markov model and proved essential mathematical results in this case, results that generalize quite easily. Orchard and Woodbury (1972) first noted the general applicability of the underlying idea, calling it the "missing information principle." Sundberg (1974) explicitly considered properties of the general likelihood equations, and Beale and Little (1975) further developed the theory for the normal model. The term EM was introduced in Dempster, Laird, and Rubin (1977), and this work exposed the full generality of the algorithm by (1) proving general results about its behavior, specifically that each iteration increases the likelihood $l(\theta | Y_{\text{obs}})$, and (2) providing a wide range of examples. Since 1977, there have been many new uses of the EM algorithm, as well as further work on its convergence properties (e.g., Wu, 1983).

Each iteration of EM consists of an E step (expectation step) and an M step (maximization step). These steps are often easy to construct conceptually, to program for calculation, and to fit into computer storage. Also each step has a direct statistical interpretation. An additional advantage of the algorithm is that it can be shown to converge reliably, in the sense that under general conditions each iteration increases the loglikelihood $l(\theta | Y_{obs})$, and if $l(\theta | Y_{obs})$ is bounded, the sequence $l(\theta^{(t)} / Y_{obs})$ converges to a stationary value of $l(\theta | Y_{obs})$. Quite generally, if the sequence $\theta^{(t)}$ converges, it converges to a local maximum or saddle point of $l(\theta | Y_{obs})$. A disadvantage of EM is that its rate of convergence can be painfully slow if a lot of data are missing: Dempster, Laird and Rubin (1977) show that convergence is linear with rate proportional to the fraction of information about θ in $l(\theta | Y)$ that is observed, in a sense made precise in Section 7.5. Furthermore, EM does *not* share with Newton-Raphson or scoring algorithms the property of yielding estimates asymptotically equivalent to ML estimates after a single iteration.

7.3. THE E STEP AND THE M STEP OF EM

The M step is particularly simple to describe: perform maximum likelihood estimation of θ just as if there were no missing data, that is, as if they had been filled in. Thus the M step of EM uses the identical computational methods as ML estimation from $l(\theta | Y)$.

The E step finds the conditional expectation of the "missing data" given the observed data and current estimated parameters, and then substitutes these expectations for the "missing data." The quotations around "missing data" are there because the missing values themselves are not necessarily being substituted by EM. The key idea of EM, which delineates it from the ad hoc idea of filling in missing values and iterating, is that "missing data" is not Y_{mis} but the functions of Y_{mis} appearing in the complete-data loglikelihood, that is, $l(\theta | Y)$.

Specifically, let $\theta^{(t)}$ be the current estimate of the parameter θ. The *E step of EM* finds the expected loglikelihood if θ were $\theta^{(t)}$:

$$Q(\theta | \theta^{(t)}) = \int l(\theta | Y) f(Y_{mis} | Y_{obs}, \theta = \theta^{(t)}) \, dY_{mis}.$$

The *M step of EM* determines $\theta^{(t+1)}$ by maximizing this expected loglikelihood:

$$Q(\theta^{(t+1)} | \theta^{(t)}) \geqslant Q(\theta | \theta^{(t)}), \qquad \text{for all } \theta.$$

EXAMPLE 7.1. *Univariate Normal Data.* Suppose y_i are iid $N(\mu, \sigma^2)$ where y_i for $i = 1, \ldots, m$ are observed, and y_i for $i = m + 1, \ldots, n$ are missing, and assume MAR. The expectation of each missing y_i given Y_{obs} and $\theta = (\mu, \sigma^2)$ is μ.

However, from Example 5.1, the loglikelihood $l(\theta | Y)$, based on all y_i, $i = 1, \ldots, n$, is linear in the sufficient statistics $\sum_1^n y_i$ *and* $\sum_1^n y_i^2$. Thus the E step of the algorithm calculates

$$E\left(\sum_1^n y_i | \theta^{(t)}, Y_{\text{obs}}\right) = \sum_1^m y_i + (n - m)\mu^{(t)} \tag{7.5}$$

$$E\left(\sum_1^n y_i^2 | \theta^{(t)}, Y_{\text{obs}}\right) = \sum_1^m y_i^2 + (n - m)[(\mu^{(t)})^2 + (\sigma^{(t)})^2] \tag{7.6}$$

for current estimates $\theta^{(t)} = (\mu^{(t)}, \sigma^{(t)})$ of the parameters. Note that simple substitution of $\mu^{(t)}$ for missing values $y_{m+1}, \ldots, y_n$ would lead to the omission of the term $(n - m)(\sigma^{(t)})^2$ in (7.6).

With no missing data, the ML estimate of μ is $\sum_1^n y_i/n$ and the ML estimate of σ^2 is $\sum_1^n y_i^2/n - (\sum_1^n y_i/n)^2$. The M step uses these same expressions with the current expectations of the sufficient statistics calculated in the E step substituted for the incompletely observed sufficient statistics. Thus the M step calculates

$$\mu^{(t+1)} = E\left(\sum_1^n y_i | \theta^{(t)}, Y_{\text{obs}}\right) \Big/ n \tag{7.7}$$

$$(\sigma^{(t+1)})^2 = E\left(\sum_1^n y_i^2 | \theta^{(t)}, Y_{\text{obs}}\right) \Big/ n - (\mu^{(t+1)})^2. \tag{7.8}$$

Setting $\mu^{(t)} = \mu^{(t+1)} = \hat{\mu}$ and $\sigma^{(t)} = \sigma^{(t+1)} = \hat{\sigma}$ in Eqs. (7.5)–(7.8), shows that these iterations converge to

$$\hat{\mu} = \sum_1^m y_i/m$$

and

$$\hat{\sigma}^2 = \sum_1^m y_i^2/m - \hat{\mu}^2,$$

the ML estimates of μ and σ^2 from Y_{obs} assuming MAR. Of course, the EM algorithm is unnecessary for this example since the explicit ML estimates $(\hat{\mu}, \hat{\sigma}^2)$ are available.

EXAMPLE 7.2. *A Multinomial Example.* This example is used in Dempster, Laird, and Rubin (1977) to introduce EM. Suppose that the data vector of observed counts $Y_{\text{obs}} = (38, 34, 125)$ is postulated to arise from a multinomial with cell probabilities $(\frac{1}{2} - \frac{1}{2}\theta, \frac{1}{4}\theta, \frac{1}{2} + \frac{1}{4}\theta)$. The objective is to find the ML estimate of θ. Define $y = (y_1, y_2, y_3, y_4)$ to be multinomial with probabilities $(\frac{1}{2} - \frac{1}{2}\theta, \frac{1}{4}\theta, \frac{1}{4}\theta, \frac{1}{2})$ where $Y_{obs} = (y_1, y_2, y_3 + y_4)$. Notice that if Y were observed, the ML estimate of θ would be immediate:

$$\frac{y_2 + y_3}{y_1 + y_2 + y_3}. \tag{7.9}$$

Also note that the loglikelihood $l(\theta | Y)$ is linear in Y, so finding the expectation of $l(\theta | Y)$ given θ and Y_{obs} involves the same calculation as finding the expectation of Y given θ and Y_{obs}, which in effect fills in estimates of the missing values:

$$E(y_1 | \theta, Y_{obs}) = 38$$

$$E(y_2 | \theta, Y_{obs}) = 34$$

$$E(y_3 | \theta, Y_{obs}) = 125(\tfrac{1}{4}\theta)/(\tfrac{1}{2} + \tfrac{1}{4}\theta)$$

$$E(y_4 | \theta, Y_{obs}) = 125(\tfrac{1}{2})/(\tfrac{1}{2} + \tfrac{1}{4}\theta).$$

Thus at the tth iteration, with estimate $\theta^{(t)}$, we have for the E step

$$y_3^{(t)} = 125(\tfrac{1}{4}\theta^{(t)})/(\tfrac{1}{2} + \tfrac{1}{4}\theta^{(t)}) \tag{7.10}$$

and for the M step, from (7.9) we have

$$\theta^{(t+1)} = (34 + y_3^{(t)})/(72 + y_3^{(t)}). \tag{7.11}$$

Iterating between (7.10) and (7.11) defines the EM algorithm for this problem. In fact, setting $\theta^{(t+1)} = \theta^{(t)} = \hat{\theta}$ and combining the two equations yields a quadratic equation in $\hat{\theta}$ and thus a closed-form solution for the ML estimate. Table 7.1 shows the convergence of EM to this solution starting from $\theta^{(0)} = \frac{1}{2}$.

EXAMPLE 7.3. *The Bivariate Normal with Missing Data on Both Variables.* A simple but nontrivial example of EM is its application to the bivariate normal with a general pattern of missing data: The first group of units have

Table 7.1 The EM Algorithm for Example 7.2

t	$\theta^{(t)}$	$\theta^{(t)} - \hat{\theta}$	$(\theta^{(t+1)} - \hat{\theta})/(\theta^{(t)} - \hat{\theta})$
0	0.500000000	0.126821498	0.1465
1	0.608247423	0.018574075	0.1346
2	0.624321051	0.002500447	0.1330
3	0.626488879	0.000332619	0.1328
4	0.626777323	0.000044176	0.1328
5	0.626815632	0.000005866	0.1328
6	0.626820719	0.000000779	·
7	0.626821395	0.000000104	·
8	0.626821484	0.000000014	·

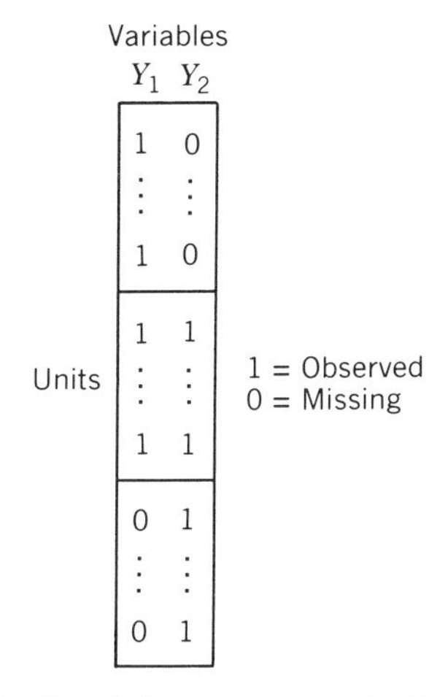

Figure 7.1. The missing-data pattern for Example 7.3.

Y_1 observed but are missing Y_2, the second group of units have both Y_1 and Y_2 observed, and the third group of units have Y_2 observed but are missing Y_1 (see Figure 7.1). We wish to calculate the ML estimates of μ and Σ, the mean and covariance matrix of Y_1 and Y_2.

Like Example 7.1, but unlike Example 7.2, filling in missing values in the E step does not work because the loglikelihood $l(\theta | Y)$ is not linear in the data, but rather is linear in the following sufficient statistics:

$$s_1 = \sum_1^n y_{i1}, \quad s_2 = \sum_1^n y_{i2}, \quad s_{11} = \sum_1^n y_{i1}^2, \quad s_{22} = \sum_1^n y_{i2}^2, \quad s_{12} = \sum_1^n y_{i1} y_{i2}, \tag{7.12}$$

which are simple functions of the sample means, variances, and covariances. The task at the E step is thus to find the conditional expectation given Y_{obs} and $\theta = (\mu, \Sigma)$ of the sums in (7.12). For the group of units with both y_{i1} and y_{i2} observed, the conditional expectations of the quantities in (7.12) equal their observed values. For the group of units with y_{i1} observed but y_{i2} missing, the expectations of y_{i1} and y_{i1}^2 equal their observed values; the expectations of y_{i2}, y_{i2}^2 and $y_{i1} y_{i2}$ are found from the regression of the y_{i2} on y_{i1}:

$$E(y_{i2} | y_{i1}, \mu, \Sigma) = \beta_{20 \cdot 1} + \beta_{21 \cdot 1} y_{i1}$$

$$E(y_{i2}^2 | y_{i1}, \mu, \Sigma) = (\beta_{20 \cdot 1} + \beta_{21 \cdot 1} y_{i1})^2 + \sigma_{22 \cdot 1}$$

$$E(y_{i2} y_{i1} | y_{i1}, \mu, \Sigma) = (\beta_{20 \cdot 1} + \beta_{21 \cdot 1} y_{i1}) y_{i1}$$

where $\beta_{20 \cdot 1}$, $\beta_{21 \cdot 1}$, and $\sigma_{22 \cdot 1}$ are functions of Σ corresponding to the regression of y_{i2} on y_{i1} (see Example 6.1 for details). For the units with y_{i2} observed and y_{i1} missing, the regression of y_{i1} on y_{i2} is used to calculate the missing contributions to the sufficient statistics. Having found the expectations of y_{i1}, y_{i2}, y_{i1}^2, y_{i2}^2, and $y_{i1}y_{i2}$ for each unit in the three groups, the expectations of the sufficient statistics in (7.12) are found as the sums of these quantities over all n units. The M step calculates the usual moment-based estimators of μ and Σ from those filled-in sufficient statistics:

$$\hat{\mu}_1 = s_1/n, \quad \hat{\mu}_2 = s_2/n,$$

$$\hat{\sigma}_1^2 = s_{11}/n - \hat{\mu}_1^2, \quad \hat{\sigma}_2^2 = s_{22}/n - \hat{\mu}_2^2, \quad \hat{\sigma}_{12} = s_{12}/n - \hat{\mu}_1\hat{\mu}_2.$$

The EM algorithm for this problem consists of performing these steps iteratively. More details of this example are considered in Chapter 8, where the EM algorithm is presented for the general multivariate normal distribution with any pattern of missing values.

7.4. THEORY OF THE EM ALGORITHM

The distribution of the complete data Y can be factored as

$$f(Y|\theta) = f(Y_{\text{obs}}, Y_{\text{mis}}|\theta) = f(Y_{\text{obs}}|\theta)f(Y_{\text{mis}}|Y_{\text{obs}}, \theta) \tag{7.13}$$

where $f(Y_{\text{obs}}|\theta)$ is the density of the observed data Y_{obs} and $f(Y_{\text{mis}}|Y_{\text{obs}}, \theta)$ is the density of the missing data given the observed data. The decomposition of the loglikelihood that corresponds to (7.13) is

$$l(\theta|Y) = l(\theta|Y_{\text{obs}}, Y_{\text{mis}}) = l(\theta|Y_{\text{obs}}) + \ln f(Y_{\text{mis}}|Y_{\text{obs}}, \theta).$$

We wish to estimate θ by maximizing the incomplete-data likelihood $l(\theta|Y_{\text{obs}})$ with respect to θ for fixed Y_{obs}; this task, however, can be difficult to accomplish directly.

Write

$$l(\theta|Y_{\text{obs}}) = l(\theta|Y) - \ln f(Y_{\text{mis}}|Y_{\text{obs}}, \theta) \tag{7.14}$$

where $l(\theta|Y_{\text{obs}})$ is the observed loglikelihood to be maximized, $l(\theta|Y)$ is the complete-data loglikelihood which is presumably relatively easy to maximize, and $\ln f(Y_{\text{mis}}|Y_{\text{obs}}, \theta)$ is the missing part of the complete-data loglikelihood.

The expectation of both sides of (7.14) over the distribution of the missing data Y_{mis}, given the observed data Y_{obs}, and a current estimate of θ, say $\theta^{(t)}$, is

$$l(\theta|Y_{\text{obs}}) = Q(\theta|\theta^{(t)}) - H(\theta|\theta^{(t)}),$$

where

$$Q(\theta|\theta^{(t)}) = \int [l(\theta|Y_{\text{obs}}, Y_{\text{mis}})] f(Y_{\text{mis}}|Y_{\text{obs}}, \theta^{(t)})\, dY_{\text{mis}}$$

and

$$H(\theta|\theta^{(t)}) = \int [\ln f(Y_{\text{mis}}|Y_{\text{obs}}, \theta)] f(Y_{\text{mis}}|Y_{\text{obs}}, \theta^{(t)})\, dY_{\text{mis}}.$$

Note that

$$H(\theta|\theta^{(t)}) \leqslant H(\theta^{(t)}|\theta^{(t)}) \tag{7.15}$$

by Jensen's inequality (see Rao, 1972, p. 47).

Consider a sequence of iterates $\theta^{(0)}, \theta^{(1)}, \ldots$, where $\theta^{(t+1)} = M(\theta^{(t)})$ for some function $M(\cdot)$. The difference in values of $l(\theta|Y_{\text{obs}})$ at successive iterates is given by

$$l(\theta^{(t+1)}|Y_{\text{obs}}) - l(\theta^{(t)}|Y_{\text{obs}}) = [Q(\theta^{(t+1)}|\theta^{(t)}) - Q(\theta^{(t)}|\theta^{(t)})] - [H(\theta^{(t+1)}|\theta^{(t)}) - H(\theta^{(t)}|\theta^{(t)})]. \tag{7.16}$$

An EM algorithm chooses $\theta^{(t+1)}$ to maximize $Q(\theta|\theta^{(t)})$ with respect to θ. More generally, a GEM (generalized EM) algorithm chooses $\theta^{(t+1)}$ so that $Q(\theta^{(t+1)}|\theta^{(t)})$ is greater than $Q(\theta^{(t)}|\theta^{(t)})$. Hence the difference of Q functions in (7.16) is positive for any EM or GEM algorithm. Furthermore, note that the difference in the H functions in (7.16) is negative by (7.15). Hence for any EM or GEM algorithm, the change from $\theta^{(t)}$ to $\theta^{(t+1)}$ increases the loglikelihood. This proves the following theorem, which is a key result of Dempster, Laird, and Rubin (1977).

Theorem 7.1. Every GEM algorithm increases $l(\theta|Y_{\text{obs}})$ at each iteration, that is,

$$l(\theta^{(t+1)}|Y_{\text{obs}}) \geqslant l(\theta^{(t)}|Y_{\text{obs}}),$$

with equality if and only if

$$Q(\theta^{(t+1)}|\theta^{(t)}) = Q(\theta^{(t)}|\theta^{(t)}).$$

Corollary 1. Suppose that for some θ^* in the parameter space of θ, $l(\theta^*|Y_{\text{obs}}) \geqslant l(\theta|Y_{\text{obs}})$ for all θ. Then for every GEM algorithm,

$$l(M(\theta^*)|Y_{\text{obs}}) = l(\theta^*|Y_{\text{obs}})$$

$$Q(M(\theta^*)|\theta^*) = Q(\theta^*|\theta^*)$$

and

$$f(Y_{\text{mis}}|Y_{\text{obs}}, M(\theta^*)) = f(Y_{\text{mis}}|Y_{\text{obs}}, \theta^*),$$

almost everywhere.

Corollary 2. Suppose that for some θ^* in the parameter space of θ, $l(\theta^*|Y_{\text{obs}}) > l(\theta|Y_{\text{obs}})$ for all θ. Then for every GEM algorithm,

$$M(\theta^*) = \theta^*.$$

Theorem 7.1 implies that $l(\theta \mid Y_{\text{obs}})$ is nondecreasing on each iteration of a GEM algorithm, and is strictly increasing on any iteration such that Q increases (i.e., $Q(\theta^{(t+1)} \mid \theta^{(t)}, Y_{\text{obs}}) > Q(\theta^{(t)} \mid \theta^{(t)}, Y_{\text{obs}})$). The corollaries imply that a ML estimate of θ is a fixed point of a GEM algorithm.

Another important result concerning EM algorithms is given by Theorem 7.2, which applies when $Q(\theta \mid \theta^{(t)})$ is maximized by setting the first derivative equal to zero.

Theorem 7.2. Suppose a sequence of EM iterates are such that

(a) $D^{10}Q(\theta^{(t+1)} \mid \theta^{(t)}) = 0$,

where D^{10} means the first derivative with respect to the first argument, that is,

$$D^{10}Q(\theta^{(t+1)} \mid \theta^{(t)}) = \frac{\partial}{\partial\theta} Q(\theta \mid \theta^{(t)})|_{\theta=\theta^{(t+1)}},$$

(b) $\theta^{(t)}$ converges to θ^*,

and

(c) $f(Y_{\text{mis}} \mid Y_{\text{obs}}, \theta)$ is smooth in θ, where smooth is defined in the proof. Then

$$D(l(\theta^* \mid Y_{\text{obs}}) = \frac{\partial}{\partial\theta} l(\theta \mid Y_{\text{obs}})|_{\theta=\theta^*} = 0,$$

so that if the $\theta^{(t)}$ converge, they converge to a stationary point.

PROOF.

$$\begin{aligned} Dl(\theta^{(t+1)} \mid Y_{\text{obs}}) &= D^{10}Q(\theta^{(t+1)} \mid \theta^{(t)}) - D^{10}H(\theta^{(t+1)} \mid \theta^{(t)}) \\ &= -D^{10}H(\theta^{(t+1)} \mid \theta^{(t)}) \\ &= \frac{\partial}{\partial\theta} \int [\ln f(Y_{\text{mis}} \mid Y_{\text{obs}}, \theta)] f(Y_{\text{mis}} \mid Y_{\text{obs}}, \theta^*)\, dY_{\text{mis}}, \end{aligned}$$

which, assuming sufficient smoothness to interchange the order of differentiation and integration, converges to

$$\int \frac{\partial}{\partial\theta} f(Y_{\text{mis}} \mid Y_{\text{obs}}, \theta^*) D,$$

which equals zero after interchanging the order of integration and differentiation.

Other EM results in Dempster, Laird, and Rubin (1977) and Wu (1983) regarding convergence include the following:

1. If $l(\theta \mid Y_{\text{obs}})$ is bounded, $l(\theta^{(t)} \mid Y_{\text{obs}})$ converges to some l^*.
2. If $f(Y \mid \theta)$ is a general (curved) exponential family and $l(\theta \mid Y_{\text{obs}})$ is bounded, then $l(\theta^{(t)} \mid Y_{\text{obs}})$ converges to a stationary value l^*.

3. If $f(Y|\theta)$ is a regular exponential family and $l(\theta|Y_{obs})$ is bounded, then $\theta^{(t)}$ converges to a stationary point θ^*.

7.5. THE MISSING INFORMATION

The observed information matrix $I(\theta|Y_{obs})$ can be found directly by differentiating the loglikelihood $l(\theta|Y_{obs})$ twice with respect to θ. Alternatively, note that differentiating (7.14) twice with respect to θ yields for any Y_{mis}

$$I(\theta|Y_{obs}) = I(\theta|Y_{obs}, Y_{mis}) + \partial^2 \ln f(Y_{mis}|Y_{obs}, \theta)/\partial\theta\,\partial\theta,$$

where $I(\theta|Y_{obs}, Y_{mis})$ is the observed information based on $Y = (Y_{obs}, Y_{mis})$ and the negative of the last term is the missing information from Y_{mis}. Taking expectations over the distribution of Y_{mis} given Y_{obs} and θ yields

$$I(\theta|Y_{obs}) = -D^{20}Q(\theta|\theta) + D^{20}H(\theta|\theta), \tag{7.17}$$

provided differentials with respect to θ can be passed through the integral signs. If we call $-D^{20}Q(\theta|\theta)$ the complete information and $-D^{20}H(\theta|\theta)$ the missing information, then (7.17) has the following appealing interpretation:

Observed information = complete information − missing information.

The rate of convergence of the EM algorithm is closely related to these quantities: The greater the proportion of missing information, the slower the rate of convergence. Specifically, Dempster, Laird and Rubin (1977) show that if the EM iterates $\theta^{(t)}$ converge to θ^*, then for $\theta^{(t)}$ near θ^*

$$|\theta^{(t+1)} - \theta^*| = \lambda|\theta^{(t)} - \theta^*|, \tag{7.18}$$

where λ is the ratio of the missing information to the complete information for scalar θ or the largest eigenvalue of the corresponding matrix for vector θ.

Louis (1982) rewrites the missing information in terms of complete-data quantities, and shows that

$$-D^{20}H(\theta|\theta) = E\{S(\theta|Y_{obs}, Y_{mis})S^{T}(\theta|Y_{obs}, Y_{mis})|Y_{obs}, \theta\} - S(\theta|Y_{obs})S^{T}(\theta|Y_{obs}),$$

where, as earlier, S denotes the score function and T denotes matrix transpose. At the ML estimate $S(\hat{\theta}|Y_{obs}) = 0$, so the last term vanishes. Equation (7.17) becomes

$$I(\hat{\theta}|Y_{obs}) = -D^{20}Q(\hat{\theta}|\hat{\theta}) - E\{S(\theta|Y_{obs}, Y_{mis})S^{T}(\theta|Y_{obs}, Y_{mis})|Y_{obs}, \theta\}|_{\theta=\hat{\theta}}, \tag{7.19}$$

which may be useful for computations.

An analogous expression to (7.17) for the expected information $J(\theta)$ is obtained by taking expectations of (7.17) over Y_{obs}. Specifically,

$$J(\theta) = J_c(\theta) + E\{D^{20}H(\theta|\theta)\}, \tag{7.20}$$

where $J_c(\theta)$ is the expected complete information based on $Y = (Y_{obs}, Y_{mis})$. Orchard and Woodbury (1972) give a slightly different form of this expression.

EXAMPLE 7.4. (*Example 7.2 continued*). For the multinomial Example 7.2, the complete-data loglikelihood is proportional to

$$y_1 \ln(1-\theta) + (y_2 + y_3)\ln\theta,$$

ignoring terms not involving θ. Differentiating with respect to θ yields

$$S(\theta|Y) = -y_1/(1-\theta) + (y_2 + y_3)/\theta;$$

$$I(\theta|Y) = y_1/(1-\theta)^2 + (y_2 + y_3)/\theta^2.$$

Hence

$$E\{I(\theta|Y)|Y_{obs}, \theta\} = y_1/(1-\theta)^2 + (y_2 + \hat{y}_3)/\theta^2$$

$$E\{S^2(\theta|Y)|Y_{obs}, \theta\} = \text{Var}\{s(\theta|Y)|Y_{obs}\} = V/\theta^2,$$

where $\hat{y}_3 = E(y_3|Y_{obs}, \theta) = (y_3 + y_4)(0.25\theta)(0.25\theta + 0.5)^{-1}$, and $V = \text{Var}(y_3|Y_{obs}, \theta) = (y_3 + y_4)(0.5)(0.25\theta)(0.25\theta + 0.5)^{-2}$. Substituting $y_1 = 38$, $y_2 = 34$, $y_3 + y_4 = 125$, and $\hat{\theta} = 0.6268215$ in these expressions yields

$$E\{I(\theta|Y)|Y_{obs}, \theta\}|_{\theta=\hat{\theta}} = 435.3,$$

$$E\{S^2(\theta|Y)|Y_{obs}, \theta\}|_{\theta=\hat{\theta}} = 57.8,$$

Hence $I(\hat{\theta}|Y_{obs}) = 435.3 - 57.8 = 377.5$, as can be verified by direct computation. Note that the ratio of missing information to complete information is $57.8/435.3 = 0.1328$, which governs the speed of convergence of EM near $\hat{\theta}$ as shown in the last column of Table 7.1.

7.6. EM THEORY FOR EXPONENTIAL FAMILIES

The EM algorithm has a particularly simple and useful interpretation when the complete data Y have a distribution from the *regular exponential family* defined by

$$f(Y|\theta) = b(Y)\exp(s(Y)\theta)/a(\theta), \tag{7.21}$$

where θ denotes a $(d \times 1)$ parameter vector, $s(Y)$ denotes a $(1 \times d)$ vector of *complete-data* sufficient statistics, and a and b are functions of θ and Y, respectively. Many complete-data problems can be modeled by a distribution of the form (7.21), which includes as special cases essentially all the examples in Part II of this book. The E step for (7.21) consists in estimating the complete-data sufficients statistics $s(Y)$ by

$$s^{(t+1)} = E(s(Y)|Y_{\text{obs}}, \theta^{(t)}). \tag{7.22}$$

The M step determines the new estimate $\theta^{(t+1)}$ of θ as the solution of the likelihood equations

$$E(s(Y)|\theta) = s^{(t)}, \tag{7.23}$$

which are simply the likelihood equations for the complete data Y with $s(Y)$ replaced by $s^{(t)}$. Equation (7.23) can often be solved for θ explicitly, or at least through existing complete-data computer programs. In such cases the computational problem is effectively confined to the E step, which involves estimating (or "imputing") the statistics $s(Y)$, using (7.22). The decomposition (7.19) of the observed information is particularly simple in this case. The complete information is $\text{Var}(s(Y)|\theta)$, and the missing information is $\text{Var}(s(Y)|Y_{\text{obs}}, \theta)$. Thus the observed information is

$$I(\theta|Y_{\text{obs}}) = \text{Var}(s(Y)|\theta) - \text{Var}(s(Y)|Y_{\text{obs}}, \theta), \tag{7.24}$$

the difference between the unconditional and conditional variance of the complete-data sufficient statistic. The ratio of the conditional to the unconditional variance determines the rate of convergence in this case.

REFERENCES

Baum, L. E., Petrie, T., Soules, G., and Weiss, N. (1970). A maximization technique occurring in the statistical analysis of probabilistic functions of Markov chains, *Ann. Math. Statist.* **41**, 164–171.

Beale, E. M. L., and Little, R. J. A. (1975). Missing values in multivariate analysis, *J. Roy. Statist. Soc.* **B37**, 129–145.

Berndt, E. B., Hall, B., Hall, R., and Hausman, J. A. (1974). Estimation and inference in nonlinear structural models, *Ann. Econ. Soc. Meas.* **3**, 653–665.

Dempster, A. P., Laird, N. M., and Rubin, D. B. (1977). Maximum likelihood estimation from incomplete data via the EM Algorithm (with discussion), *J. Roy. Statist. Soc.* **B39**, 1–38.

Hartley, H. O. (1958). Maximum likelihood estimation from incomplete data, *Biometrics* **14**, 174–194.

Louis, T. A. (1982). Finding the observed information when using the EM algorithm, *J. Roy. Statist. Soc.* **B44**, 226–233.

McKendrick, A. G. (1926). Applications of mathematics to medical problems, *Proc. Edinburgh Math. Soc.* **44**, 98–130.

Orchard, T., and Woodbury, M. A. (1972). A missing information principle: theory and applications. *Proceedings of the 6th Berkeley Symposium on Mathematical Statistics and Probability*, **1**, 697–715.

Rao, C. R. (1972). *Linear Statistical Inference and Its Applications*. New York: Wiley.

Sundberg, R. (1974). Maximum likelihood theory for incomplete data from an exponential family, *Scand. J. Statist.* **1**, 49–58.

Wu, C. F. J. (1983). On the convergence properties of the EM algorithm, *Ann. Statist.* **11**, 95–103.

PROBLEMS

1. Show that for a scalar parameter the Newton–Raphson algorithm converges in one step if the loglikelihood is quadratic.

2. Describe in words the function of the E and M steps of the EM algorithm.

3. Prove that the loglikelihood in Example 7.3 is linear in the statistics in Eq. (7.12).

4. Show how Corollaries 1 and 2 follow from Theorem 1.

5. Review results concerning the convergence of EM.

6. Show that (7.22) and (7.23) are the E and M steps for the regular exponential family (7.21).

7. Suppose $Y = (y_1, \ldots, y_n)^{\mathrm{T}}$ are independent gamma random variables with unknown index k and mean $\mu_i = g(\Sigma\beta_j x_{ij})$, where g is a known function, $\beta = (\beta_1, \ldots, \beta_J)$ are unknown regression coefficients, and x_{i1}, ..., x_{iJ} are the values of covariates X_1, ..., X_J for case i. For what choice of g does y belong to the regular $J + 1$ parameter exponential family, and what are the natural parameters and complete-data sufficient statistics?

8. Suppose values y_i in Problem 7 are missing if and only if $y_i > c$, for some unknown censoring point c. Explore the E step of the EM algorithm for estimating (a) β_1, ..., β_J when k is assumed known; (b) $\beta_1, \ldots, \beta_J$ and k, when k requires estimation.

9. By hand calculation, carry out the multivariate normal EM algorithm for the data set in Table 6.1, with initial estimates based on the complete observations. Hence verify that for this pattern of data and choice of starting values the algorithm converges after one iteration (i.e., subsequent iterations lead to the same answer as the first iteration). Why does Eq. (7.18) not apply in this case? (Hint: consider Corollary 2 of Theorem 7.1 with $\theta^{(t)} = \theta^*$).

10. Write down the loglikelihood of θ for the observed data in Example 7.2. Show directly by differentiating this function that $I(\theta | Y_{\text{obs}}) = 435.3$, as found in Example 7.4.

11. Write down the large sample variance of the ML estimate of θ in Example 7.2, and compare it with the variance of the ML estimate when the first and third counts (namely, 38 and 125) are combined, yielding counts (163, 34) from a binomial distribution with probabilities $(1 - \theta/4, \theta/4)$.

12. For the censored exponential sample in the second part of Example 5.14, suppose $y_1, \ldots, y_m$ are observed and $y_{m+1}, \ldots, y_n$ are censored at c. Show that the complete data sufficient statistic for this problem is $s(Y) = \sum_{i=1}^{n} y_i$ and the natural parameter is $\phi = 1/\theta$, the reciprocal of the mean. Find the observed information for ϕ by computing the unconditional and conditional variance of $s(Y)$ and substracting, as in (7.24). Hence find the proportion of missing information from the censoring, and the large sample variances of $\hat{\phi} - \phi$ *and* $\hat{\theta} - \theta$.

CHAPTER 8

Maximum Likelihood Estimation for Multivariate Normal Examples, Ignoring the Missing-Data Mechanism

8.1. INTRODUCTION

In this chapter we apply the EM algorithm to a variety of common problems involving incomplete data on multivariate normally distributed variables: estimation of the mean vector and covariance matrix; estimation of these quantities when there are restrictions on the covariance matrix; multiple linear regression, including ANOVA and multivariate regression; repeated measures models, including random coefficient regression models where the coefficients themselves are regarded as missing data; and selected time series models. The analysis of categorical data with partially observed or missing data is deferred until the chapter on contingency tables, Chapter 9, and the analysis of mixed normal and nonnormal data is considered in Chapter 10. All examples assume MAR.

8.2. ESTIMATING A MEAN VECTOR AND COVARIANCE MATRIX

8.2.1. The EM Algorithm for Incomplete Multivariate Normal Samples

Many multivariate statistical analyses, including multiple linear regression, principal component analysis, discriminant analysis, and canonical correlation analysis are based on the initial summary of the data matrix into the sample mean and covariance matrix of the variables. Thus the efficient estimation of these quantities for an arbitrary pattern of missing values is a particularly important problem. In this section we discuss ML estimation of the mean and covariance matrix from an incomplete multivariate normal sample, assuming the data are missing at random. Although the assumption of multivariate normality may appear restrictive, the methods discussed here can provide consistent estimates under weaker assumptions about the underlying distribution. Furthermore, the normality will be relaxed somewhat when we consider linear regression in Section 8.4. Robust ML estimation of the mean and covariance matrix under multivariate t and contaminated multivariate normal models is considered in Section 10.5.

Suppose that we measure K variables $(Y_1, Y_2, \ldots, Y_K)$ that have a K-variate normal distribution with mean $\mu = (\mu_1, \ldots, \mu_K)$ and covariance matrix $\Sigma = (\sigma_{jk})$. We write $Y = (Y_{\text{obs}}, Y_{\text{mis}})$, where Y represents a random sample of size n on $(Y_1, \ldots, Y_K)$, Y_{obs} the set of observed values, and Y_{mis} the missing data. We write

$$Y_{\text{obs}} = (y_{\text{obs},1}, y_{\text{obs},2}, \ldots, y_{\text{obs},n}),$$

where $y_{\text{obs},i}$ represents the set of variables observed for observation i, $i = 1, \ldots, n$.

To derive the EM algorithm, we note that the hypothetical complete data Y belong to the regular exponential family (7.21) with sufficient statistics

$$S = \left(\sum_{i=1}^{n} y_{ij}, \quad j = 1, \ldots, K \quad \text{and} \quad \sum_{i=1}^{n} y_{ij} y_{ik}, \quad j, k = 1, \ldots, K \right).$$

At the tth iteration, let $\theta^{(t)} = (\mu^{(t)}, \Sigma^{(t)})$ denote current estimates of the parameters. The E step of the algorithm consists in calculating

$$\begin{aligned} E\left(\sum_{i=1}^{n} y_{ij} \mid Y_{\text{obs}}, \theta^{(t)} \right) &= \sum_{i=1}^{n} y_{ij}^{(t)}, \qquad j = 1, \ldots, K \\ E\left(\sum_{i=1}^{n} y_{ij} y_{ik} \mid Y_{\text{obs}}, \theta^{(t)} \right) &= \sum_{i=1}^{n} (y_{ij}^{(t)} y_{ik}^{(t)} + c_{jki}^{(t)}), \qquad j, k = 1, \ldots, K, \end{aligned} \tag{8.1}$$

where

$$y_{ij}^{(t)} = \begin{cases} y_{ij}, & \text{if } y_{ij} \text{ is observed;} \\ E(y_{ij} \mid y_{\text{obs},i}, \theta^{(t)}), & \text{if } y_{ij} \text{ is missing,} \end{cases}$$

and

$$c_{jki}^{(t)} = \begin{cases} 0 & \text{if } y_{ij} \text{ or } y_{ik} \text{ are observed;} \\ \text{Cov}(y_{ij}, y_{ik} \mid y_{\text{obs},i}, \theta^{(t)}), & \text{if } y_{ij} \text{ and } y_{ik} \text{ are missing.} \end{cases}$$

Missing values y_{ij} are thus replaced by the conditional mean of y_{ij} given the set of values, $y_{\text{obs},i}$ observed for that observation. These conditional means and the nonzero conditional covariances are easily found from the current parameter estimates by sweeping the augmented covariance matrix so that the variables $y_{\text{obs},i}$ are predictors in the regression equation and the remaining variables are outcome variables. The sweep operator is described in Section 6.5.

The M step of the EM algorithm is straightforward. The new estimates $\theta^{(t+1)}$ of the parameters are estimated from the estimated complete-data sufficient statistics. That is,

$$\mu_j^{(t+1)} = n^{-1} \sum_{i=1}^{n} y_{ij}^{(t)}, \qquad j = 1, \ldots, K;$$

$$\sigma_{jk}^{(t+1)} = n^{-1} E\left(\sum_{i=1}^{n} y_{ij} y_{ik} \mid Y_{\text{obs}} \right) - \mu_j^{(t+1)} \mu_k^{(t+1)} \tag{8.2}$$

$$= n^{-1} \sum_{i=1}^{n} [(y_{ij}^{(t)} - \mu_j^{(t+1)})(y_{ik}^{(t)} - \mu_k^{(t+1)}) + c_{jki}^{(t)}], \qquad j, k = 1, \ldots, K.$$

Beale and Little (1975) suggest replacing the factor n^{-1} in the estimate of σ_{jk} by $(n-1)^{-1}$, which parallels the correction for degrees of freedom in the complete data case.

It remains to suggest initial values of the parameters. Four straightforward possibilities are (1) to use the complete cases solution of Section 3.2; (2) to use one of the available case solutions of Section 3.3; (3) to form the sample mean and covariance matrix of the data filled in by one of the imputation methods of Section 3.4; and (4) to form means and variances from observed values of each variable and set all correlations equal to zero. Option 1 provides consistent estimates of the parameters if the data are MCAR and there are at least $K + 1$ complete observations. Option 2 makes use of all the available data but can yield an estimated covariance matrix that is not positive definite, leading to problems in the first iteration. Options 3 and 4 generally yield inconsistent estimates of the covariance matrix, but estimates that are positive semidefinite and hence usually workable as initial estimates. A computer program for general use should have several alternative initializations of the parameters available so that a suitable choice can be made.

The link between ML estimation and an efficient form of imputation for the missing values is clear from the EM algorithm. The E step imputes the best linear predictors of the missing values, using current estimates of the parameters. It also calculates the adjustments c_{jk} to the estimated covariance matrix needed to allow for imputation of the missing values.

The EM algorithm given here was first described by Orchard and Woodbury (1972). Earlier, the scoring algorithm for this problem had been described by Trawinski and Bargmann (1964); the iterating equations are presented in an elegant form by Hartley and Hocking (1971). An important difference between the scoring algorithm and the EM algorithm is that the former algorithm requires inversion of the information matrix of μ and Σ at each iteration. After convergence this matrix provides an estimate of the asymptotic covariance matrix of the ML estimates, which is not directly obtained from the EM algorithm. The inversion of the information matrix of θ at each iteration, however, can be expensive because this is a large matrix if the number of variables is large. For the K-variable case, the information matrix of θ has $\frac{1}{2}K(K+1)$ rows and columns, and when $K = 30$ it has over 100,000 elements! With EM, an asymptotic covariance matrix of θ can be obtained using just one inversion of the information matrix evaluated at the final ML estimate of θ.

Three versions of the EM algorithm can be defined. The first stores the raw data (Beale and Little, 1975). The second stores the sums, sums of squares, and sums of cross products for each pattern of missing data (Dempster, Laird, and Rubin, 1977). Because the version that takes less storage and computation is to be preferred, a preferable option is a third, which mixes the two previous versions, storing raw data for those patterns with fewer than $(K+1)/2$ units and storing sufficient statistics otherwise.

8.2.2. Estimated Asymptotic Covariance Matrix of $(\theta-\hat{\theta})$ Based on the Information Matrix

If the data are MCAR, the expected information matrix of $\theta = (\mu, \Sigma)$ represented as a vector has the form

$$J(\theta) = \begin{bmatrix} J(\mu) & 0 \\ 0 & J(\Sigma) \end{bmatrix}.$$

Here, the (j,k)th element of $J(\mu)$, corresponding to row μ_j, column μ_k, is

$$\sum_{i=1}^{n} \psi_{jki},$$

where

$$\psi_{jki} = \begin{cases} (j,k)\text{th element of } \Sigma_{\text{obs},i}^{-1}, & \text{if both } x_{ij} \text{ and } x_{ik} \text{ present,} \\ 0, & \text{otherwise,} \end{cases}$$

and $\Sigma_{\text{obs},i}$ is the covariance matrix of the variables present in observation i. The (lm, rs)th element of $J(\Sigma)$, corresponding to row σ_{lm}, column σ_{rs}, is

$$\tfrac{1}{4}(2-\delta_{lm})(2-\delta_{rs})\sum_{i=1}^{n}(\psi_{lri}\psi_{msi}+\psi_{lsi}\psi_{mri}),$$

where $\delta_{lm}=1$ if $l=m$, 0 if $l\neq m$. As noted earlier, the inverse of $J(\hat{\theta})$ supplies an estimated covariance matrix for the ML estimate $\hat{\theta}$. The matrix $J(\theta)$ is estimated and inverted at each step of the scoring algorithm. Note that the expected information matrix is block diagonal with respect to the means and the covariances. Hence if these asymptotic variances are required for ML estimates of means or linear combinations of means, then it is only necessary to calculate and invert the information matrix $J(\mu)$ corresponding to the means, which has relatively small dimension.

The observed information matrix, which is calculated and inverted at each iteration of the Newton-Raphson algorithm, is not block diagonal with respect to μ and Σ, so this simplification does not occur if standard errors are based on this matrix. On the other hand, the standard errors based on the observed information matrix are more conditional and thus valid when the data are MAR but not MCAR, and hence should be preferable to those based on $J(\theta)$ in applications.

EM uses neither the observed nor the expected information matrix; hence if either is used as a basis for standard errors it must be calculated and inverted after the ML estimates are obtained.

8.3. ESTIMATION WITH A RESTRICTED COVARIANCE MATRIX

In Section 8.2 there were no restrictions on the parameters of the multivariate normal, θ being free to vary anywhere in its natural parameter space. Some statistical models, however, place restrictions on θ. ML estimation with incomplete data from such restricted models can be handled easily by EM whenever ML estimation with complete data is simple. The reason is that the E step of EM takes the same form whether θ is restricted or not; the only alteration in EM with a restriction on θ is to modify the M step to be ML for the restricted model.

For some kinds of restrictions on θ noniterative ML estimates do not exist even with complete data. In some of these cases EM can be used to compute iterative ML estimates by *creating* fully missing variables in such a way that the M step is noniterative. Two examples illustrate this idea. Both examples can be modified to handle missing data among the observed variables.

EXAMPLE 8.1. *Patterned Covariance Matrices.* Some patterned covariance matrices that do not have explicit ML estimates can be viewed as submatrices of larger patterned covariance matrices that do have explicit ML estimates. In such a case the smaller covariance matrix, say Σ_{11}, can be viewed as the covariance matrix for observed variables and the larger covariance matrix, say Σ, can be viewed as the covariance matrix for both observed and missing variables. In such a case the EM algorithm can be used to calculate the desired ML estimates for the original problem, as described by Rubin and Szatrowski (1982).

As an illustration, consider the 3×3 stationary covariance pattern Σ_{11} and the 4×4 circular symmetry pattern Σ:

$$\Sigma_{11} = \begin{bmatrix} \theta_1 & \theta_2 & \theta_3 \\ \theta_2 & \theta_1 & \theta_2 \\ \theta_3 & \theta_2 & \theta_1 \end{bmatrix}, \qquad \Sigma = \left[\begin{array}{ccc|c} \theta_1 & \theta_2 & \theta_3 & \theta_2 \\ \theta_2 & \theta_1 & \theta_2 & \theta_3 \\ \theta_3 & \theta_2 & \theta_1 & \theta_2 \\ \hline \theta_2 & \theta_3 & \theta_2 & \theta_1 \end{array}\right] = \left[\begin{array}{c|c} \Sigma_{11} & \Sigma_{12} \\ \hline \Sigma_{21} & \Sigma_{22} \end{array}\right].$$

Suppose that we have a random sample $y_1, \ldots, y_n$ from a multivariate normal distribution, $N_3(0, \Sigma_{11})$. These observations can be viewed as the first three of four components from a random sample $(y_1, z_1), \ldots, (y_n, z_n)$ from a multivariate normal distribution $N_4(0, \Sigma)$, where the first three components of each (y_i, z_i) are observed, and the last component, z_i, is missing. The y_i are the observed data, and the (y_i, z_i) are the complete data, both observed and missing. Let $C = \sum (y_i, z_i)^{\mathrm{T}}(y_i, z_i)/n$ and $C_{11} = \sum (y_i^{\mathrm{T}} y_i)/n$. The matrix C is the complete-data sufficient statistic and C_{11} is the observed sufficient statistic. The explicit maximum likelihood estimate of Σ is obtained from complete data, C, by simple averaging (Szatrowski, 1978), whence the M step of EM at iteration t is given by

$$\begin{gathered} \theta_1^{(t+1)} = \tfrac{1}{4}\left(\sum c_{kk}^{(t)}\right), \qquad \theta_2^{(t+1)} = \tfrac{1}{4}(c_{12}^{(t)} + c_{23}^{(t)} + c_{34}^{(t)} + c_{14}^{(t)}), \\ \theta_3^{(t+1)} = \tfrac{1}{2}(c_{13}^{(t)} + c_{24}^{(t)}), \end{gathered} \tag{8.3}$$

where $c_{kj}^{(t)}$ is the (k, j)th element of $C^{(t)}$, the expected value of C from the E step at iteration t. These estimates of θ_1, θ_2, and θ_3 yield a new estimate of Σ for iteration $t + 1$.

Since there is only one pattern of incomplete data (y_i observed and z_i missing), the E step of the EM algorithm involves calculating the expected value of C given the observed sufficient statistic C_{11} and the current estimate $\Sigma^{(t)}$ of Σ, namely, $C^{(t)} = E(C | C_{11}, \Sigma^{(t)})$. First, find the regression parameters of the conditional distribution of z_i given y_i by sweeping Y from the current estimate of Σ, $\Sigma^{(t)}$, to obtain

$$\begin{bmatrix} \Sigma_{11}^{(t)-1} & \Sigma_{11}^{(t)-1}\Sigma_{12}^{(t)} \\ \Sigma_{21}^{(t)}\Sigma_{11}^{(t)-1} & \Sigma_{22}^{(t)} - \Sigma_{21}^{(t)}\Sigma_{11}^{(t)-1}\Sigma_{12}^{(t)} \end{bmatrix} = \mathrm{SWP}(1, 2, 3)\begin{bmatrix} \Sigma_{11}^{(t)} & \Sigma_{12}^{(t)} \\ \Sigma_{21}^{(t)} & \Sigma_{22}^{(t)} \end{bmatrix}.$$

The expected value of z_i given the observed data and $\Sigma = \Sigma^{(t)}$ is $y_i \Sigma_{11}^{(t)-1}\Sigma_{12}^{(t)}$, so that the expected value of $C_{12} = \sum_i y_i^{\mathrm{T}} z_i / n$ given C_{11} and $\Sigma^{(t)}$ is $C_{11}\Sigma_{11}^{(t)-1}\Sigma_{12}^{(t)}$. The expected value of $z_i^{\mathrm{T}} z_i$ given the observed data and $\Sigma = \Sigma^{(t)}$ is

$$(\Sigma_{21}^{(t)}\Sigma_{11}^{(t)-1}y_i^{\mathrm{T}})(y_i\Sigma_{11}^{(t)-1}\Sigma_{12}^{(t)}) + \Sigma_{22}^{(t)} - \Sigma_{21}^{(t)}\Sigma_{11}^{(t)-1}\Sigma_{12}^{(t)},$$

so that the expected value of $C_{22} = \sum_1^n z_i^{\mathrm{T}} z_i / n$ is

$$\Sigma_{22}^{(t)} - \Sigma_{21}^{(t)}\Sigma_{11}^{(t)-1}\Sigma_{12}^{(t)} + \Sigma_{21}^{(t)}\Sigma_{11}^{-1}C_{11}\Sigma_{11}^{(t)-1}\Sigma_{12}^{(t)}.$$

These calculations are summarized as:

$$E(C|C_{11}, \Sigma^{(t)}) = \begin{bmatrix} C_{11} & C_{11}\Sigma_{11}^{(t)-1}\Sigma_{12}^{(t)} \\ \Sigma_{21}^{(t)}\Sigma_{11}^{(t)-1}C_{11} & \{\Sigma_{22}^{(t)} - \Sigma_{21}^{(t)}(\Sigma_{11}^{(t)-1} - \Sigma_{11}^{(t)-1}C_{11}\Sigma_{11}^{(t)-1})\Sigma_{12}^{(t)}\} \end{bmatrix}. \tag{8.4}$$

This new value of C is used in (8.3) to calculate new estimates of θ_1, θ_2, and θ_3 and thus $\Sigma^{(t+1)}$.

An advantage of the EM algorithm is its ability to handle simultaneously both missing values in the data matrix and patterned covariance matrices, both of which occur frequently in educational testing examples. In some of these examples unrestricted covariance matrices do not have unique ML estimates because of the missing data, and the patterned structure is easily justified from theoretical considerations and from empirical evidence on related data (Holland and Wightman, 1982; Rubin and Szatrowski, 1982). When there is more than one pattern of incomplete data, the E step computes expected sufficient statistics for each of the patterns rather than just one pattern as in (8.4).

EXAMPLE 8.2. *Factor Analysis.* Let Y be an $n \times K$ observed data matrix and Z be an $n \times q$ unobserved "factor-score matrix," $q < K$. The rows of (Y, Z) are independently and identically distributed. The marginal distribution of each row of Z, the factors, is normal with means $(0, \ldots, 0)$, variances $(1, \ldots, 1)$ and correlation matrix B. The conditional distribution of the ith row of Y, y_i, given the ith row of Z, z_i, is normal with mean $\alpha + z_i\beta$ and residual covariance matrix $\tau^2 = \mathrm{diag}(\tau_1^2, \ldots, \tau_K^2)$; this assumption of the conditional independence of the variables given the factors is crucial. The regression coefficient matrix β is commonly called the factor-loading matrix and the residual variances in τ^2 are commonly called the uniquenesses.

The parameters to be estimated in general consist of α, β, τ^2, and B. Since the marginal distribution of each row of Y is normal with mean α and covariance matrix $\tau^2 + \beta^{\mathrm{T}}B\beta$, the ML estimate of α is $\bar{Y}$. Thus for purposes of ML estimation, we may replace $(y_{ij} - \alpha_j)$ by $(y_{ij} - \bar{y}_j)$ and consider the parameters to be (β, τ^2, B). For notational simplicity we simply suppose $\bar{y}_j = 0$ (i.e., the observed variables have been centered at the sample means). Consequently, the marginal distribution of the observed y given (β, τ^2, B) is normal with mean 0 and covariance matrix $\tau^2 + \beta^T B\beta$, a patterned covariance matrix of a special type. The general results of Example 8.1 can be applied to derive an EM algorithm for ML factor analysis. In particular, the E step is given by Eq. (8.4), with the first block of variables corresponding to the observed variables, Y, and the second block of variables corresponding to the missing factors, Z. Rubin and Thayer (1982, 1983) provide details of the M step for three cases defined by restrictions on the parameters:

Case 1: $B = I$ (orthogonal factors) and unrestricted β
Case 2: $B = I$ and *a priori* zeroes in β
Case 3: B free to be estimated and *a priori* zeroes in β

Case 1 is sometimes referred to as exploratory factor analysis, and cases 2 and 3 are sometimes referred to as confirmatory factor analysis.

In case 1, ML estimates of β and τ^2 are found simply by sweeping Z from the current estimate of the (Y, Z) cross-products matrix found in the previous E step:

$$\left[\begin{array}{ccc|c} \tau_1^{(t)2} & \cdots & 0 & \\ \vdots & & \vdots & \beta^{(t)\mathrm{T}} \\ 0 & \cdots & \tau_k^{(t)2} & \\ \hline & \beta^{(t)} & & -(C_{22}^{(t)})^{-1} \end{array}\right] = \mathrm{SWP}[Z]\begin{bmatrix} C_{11} & C_{12}^{(t)} \\ C_{21}^{(t)} & C_{22}^{(t)} \end{bmatrix}. \tag{8.5}$$

In case 2 ML estimates of the regression coefficients and residual variances for the jth Y variable (β_j, τ_j^2) are found by sweeping out only the factors with nonzero coefficients in β_j. Thus in (8.5) a different set of Z variables must be swept from C for each group of Y variables with a distinct pattern of *a priori* zeroes in β. Rubin and Thayer (1982) illustrate the computations with a nine-variable, four-factor example. Case 3 extends case 2 by requiring the estimation of B. The ML estimate of the covariance matrix of Z is simply the current expected cross-product matrix of Z, $C_{22}^{(t)}$. When the restriction that the variances of the factors are 1 is interpreted to mean that there exist infinite data sets on the marginal distribution of each factor with sample variance 1, the ML estimate of B is $C_{22}^{(t)}$ normed to be a correlation matrix.

As in Example 8.1, the EM algorithm for factor analysis can handle missing data in the Y variables with only minor modification; specifically, the E step must calculate expected sufficient statistics for each pattern of incomplete data, rather than just the single pattern with Y_i completely observed.

EXAMPLE 8.3. *Variance Components.* A large collection of patterned covariance matrices arises from variance components models, also called random effects or mixed effects analysis of variance models. The EM al-

Table 8.1 Data for Example 8.3

Bull(i)	Percentages of Conception to Services for Successive Samples	n_i	X_i
1	46, 31, 37, 62, 30	5	206
2	70, 59	2	129
3	52, 44, 57, 40, 67, 64, 70	7	394
4	47, 21, 70, 46, 14	5	198
5	42, 64, 50, 69, 77, 81, 87	7	470
6	35, 68, 59, 38, 57, 76, 57, 29, 60	9	479
Total		35	1876

gorithm can be used to obtain ML estimates of variance components and more generally covariance components (Dempster, Laird, and Rubin, 1977; Dempster, Rubin, and Tsutakawa, 1981). The following example is taken from Snedecor and Cochran (1967, p. 290).

In a study of artificial insemination of cows, semen samples from a sample of K = six bulls were tested for their ability to produce conceptions, where the number, n_i, of semen samples tested from bulls varied from bull to bull. The data are given in Table 8.1. Interest focuses on the variability of the bull effects; that is, if an infinite number of samples had been taken from each bull, the variance of the six means would be calculated and used to estimate the variance of the bull effects in the population. Thus, with the actual data, there is one component of variability due to sampling bulls from a population of bulls, which are of primary interest, and another to sampling within each bull.

A common normal model for such data is

$$y_{ij} = \alpha_i + e_{ij} \tag{8.6}$$

where

$$\alpha_i \overset{\text{iid}}{\sim} N(\mu, \sigma_\alpha^2) \quad \text{are the bull effects}$$

$$e_{ij} \overset{\text{iid}}{\sim} N(0, \sigma_e^2) \quad \text{are the within-bull effects.}$$

Integrating over the α_i, the y_{ij} are jointly normal with common mean μ, common variance $\sigma_e^2 + \sigma_\alpha^2$, and covariance σ_α^2 within the same bull and 0 between bulls. That is,

$$\operatorname{Corr}(y_{ij}, y_{i'j'}) = \begin{cases} \rho = [1 + \sigma_e^2/\sigma_\alpha^2]^{-1}, & \text{if } i = i', j \neq j', \\ 0, & \text{if } i \neq i', \end{cases}$$

where ρ is commonly called the intraclass correlation.

Treating the unobserved random variables $\alpha_1, \ldots, \alpha_6$ as missing data (with all y_{ij} observed) leads to an EM algorithm for obtaining ML estimates of the parameters $\theta = (\mu, \sigma_\alpha^2, \sigma_e^2)$. Specifically, the complete-data likelihood has two factors, the first corresponding to the distribution of y_{ij} given α_i and θ, and the second to the distribution of α_i given θ:

$$\prod_{i,j} \frac{1}{\sqrt{2\pi}\sigma_e} \exp\left[\frac{-\frac{1}{2}(y_{ij} - \alpha_i)^2}{\sigma_e^2}\right] \prod_i \frac{1}{\sqrt{2\pi}\sigma_\alpha} \exp\left[\frac{-\frac{1}{2}(\alpha_i - \mu)^2}{\sigma_\alpha^2}\right].$$

The loglikelihood is linear in the following complete-data sufficient statistics:

$$\begin{aligned} T_1 &= \sum \alpha_i \\ T_2 &= \sum \alpha_i^2 \\ T_3 &= \sum_{i,j} (y_{ij} - \alpha_i)^2 = \sum_{i,j} (y_{ij} - \bar{y}_i)^2 + \sum_i n_i(\bar{y}_i - \alpha_i)^2. \end{aligned}$$

The ML estimates based on complete data are:

$$\begin{aligned} \hat{\mu} &= \frac{T_1}{K}, \\ \hat{\sigma}_\alpha^2 &= \frac{T_2}{K} - \hat{\mu}^2, \\ \hat{\sigma}_e^2 &= \frac{T_3}{\sum_i n_i}. \end{aligned} \tag{8.7}$$

These equations define the M step of EM. The E step of EM is defined by taking the expectations of T_1, T_2, T_3 given current estimates of θ and the observed data y_{ij}, $i = 1, \ldots, K, j = 1, \ldots, n_i$. These follow by applying Bayes's theorem to the joint distribution of the α_i and the y_{ij} to obtain the conditional distribution of the α_i given the y_{ij}:

$$(\alpha_i | \{y_{ij}\}, \theta) \overset{\text{indep}}{\sim} N(w_i\mu + (1 - w_i)\bar{y}_i, v_i)$$

where

$$\begin{aligned} w_i &= \sigma_\alpha^{-2} v_i, \\ v_i &= (\sigma_\alpha^{-2} + n_i\sigma_e^{-2})^{-1}. \end{aligned}$$

Hence

$$\begin{aligned} T_1^{(t+1)} &= \sum [w_i^{(t)}\mu^{(t)} + (1 - w_i^{(t)})\bar{y}_i] \\ T_2^{(t+1)} &= \sum [w_i^{(t)}\mu^{(t)} + (1 - w_i^{(t)})\bar{y}_i]^2 + \sum v_i^{(t)}, \\ T_3^{(t+1)} &= \sum_{i,j} (y_{ij} - \bar{y}_i)^2 + \sum_i n_i[w_i^{(t)2}(\mu^{(t)} - \bar{y}_i)^2 + v_i^{(t)}]. \end{aligned} \tag{8.8}$$

ML estimates from this algorithm are $\hat{\mu} = 53.3184$, $\hat{\sigma}_\alpha^2 = 54.8223$ and $\hat{\sigma}_e^2 = 249.2235$. The latter two estimates can be compared with $\tilde{\sigma}_\alpha^2 = 53.8740$, $\tilde{\sigma}_e^2 = 248.1876$, obtained by equating observed and expected mean squares from a random effects analysis of variance (e.g., see Brownlee, 1965, section 10.4). Far more complex variance components models can be fit using EM including those with multivariate y_i, α_i, and X variables; see, for example, Dempster, Rubin, and Tsutakawa (1981) and Laird and Ware (1982).

8.4. MULTIPLE LINEAR REGRESSION

8.4.1. Linear Regression with Missing Values Confined to the Dependent Variable

When a scalar outcome variable Y is regressed on p predictor variables $X_1, \ldots, X_p$ and missing values are confined to Y, the incomplete observations do not contain information about the regression parameters, $\theta_{Y \cdot X} = (\beta_{Y \cdot X}, \sigma_{Y \cdot X}^2)$, if $\theta_{Y \cdot X}$ is distinct, as when the X's are regarded as fixed constants. Nevertheless, the EM algorithm can be applied to all observations and will obtain iteratively the same ML estimates as would have been obtained noniteratively using only the complete observations. In some cases it may be easier to find these ML estimates iteratively by EM than noniteratively.

EXAMPLE 8.4. *Missing Outcomes in ANOVA.* In designed experiments the set of values of $(X_1, \ldots, X_p)$ is chosen to make the least squares estimation of parameters computationally simple. When Y given $(X_1, \ldots, X_p)$ is normal, least squares computations yield ML estimates. When values of Y, say y_i, $i = 1, \ldots, m_0$, are missing, the remaining complete observations no longer have the balance in the original design with the result that ML (least squares) estimation is more complicated. For a variety of reasons given in Chapter 2, it can be desirable to retain all observations and treat the problem as one with missing data.

When EM is applied to this problem, the M step corresponds to the least squares analysis on the original design and the E step involves finding the expected values and expected squared values of the missing y_i's given the current estimated parameters $\theta_{Y \cdot X}^{(t)} = (\beta_{Y \cdot X}^{(t)}, \sigma_{Y \cdot X}^{(t)2})$:

$$E(y_i | X, Y_{\text{obs}}, \theta_{Y \cdot X}^{(t)}) = \begin{cases} y_i, & \text{if } y_i \text{ is observed } (i = m_0 + 1, \ldots, n), \\ \beta_{Y \cdot X}^{(t)} x_i, & \text{if } y_i \text{ is missing } (i = 1, \ldots, m_0). \end{cases}$$

$$E(y_i^2 | X, Y_{\text{obs}}, \theta_{Y \cdot X}^{(t)}) = \begin{cases} y_i^2, & \text{if } y_i \text{ is observed,} \\ (\beta_{Y \cdot X}^{(t)} x_i)^2 + \sigma_{Y \cdot X}^{(t)2}, & \text{if } y_i \text{ is missing,} \end{cases}$$

where X is the $(n \times p)$ matrix of X values. Let Y be the $(n \times 1)$ vector of Y values, and $Y^{(t)}$ the vector Y with missing components y_i replaced by estimates from the E step at iteration t. The M step calculates

$$\beta_{Y \cdot X}^{(t+1)} = (X^{\mathrm{T}}X)^{-1}X^{\mathrm{T}}Y^{(t)}, \tag{8.9}$$

$$\sigma_{Y \cdot X}^{(t+1)2} = n^{-1}\left[\sum_{m_0+1}^{n} (y_i - \beta_{Y \cdot X}^{(t)}X_i)^2 + m_0\sigma_{Y \cdot X}^{(t)2}\right]. \tag{8.10}$$

The algorithm can be simplified by noting that (8.9) does not involve $\sigma_{Y \cdot X}^{(t)2}$, and that at convergence we have

$$\sigma_{Y \cdot X}^{(t+1)2} = \sigma_{Y \cdot X}^{(t)2} = \hat{\sigma}_{Y \cdot X}^2,$$

so from (8.10)

$$\hat{\sigma}_{Y \cdot X}^2 = \frac{1}{n}\sum_{m_0+1}^{n} (y_i - \hat{\beta}_{Y \cdot X}x_i)^2 + \frac{m_0}{n}\hat{\sigma}_{Y \cdot X}^2$$

or

$$\hat{\sigma}_{Y \cdot X}^2 = \frac{1}{n - m_0}\sum_{m_0+1}^{n} (y_i - \hat{\beta}_{Y \cdot X}x_i)^2. \tag{8.11}$$

Consequently, the EM iterations can omit the M-step estimation of $\sigma_{Y \cdot X}^2$ and the E-step estimation of $E(y_i^2|\text{data}, \theta_{Y \cdot X}^{(t)})$, and find $\hat{\beta}_{Y \cdot X}$ by iteration. After convergence, we can calculate $\hat{\sigma}_{Y \cdot X}^2$ directly from (8.11). These iterations, which fill in the missing data, reestimate the missing values from the ANOVA, and so forth, comprise the algorithm of Healy and Westmacott (1956) mentioned in Section 2.4.3.

8.4.2. Linear Regression with Missing Values in the Predictor Variables

In general, there can be missing values in the predictor variables as well as in the outcome variable. For the moment, assume joint multivariate normality for $(Y, X_1, \ldots, X_p)$. Then we can find ML estimates for the regression of Y on $X_1, \ldots, X_p$ directly from the EM algorithm for multivariate normal data discussed in the previous section. Let

$$\theta = \begin{bmatrix} -1 & \mu_1 & \cdots & \mu_{p+1} \\ \mu_1 & \sigma_{11} & \cdots & \sigma_{1,p+1} \\ \vdots & \vdots & & \vdots \\ \mu_{p+1} & \sigma_{1,p+1} & \cdots & \sigma_{p+1,p+1} \end{bmatrix}$$

denote the augmented covariance matrix corresponding to the variables $X_1, \ldots, X_p$ and $X_{p+1} \equiv Y$. The intercept, slopes, and residual variance for the regression of Y on $X_1, \ldots, X_p$ are found in the last column of the matrix SWP$[1,\ldots,p]\theta$, where the constant term and the predictor variables have been swept out of the matrix θ. Hence if $\hat{\theta}$ is the ML estimate of θ found by the methods of Section 8.2, then ML estimates of the intercept, slopes, and residual variance are found from the last column of SWP$[1,\ldots,p]\hat{\theta}$.

Let $\hat{\beta}_{Y\cdot X}$ and $\hat{\sigma}^2_{Y\cdot X}$ be the ML estimates of the regression coefficient of Y on X and residual variance of Y given X as found by the EM algorithm just described. These estimates are ML under conditions more general than the multivariate normality of Y and $(X_1,\ldots,X_p)$. Specifically, suppose we partition $(X_1,\ldots,X_p)$ as $(X_{(1)}, X_{(0)})$ where the variables in $X_{(1)}$ are more observed than both Y and the variables in $X_{(0)}$ in the sense of Section 6.6 that any unit with any observation on Y or $X_{(0)}$ has all variables in $X_{(1)}$ observed. A particularly simple case occurs when $X = (X_1,\ldots,X_p)$ is fully observed so that $X_{(1)} = X$: see Figure 6.1 for the general case, where Y_1 corresponds to $(Y, X_{(0)})$ and Y_3 corresponds to $X_{(1)}$ and Y_2 is null. Then if the conditional distribution of $(Y, X_{(0)})$ given $X_{(1)}$ is multivariate normal, $\hat{\mu}_{Y\cdot X}$ and $\hat{\sigma}_{YY\cdot X}$ are ML. See Chapter 6 for details.

This assumption is much less stringent than multivariate normality for $X_1,\ldots,X_{p+1}$ since it allows the predictors in $X_{(1)}$ to be categorical variables, as in dummy variable regression, and also allows interactions and polynomials in the completely observed predictors to be introduced into the regression without affecting the propriety of the incomplete-data procedure.

Unfortunately the $(p \times p)$ submatrix of the first p rows and columns of SWP$[1,\ldots,p]\hat{\theta}$ does not provide the covariance matrix of the estimated regression coefficients, as is the case with complete data. The asymptotic covariance matrix of the estimated slopes based on the usual large sample approximation generally involves the inversion of the full information matrix of the means, variances, and covariances, which is displayed in Section 8.2.2. However, an approximate method for estimating this covariance matrix suggested by Beale and Little (1975) performed well in simulations reported by Little (1979).

The method is to let

$$\mathrm{Var}(\hat{\beta}_{Y\cdot X}) = S_W^{-1}\hat{\sigma}^2_{Y\cdot X},$$

where S_W is a $(p \times p)$ weighted sum of squares and cross-products matrix with (j,k)th element

$$s_{Wjk} = \sum_{i=1}^{n} w_i(\hat{x}_{ij} - \tilde{x}_j)(\hat{x}_{ik} - \tilde{x}_k),$$

where $\hat{x}_{ij}$ and $\hat{x}_{ik}$ are observed or estimated values of x_{ij} and x_{ik}, respectively, from the last iteration of the EM algorithm, $\tilde{x}_j = \sum_i w_i\hat{x}_{ij}/\sum_i w_i$ is a weighted mean, and w_i is defined as

$$w_i = \begin{cases} \hat{\sigma}^2_{Y \cdot X}/\hat{\sigma}^2_{Y \cdot X\text{obs}, i}, & \text{if } y_i \text{ is present,} \\ 0, & \text{if } y_i \text{ is missing,} \end{cases}$$

where $\hat{\sigma}^2_{Y \cdot X\text{obs}, i}$ is the ML estimate of the variance of y given the independent variables observed in unit i.

ML estimation for multivariate linear regression can be achieved by applying the algorithm of Section 8.2.1 and then sweeping the independent variables in the resulting augmented covariance matrix. Specifically, if the dependent variables are $Y_1, \ldots, Y_K$ and the independent variables are $X_1, \ldots, X_p$, then the augmented covariance matrix of the combined set of variables $(X_1, \ldots, X_p, Y_1, \ldots, Y_K)$ is estimated using the multivariate normal EM algorithm, and then the variables $X_1, \ldots, X_p$ are swept in the matrix. The resulting matrix contains the ML estimates of the $(p \times K)$ matrix of regression coefficients of Y on X and the $(K \times K)$ residual covariance matrix of Y given X.

EXAMPLE 8.5. *Missing Values in Many Outcome Variables.* Regression analysis with many outcome variables, several predictor variables, and missing values largely confined to the outcomes is now illustrated using data on age of attainment of developmental behaviors in the first year of life (Reinisch et. al., 1985). A sample of 4653 full-term babies was the source of data on the date of attainment of 10 behavioral milestones (e.g., first smile) as well as nine developmentally relevant covariates. Sex of infant was recorded for all babies, the covariates were recorded for nearly all babies, but the milestones suffered from a substantial number of missing values, since they were recorded in diaries kept by the mothers.

ML estimates of means, variances, and correlations among all variables were obtained using the EM algorithm of Section 8.2. The 10 milestones were then regressed on gender, by sweeping gender from the matrix of ML estimates, to obtain estimates for parameters representing male and female mean ages of milestone attainment and female–male differences in mean age of attainment. Standard errors for parameters were obtained by substituting ML estimates of parameters into standard complete-data expressions for standard errors with the sample size being the number of observations of the milestones. In other words, ML estimates of means, variances, and covariances were input into a standard complete-data regression program with $n = 4653$, and the only modification made to the output was to adjust standard errors and test statistics for the actual number of observations of the dependent variable in the regression. This procedure is consistent with the suggestion of Beale and Little (1975) since gender is never missing. The results are summarized in Table 8.2. Also reported in the rightmost columns of Table 8.2 are the results

Table 8.2 Sex Differences in the Attainment of Developmental Milestones (in Age)

Milestone	Number of Babies with Milestone Data Reported	Male Mean	Female Mean	Comparisons			
				Unadjusted		Adjusted[a]	
				Diff	T	Diff	T
Lifts head	3273	32.1	34.8	2.7	3.16	2.2	2.51
Smiles	3229	49.1	48.4	0.7	0.87	1.4	1.69
Holds head up	2618	118.5	116.9	1.6	1.18	2.0	1.27
Reaches for objects	2593	137.9	138.8	0.9	0.66	0.6	0.46
Sits without support	3036	230.0	227.5	2.5	1.69	2.8	1.91
Stands with support	3840	279.8	285.4	5.6	3.25	5.4	3.05
Crawls independently	3391	306.5	313.9	7.4	3.72	7.1	3.55
Walks with support	3689	324.6	325.6	1.0	1.71	2.0	1.35
Stands without support	2128	337.7	339.8	2.1	1.28	1.7	1.04
Walks without support	1985	361.2	361.1	0.1	0.04	0.2	0.2

[a] For covariates: birth weight, length of gestation, socioeconomic status, score for predisposition for pregnancy and birth complications, pregnancy complications, delivery complications, neonatal physical examination score, neonatal neurological examination score, neonatal nonmaturity score.

of analyses that adjusted for the nine covariates. The estimates were obtained by sweeping the matrix of ML estimates of means, variances, and covariances on gender and the nine covariates. The standard errors were obtained by using the sample size relevant to each outcome variable since the covariates were essentially fully observed.

A final set of analyses involved analogous regressions using differences in outcome variables (i.e., gaps in milestone attainments) as the dependent variables. Again, for standard errors standard complete-data regression methods were used with the modification that the sample size for a dependent variable was the number of babies for whom both milestones comprising the dependent variable were observed. These sample sizes varied between a low of 1262 for the difference between "walks without support" and "stands without support" and a high of 3524 for the difference between "stands with support" and "sits without support."

8.5. A GENERAL REPEATED MEASURES MODEL WITH MISSING DATA

Missing data often occur in longitudinal studies, where subjects are observed at different times and/or under different experimental conditions. Normal models for such data often combine special covariance structures such as those discussed in Section 8.3 with mean structures that relate the mean of the repeated measures to design variables. The following general repeated measures model is given in Jennrich and Schluchter (1987) and builds on earlier work by Ware (1985) and Helms (1984).

Suppose that the hypothetical complete data for case i consists of K measurements $y_i = (y_{i1}, \ldots, y_{iK})$ on an outcome variable Y, and that y_i are assumed to be independent multivariate normal random variables with mean μ_i and covariance matrix Σ. We write

$$\mu_i^{\mathrm{T}} = X_i \beta, \qquad \Sigma = \Sigma(\psi), \tag{8.12}$$

where X_i is a known $(K \times m)$ design matrix for case i, β is a $(m \times 1)$ vector of unknown regression coefficients, and the elements of Σ are known functions of a set of v unknown parameters ψ. The model thus incorporates a mean structure, defined by the set of design matrices $\{X_i\}$, and a covariance structure, defined by the form of the covariance matrix Σ. The observed data consist of the design matrices $\{X_i\}$ and $\{y_{\mathrm{obs},i}, i = 1, \ldots, n\}$, where $y_{\mathrm{obs},i}$ is the observed part of the vector y_i. Missing values of y_i are assumed to be MAR. The complete-data loglikelihood is linear in the quantities $\{y_i, y_i^{\mathrm{T}} y_i, i = 1, \ldots, n\}$. Hence the E step consists in calculating the means of y_i and $y_i^{\mathrm{T}} y_i$ given $y_{\mathrm{obs},i}$, X_i, and current estimates of β and Σ. These calculations involve sweep operations on the current estimate of Σ analogous to those in the multivariate normal model of Section 8.2.1. The M step for the model is itself iterative except in special cases, and thus a primary attraction of EM, simplicity of the M step, is lost. Jennrich and Schluchter (1987) present a GEM algorithm (see Section 7.4) and also discuss scoring and Newton–Ralphson algorithms that can be attractive when Σ depends on a modest number of parameters, ψ.

A large number of situations can be modeled by combining different choices of mean and covariance structures. Jennrich and Schluchter discuss the following covariance structures:

Independence: $\Sigma = \psi I$, ψ scalar, $I = (K \times K)$ identity matrix

Compound symmetry: $\Sigma = \psi_1 U + \psi_2 I$, ψ_1 and ψ_2 scalar, $U = (K \times K)$ matrix of ones, $I = (K \times K)$ identity matrix

Autoregressive, lag 1: $\Sigma = (\sigma_{jk})$, $\sigma_{jk} = \psi_1 \psi_2^{|j-k|}$, ψ_1, ψ_2 scalars

Banded: $\Sigma = (\sigma_{jk})$, $\sigma_{jk} = \psi_r$, where $r = |j - k| + 1$, $r = 1, \ldots, K$

Factor analytic: $\Sigma = \Gamma\Gamma^{\mathrm{T}} + \psi$, $\Gamma = (K \times q)$ matrix of unknown factor loadings, and ψ = diagonal matrix of specific variances

Random Effects: $\Sigma = Z\psi Z^{\mathrm{T}} + \sigma^2 I$, $Z = (K \times q)$ known matrix, $\psi = (q \times q)$ unknown dispersion matrix, σ^2 scalar, I the $q \times q$ identity matrix

Unstructured: $\Sigma = (\sigma_{jk})$, $\psi_1 = \sigma_{11}$, $\psi_2 = \sigma_{12}$, $\psi_3 = \sigma_{22}, \ldots, \psi_v = \sigma_{KK}$.

The mean structure is also very flexible. If $X_i = I$, the $(K \times K)$ identity matrix, then $\mu_i = \beta^{\mathrm{T}}$ for all i. This constant mean structure, combined with the unstructured, factor analytic and compound symmetry covariance structures, yields the models of Section 8.2, Examples 8.2 and 8.3, respectively. Between-subject and within-subject effects are readily modeled through other choices of X_i, as in the next example.

EXAMPLE 8.6. *Growth Curve Models with Missing Data.* Potthoff and Roy (1964) present the growth data in Table 8.3 for 11 girls and 16 boys. For each subject the distance from the center of the pituitary to the maxillary fissure was recorded at the ages of 8, 10, 12, and 14. Jennrich and Schluchter (1987) fit eight repeated measures models to these data. We fit the same models to the data obtained by deleting the nine values in parentheses in Table 8.3. The deletion mechanism is designed to be MAR but not MCAR. Specifically, for each gender, values at age 10 are deleted for cases with low values at age 8. Table 8.4 summarizes the models, giving values of minus twice the loglikelihood (-2λ) and the likelihood ratio chi-squared (χ^2) for comparing models. The last column gives values for the latter statistic from the complete data before deletion, as given in Jennrich and Schluchter (1987).

For the ith subject, let y_i denote the four distance measurements, and let

Table 8.3 Growth Data for 11 Girls and 16 Boys

Individual	Age (in years)				Individual	Age (in years)			
Girl	8	10	12	14	Boy	8	10	12	14
1	21	20	21.5	23	1	26	25	29	31
2	21	21.5	24	25.5	2	21.5	(22.5)	23	26.5
3	20.5	(24)	24.5	26	3	23	22.5	24	27.5
4	23.5	24.5	25	26.5	4	25.5	27.5	26.5	27
5	21.5	23	22.5	23.5	5	20	(23.5)	22.5	26
6	20	(21)	21	22.5	6	24.5	25.5	27	28.5
7	21.5	22.5	23	25	7	22	22	24.5	26.5
8	23	23	23.5	24	8	24	21.5	24.5	25.5
9	20	(21)	22	21.5	9	23	20.5	31	26.0
10	16.5	(19)	19	19.5	10	27.5	28	31	31.5
11	24.5	25	28	28	11	23	23	23.5	25
					12	21.5	(23.5)	24	28
					13	17	(24.5)	26	29.5
					14	22.5	25.5	25.5	26
					15	23	24.5	26	30
					16	22	(21.5)	23.5	(25)

Source: Potthoff and Roy (1964) as reported by Jennrich and Schluchter (1987). Values in parentheses are treated as missing in Example 8.7.

Table 8.4 Summary of Models Fit in Example 8.7

Model Number	Description	Number of Parameters	-2λ	Comparison Model	χ^2	df	Complete Data[a] χ^2
1	Eight separate means, unstructured covariance matrix	18	386.956	—		—	—
2	Two lines, unequal slopes, unstructured covariance matrix	14	393.288	1	6.332	4	[2.968]
3	Two lines, common slope, unstructured covariance matrix	13	397.400	2	4.112	1	[6.676]
4	Two lines, unequal slopes, banded structure	8	398.030	2	4.742	6	[5.166]
5	Two lines, unequal slopes, AR(1) structure	6	409.524	2	16.236	8	[21.204]
6	Two lines, unequal slopes, random slopes and intercepts	8	400.452	2	7.164	6	[8.329]
7	Two lines, unequal slopes, random intercepts (compound symmetry)	6	401.312	2	8.024	8	[9.162]
8	Two lines, unequal slopes, uncorrelated observations	5	441.582	7	40.270	1	[50.833]

[a] From Jennrich and Schluchter (1987).

x_i be a design variable equal to 1 if the child is a boy and 0 if the child is a girl. Model 1 specifies a distinct mean for each of the gender by age groups, and assumes that the (4×4) covariance matrix is unstructured. The design matrix for subject i can be written as

$$X_i = \begin{bmatrix} 1 & x_i & 0 & 0 & 0 & 0 & 0 & 0 \\ 0 & 0 & 1 & x_i & 0 & 0 & 0 & 0 \\ 0 & 0 & 0 & 0 & 1 & x_i & 0 & 0 \\ 0 & 0 & 0 & 0 & 0 & 0 & 1 & x_i \end{bmatrix}.$$

With no missing data, the ML estimate of β is the vector of eight sample means and the ML estimate of Σ is S/n, where S is the pooled within-groups sum of squares and cross-products matrix.

This unrestricted model, model 1 in Table 8.4, was fitted to the incomplete data of Table 8.3. Seven other models were also fitted to those data. Plots suggest a linear relationship between mean distance and age, with different intercepts and slopes for girls and boys. The mean structure for this model can be written as

$$\mu_i^{\mathrm{T}} = X_i\beta = \begin{bmatrix} 1 & x_i & -3 & -3x_i \\ 1 & x_i & -1 & -x_i \\ 1 & x_i & 1 & x_i \\ 1 & x_i & 3 & 3x_i \end{bmatrix} \begin{bmatrix} \beta_1 \\ \beta_2 \\ \beta_3 \\ \beta_4 \end{bmatrix}, \tag{8.13}$$

where β_1 and $\beta_1 + \beta_2$ represent overall means and β_3 and $\beta_3 + \beta_4$ represent slopes for girls and boys, respectively. Model 2 fits this mean structure and an unstructured Σ.

The likelihood ratio statistic comparing model 2 with model 1 is $\chi^2 = 6.332$ on four degrees of freedom, indicating a fairly satisfactory fit for model 2. Model 3 is obtained from model 2 by setting $\beta_4 = 0$, that is, dropping the last column of X_i. It constrains the regression lines of distance against age to have common slope in the two gender groups. Compared with model 2, model 3 yields a likelihood ratio of 4.112 on one degree of freedom, indicating significant lack of fit. Hence the mean structure of model 2 is preferred.

The remaining models in Table 8.4 have the mean structure of model 2, but place constraints on Σ. The autoregressive (model 5) and independence (model 8) covariance structures do not fit the data, judging from the chi-squared statistics. The banded structure (model 4) and two random effects structures (models 6 and 7) fit the data well. Of these, model 7 is preferred on grounds of parsimony. The model can be interpreted as a random effects model with a fixed slope for each gender group and a random intercept that varies across subjects about common gender means. Further analysis would display the parameter estimates for this preferred model.

8.6. TIME SERIES MODELS

8.6.1. Introduction

We confine our limited discussion of time series modeling with missing data to parametric time-domain models with normal disturbances, since these models are most amenable to the ML techniques developed in Chapters 5 and 7. Two classes of models of this type appear particularly important in applications: the autoregressive–moving average (ARMA) models developed by Box and Jenkins (1976), and general state space or Kalman filter models, initiated in the engineering literature (Kalman, 1960) and currently enjoying considerable development in the econometrics and statistics literature on time series (Harvey, 1981). As discussed in the next section, autoregressive models are relatively easy to fit to incomplete time series data, with the aid of the EM algorithm. Box–Jenkins models with moving average components are less easily handled, but ML estimation can be achieved by recasting the models as general state space models, as discussed in Harvey and Phillips (1979) and Jones (1980). The details of this transformation are omitted here; however, ML estimation for general state space models from incomplete data is outlined in Section 8.6.3, following the approach of Shumway and Stoffer (1982).

8.6.2. Autoregressive Models for Univariate Time Series with Missing Values

Let $Y = (y_0, y_1, \ldots, y_T)$ denote a completely observed univariate time series with $T + 1$ observations. The autoregressive model of lag p (ARp) assumes that y_i, the value at time i, is related to values at p previous time points by the model

$$(y_i | y_1, y_2, \ldots, y_{i-1}, \theta) \sim N(\alpha + \beta_1 y_{i-1} + \cdots + \beta_p y_{i-p}, \sigma^2), \tag{8.14}$$

where $\theta = (\alpha, \beta_1, \beta_2, \ldots, \beta_p, \sigma^2)$, α is a constant term, $\beta_1, \beta_2, \ldots, \beta_p$ are unknown regression coefficients, and σ^2 is an unknown error variance. Least squares estimates of α, β_1, β_2, $\ldots$, β_p and σ^2 can be found by regressing y_i on $x_i = (y_{i-1}, y_{i-2}, \ldots, y_{i-p})$, using observations $i = p$, $p + 1$, $\ldots$, T. These estimates are only approximately ML because the contribution of the marginal distribution of y_0, y_1, $\ldots$, y_{p-1} to the likelihood is ignored, which is justified when p is small compared with T.

If some observations in the series are missing, one might consider applying the methods of Section 8.4 for regression with missing values. This approach may yield useful rough approximations, but the procedure is not ML even assuming the marginal distribution of y_0, y_1, $\ldots$, y_p can be ignored, since (1) missing values y_i $(i \geqslant p)$ appear as dependent and independent variables in the regressions, and (2) the model (8.14) induces a special structure on the mean vector and covariance matrix of Y that is not used in the analysis. Thus special EM algorithms are required to estimate the ARp model from incomplete time series. The algorithms are relatively easy to implement, although not trivial to describe. We confine attention here to the $p = 1$ case.

EXAMPLE 8.7. *The AR1 Model for Time Series with Missing Values.* Setting $p = 1$ in Eq. (8.14), we obtain the model

$$(y_i | y_1, \ldots, y_{i-1}, \theta) \overset{\text{ind}}{\sim} N(\alpha + \beta y_{i-1}, \sigma^2). \tag{8.15}$$

The AR1 series is *stationary*, yielding a constant marginal distribution of y_i over time, only if $|\beta| < 1$. The joint distribution of the y_i then has constant marginal mean $\mu \equiv E(y_i) = \alpha(1 - \beta)^{-1}$, variance $\operatorname{Var}(y_i) = \sigma^2(1 - \beta^2)^{-1}$, and covariances $\operatorname{Cov}(y_i, y_{i+k}) = \beta^k \sigma^2 (1 - \beta^2)^{-1}$ for $k \geqslant 1$. Ignoring the contribution of the marginal distribution of y_0, the complete-data loglikelihood for Y is $l(\alpha, \beta, \sigma^2) | y) = -0.5\sigma^{-2} \sum_{i=1}^{T} (y_i - \alpha - \beta y_{i-1})^2 - 0.5T \log \sigma^2$, which is equivalent to the loglikelihood for the normal linear regression of y_i on $x_i = y_{i-1}$, with data $\{(y_i, x_i), i = 1, \ldots, T\}$. The complete-data sufficient statistics are $s = (s_1, s_2, s_3, s_4, s_5)$, where

$$s_1 = \sum_{i=1}^{T} y_i, \quad s_2 = \sum_{i=1}^{T} y_{i-1}, \quad s_3 = \sum_{i=1}^{T} y_i^2, \quad s_4 = \sum_{i=1}^{T} y_{i-1}^2, \quad s_5 = \sum_{i=1}^{T} y_i y_{i-1}.$$

ML estimates of $\theta = (\alpha, \beta, \sigma)$ are $\hat{\theta} = (\hat{\alpha}, \hat{\beta}, \hat{\sigma})$, where

$$\begin{aligned} \hat{\alpha} &= (s_1 - \hat{\beta} s_2) T^{-1}, \\ \hat{\beta} &= (s_5 - T^{-1} s_1 s_2)(s_4 - T^{-1} s_2^2)^{-1}, \\ \hat{\sigma}^2 &= \{s_3 - s_1^2 T^{-1} - \hat{\beta}^2 (s_4 - s_2^2 T^{-1})\}/T. \end{aligned} \tag{8.16}$$

Now suppose some observations are missing, and the data are MAR. ML estimates of θ, still ignoring the contribution of the marginal distribution of y_0 to the likelihood, can be obtained by the EM algorithm. Let $\theta^{(t)} = (\alpha^{(t)}, \beta^{(t)}, \sigma^{(t)})$ be estimates of θ at iteration t. The M step of the algorithm calculates $\theta^{(t+1)}$ from Eq. (8.16) with complete-data sufficient statistics s replaced by estimates $s^{(t)}$ from the E step.

The E step computes $s^{(t)} = (s_1^{(t)}, s_2^{(t)}, s_3^{(t)}, s_4^{(t)}, s_5^{(t)})$, where

$$s_1^{(t)} = \sum_{i=1}^{T} \hat{y}_i^{(t)}, \quad s_2^{(t)} = \sum_{i=1}^{T} \hat{y}_{i-1}^{(t)}, \quad s_3^{(t)} = \sum_{i=1}^{T} \{(\hat{y}_i^{(t)})^2 + c_{ii}^{(t)}\},$$

$$s_4^{(t)} = \sum_{i=1}^{T} \{(\hat{y}_{i-1}^{(t)})^2 + c_{i-1,i-1}^{(t)}\}, \quad s_5^{(t)} = \sum_{i=1}^{T} \{y_{i-1}^{(t)} \hat{y}_i^{(t)} + c_{i-1,i}^{(t)}\},$$

and

$$\hat{y}_i^{(t)} = \begin{cases} y_i, & \text{if } y_i \text{ is present,} \\ E\{y_i | Y_{\text{obs}}, \theta^{(t)}\}, & \text{if } y_i \text{ is missing,} \end{cases}$$

$$c_{ij}^{(t)} = \begin{cases} 0, & \text{if } y_i \text{ or } y_j \text{ is present,} \\ \text{Cov}(y_i, y_j | Y_{\text{obs}}, \theta^{(t)}), & \text{if } y_i \text{ and } y_j \text{ are missing.} \end{cases}$$

The E step involves standard sweep operations on the covariance matrix of the observations. However, this $(T \times T)$ matrix is usually large, so it is desirable to exploit properties of the AR1 model to simplify the E step computations. Suppose $Y_{\text{mis}}^* = (y_{j+1}, y_{j+2}, \ldots, y_{k-1})$ is a sequence of missing values, bounded by present observations, y_j and y_k. Then (1) Y_{mis}^* is independent of the other missing values, given Y_{obs} and θ, and (2) the distribution of Y_{mis}^* given Y_{obs} and θ depends on Y_{obs} only through the bounding observations y_j and y_k. The latter distribution is multivariate normal, with constant covariance matrix, and means that are weighted averages of $\mu = \alpha(1 - \beta)^{-1}$, y_j and y_k. The weights and covariance matrix depend only on the number of missing values in the sequence and can be found from the current estimate of the covariance matrix of $(y_j, y_{j+1}, \ldots, y_k)$ by sweeping on elements corresponding to the observed variables y_j and y_k.

In particular, suppose y_j and y_{j+2} are present and y_{j+1} is missing. The covariance matrix of y_j, y_{j+1} and y_{j+2} is

$$A = \frac{\sigma^2}{1 - \beta^2} \begin{bmatrix} 1 & \beta & \beta^2 \\ \beta & 1 & \beta \\ \beta^2 & \beta & 1 \end{bmatrix}.$$

Sweeping on y_j and y_{j+2} yields

$$\text{SWP}[j, j+2]A = \frac{1}{1 + \beta^2} \begin{bmatrix} -\sigma^{-2} & \beta & -\beta^2\sigma^{-2} \\ \beta & \sigma^2 & \beta \\ -\beta^2\sigma^{-2} & \beta & -\sigma^{-2} \end{bmatrix}. \tag{8.17}$$

Hence from stationarity and (8.17),

$$\begin{aligned} E\{y_{j+1} | y_j, y_{j+2}, \theta\} &= \mu + \beta(1 + \beta^2)^{-1}(y_{j+2} - \mu) + \beta(1 + \beta^2)^{-1}(y_j - \mu) \\ &= \mu\left\{1 - \frac{2\beta}{1 + \beta^2}\right\} + \frac{\beta}{1 + \beta^2}\{y_j + y_{j+2}\}, \end{aligned}$$

$$\text{Var}(y_{j+1} | y_j, y_{j+2}, \theta) = \sigma^2(1 + \beta^2)^{-1}.$$

Substituting $\theta = \theta^{(t)}$ in these expressions yields $\hat{y}_{j+1}^{(t)}$ and $c_{j+1,j+1}^{(t)}$ for the E step. Note that $\hat{y}_{j+1}^{(t)}$ supplies a prediction for the missing value at the final iteration of the algorithm.

8.6.3. Kalman Filter Models

Shumway and Stoffer (1982) consider the Kalman filter model

$$\begin{aligned} (y_i|M_i, z_i, \theta) &\overset{\text{indep}}{\sim} N(z_i M_i, B), \\ (z_0|\theta) &\sim N(\mu, \Sigma), \\ (z_i|z_1, \ldots, z_{i-1}, \theta) &\sim N(z_{i-1}\phi, Q), \end{aligned} \tag{8.18}$$

where y_i is a $(1 \times q)$ vector of observed variables at time i, M_i is a known $(p \times q)$ design matrix that relates the mean of y_i to an unobserved $(1 \times p)$ stochastic vector z_i, and $\theta = (B, \mu, \Sigma, \phi, Q)$ represents the unknown parameters, where B, Σ, and Q are covariance matrices, μ is the mean of z_0, and ϕ is a $(p \times p)$ matrix of autoregression coefficients of z_i on z_{i-1}. The random unobserved series z_i, which is modeled as a first-order multivariate autoregressive process, is of primary interest.

This model can be envisioned as a kind of random effects model for time series, where the effect vector z_i has correlation structure over time. The primary aim is to predict the unobserved series z_i for $1 = 1, 2, \ldots, n$ (smoothing) and for $i = n + 1, n + 2, \ldots$ (forecasting), using the observed series $y_1, y_2, \ldots, y_n$. If the parameter θ were known, the optimal estimates of z_i would be their conditional means, given the parameters θ and the data Y. These quantities are called Kalman smoothing estimators, and the set of recursive formulas used to derive them are called the Kalman filter. In practice θ is unknown, and the forecasting and smoothing procedures involve ML estimation of θ, and then application of the Kalman filter with θ replaced by the ML estimate $\hat{\theta}$.

The same process applies when data Y are incomplete, with Y replaced by its observed component, say Y_{obs}. ML estimates of Q can be derived by Newton–Raphson techniques (Gupta and Mehra, 1974; Ledolter, 1979; Goodrich and Caines, 1979). However, the EM algorithm provides a convenient alternative method, with the missing component Y_{mis} of Y *and* z_1, $z_2, \ldots, z_n$ treated as missing data. An attractive feature of this approach is that the E step of the algorithm includes the calculation of the expected value of z_i given Y_{obs} and current estimates of θ, which is the same process as Kalman smoothing described above. Details of the E step are given in Shumway and Stoffer (1982). The M step is relatively straightforward. Estimates of ϕ and Q are obtained by autoregression applied to the expected values of the complete-data sufficient statistics

$$\sum_{i=1}^{n} z_i, \quad \sum_{i=1}^{n} z_i^{\mathrm{T}} z_i, \quad \sum_{i=1}^{n} z_{i-1}, \quad \sum_{i=1}^{n} z_{i-1}^{\mathrm{T}} z_{i-1}, \quad \text{and} \quad \sum_{i=1}^{n} z_{i-1}^{\mathrm{T}} z_i$$

from the E step; B is estimated by the expected value of the residual covariance

Table 8.5 Data Set for Example 8.9 and Predictions from EM Algorithm—Physician Expenditures (in millions)

	SSA	HCFA	Predictions from EM Algorithm	
Year, i	y_{i1}	y_{i2}	$E(x_i \mid \text{data}, \theta)$	$\text{Var}^{1/2}(x_i \mid \text{data}, \theta)$
1949	2,633	—	2,541	178
1950	2,747	—	2,711	185
1951	2,868	—	2,864	186
1952	3,042	—	3,045	186
1953	3,278	—	3,269	186
1954	3,574	—	3,519	186
1955	3,689	—	3,736	186
1956	4,067	—	4,063	186
1957	4,419	—	4,433	186
1958	4,910	—	4,876	186
1959	5,481	—	5,331	186
1960	5,684	—	5,644	186
1961	5,895	—	5,972	186
1962	6,498	—	6,477	186
1963	6,891	—	7,032	185
1964	8,065	—	7,866	179
1965	8,745	8,474	8,521	110
1966	9,156	9,175	9,198	108
1967	10,287	10,142	10,160	108
1968	11,099	11,104	11,159	108
1969	12,629	12,648	12,645	108
1970	14,306	14,340	14,289	108
1971	15,835	15,918	15,835	108
1972	16,916	17,162	17,171	108
1973	18,200	19,278	19,106	109
1974	—	21,568	21,675	119
1975	—	25,181	25,027	120
1976	—	27,931	27,932	129
1977	—	—	31,178	355
1978	—	—	34,801	512
1979	—	—	38,846	657
1980	—	—	43,361	802
1981	—	—	48,400	952

Source: Meltzer et. al. (1980) as reported in Shumway and Stoffer (1982), Tables I and III.

matrix $n^{-1}\sum_{i=1}^{n}(y_i - z_iM)^{\mathrm{T}}(y_i - z_iM)$. Finally μ is estimated as the expected value of z_0, and Σ is set from external considerations. We now provide a specific example of this very general model.

EXAMPLE 8.8. *A Bivariate Time Series Measuring an Underlying Series with Error.* Table 8.5, taken from Meltzer et al. (1980), shows two incomplete time series of total expenditures for physician services, measured by Social Security Administration (SSA), yielding Y_1, and Health Care Financing Administration (HCFA), yielding Y_2. Shumway and Stoffer (1982) analyze the data using the model

$$(y_{ij}|z_i,\theta) \overset{\text{indep}}{\sim} N(z_i, B_j),$$

$$(z_i|z_1,\ldots,z_{i-1},\theta) \sim N(z_{i-1}\phi, Q),$$

where y_{ij} is the total expenditure amount at time i for SSA ($j = 1$) and HCFA ($j = 2$), z_i is the underlying true expenditure, assumed to form an AR1 series over time with coefficient ϕ and residual variance Q, B_j is the measurement variance of y_{ij} ($j = 1, 2$), and $\theta = (B_1, B_2, \phi, Q)$. Unlike Example 8.7, the AR1 series for z_i is not assumed stationary, the parameter ϕ being an inflation factor modeling exponential growth; the assumption that ϕ is constant over time is probably an oversimplification.

Successive EM iterates are displayed in Table 8.6; initial values were simply guessed by examining portions of the completely observed series. The last columns of Table 8.5 show smoothed estimates of z_i from the final iteration of the EM algorithm for years 1949–1976, and predictions for the five years 1977–1981, together with their standard errors. The predictions for 1977–

Table 8.6 Summary of Successive EM Iterates for the Maximum Likelihood Estimators in Example 8.8

t	$\mu(t)$	$\phi(t)$	$q(t)$	$B_1(t)$	$B_2(t)$	$-2\ln L$
1	2500	1.100	10,000	10,000	10,000	885
2	2417	1.114	49,837	41,583	24,105	680
3	2396	1.116	78,153	54,666	25,486	675
4	2383	1.116	93,513	59,958	25,580	675
5	2374	1.116	100,571	62,483	25,384	674
10	2342	1.116	105,152	65,725	23,920	674
20	2279	1.116	104,814	67,760	20,971	672
40	2277	1.116	105,115	68,636	19,394	671
50	2276	1.116	105,097	68,663	19,354	671
75	2277	1.116	105,115	68,675	19,329	671

Source: Shumway and Stoffer (1982), Table II.

1981 have standard errors ranging from 355 for 1977 to 952 for 1982, reflecting considerable uncertainty.

REFERENCES

Beale, E. M. L., and Little, R. J. A. (1975). Missing values in multivariate analysis, *J. Roy. Statist. Soc.* **B37**, 129–146.

Bentler, P. M., and Tanaka, J. S. (1983) Problems with EM for ML factor analysis, *Psychometrika* **48**, 247–253.

Box, G. E. P., and Jenkins, G. M. (1976). *Time Series Analysis: Forecasting and Control.* San Francisco: Holden-Day.

Brownlee, K. A. (1965). *Statistical Theory and Methodology in Science and Engineering.* New York: Wiley.

Dempster, A. P., Laird, N. M., and Rubin, D. B. (1977). Maximum likelihood estimation from incomplete data via the EM algorithm, (with discussion), *J. Roy. Statist. Soc.* **B39**, 1–38.

Dempster, A. P., Rubin, D. B., and Tsutakawa, R. K. (1981). Estimation in covariance component models, *J. Am. Statist. Assoc.* **76**, 341–353.

Goodrich, R. L., and Caines, P. E. (1979). Linear system identification from nonstationary cross-sectional data, *IEEE Trans. Aut. Control,* **AC-24**, 403–411.

Gupta, N. K., and Mehra, R. K. (1974). Computational aspects of maximum likelihood estimation and reduction in sensitivity function calculations, *IEEE Trans. Aut. Control* **AC-19**, 774–783.

Hartley, H. O., and Hocking, R. R. (1971). The analysis of incomplete data, *Biometrics* **14**, 174–194.

Harvey, A. C. (1981). *Time Series Models*, New York: Wiley.

Harvey, A. C., and Phillips, G. D. A. (1979). Maximum likelihood estimation of regression models with autoregressive-moving average disturbances, *Biometrika* **66**, 49–58.

Healy, M. J. R., and Westmacott, M. (1956). Missing values in experiments analyzed on automatic computers, *Appl. Statist.* **5**, 203–206.

Helms, R. W. (1984). Linear models with linear covariance structure for incomplete longitudinal data, *Biometrics Section, American Statistical Association 1984.*

Holland, P. W., and Wightman, L. E. (1982). Section pre-equating: A preliminary investigation, in *Test Equating* (P. W. Holland and D. B. Rubin, Eds.). New York: Academic Press.

Jennrich, R. I., and Schluchter, M. D. (1987). Incomplete repeated measures models with structured covariance matrices, to appear in *Biometrics* **42**.

Jones, R. H. (1980). Maximum likelihood fitting of ARMA models to time series with missing observations, *Technometrics* **22**, 389–395.

Kalman, R. E. (1960). A new approach to linear filtering and prediction problems, *Trans. ASME J. Basic Eng.* **82**, 34–35.

Laird, N. M., and Ware, J. H. (1982). Random-effects models for longitudinal data, *Biometrics* **38**, 963–974.

Ledolter, J. (1979). A recursive approach to parameter estimation in regression and time series problems, *Commun. Statist. Theory Meth.* **A8**, 1227–1245.

Little, R. J. A. (1979). Maximum likelihood inference for multiple regression with missing values: A simulation study, *J. Roy. Statist. Soc.* **B41**, 76–87.

Meltzer, A., Goodman, C., Langwell, K., Cosler, J., Baghelai, C., and Bobula, J. (1980). Develop physician and physician extender data bases, *G-155, Final Report*, Applied Management Sciences, Inc., Silver Springs, MD.

Orchard, T., and Woodbury, M. A. (1972). A missing information principle: theory and applications, *Proceedings of the 6th Berkeley Symposium on Mathematical Statistics and Probability* **1**, 697–715.

Potthoff, R. F., and Roy, S. N. (1964). A generalized multivariate analysis of variance model useful especially for growth curve problems, *Biometrika* **51**, 313–326.

Reinisch, J. M., Rosenblum, L. A., Rubin, D. B., and Schulzinger, M. F. (1985). Sex differences in behavioral milestones during the first year of life, Kinsey Institute report.

Rubin, D. B., and Szatrowski, T. H. (1982). Finding maximum likelihood estimates for patterned covariance matrices by the EM algorithm, *Biometrika* **69**, 657–660.

Rubin, D. B., and Thayer, D. T. (1982). EM algorithms for factor analysis, *Psychometrika* **47**, 69–76.

Rubin, D. B., and Thayer, D. T. (1983). More on EM for ML factor analysis, *Psychometrika* **48**, 253–257.

Shumway, R. H., and Stoffer, D. S. (1982). An approach to time series smoothing and forecasting using the EM algorithm, *J. Time Series Anal.* **3**, 253–264.

Snedecor, G. W., and Cochran, W. G. (1967). *Statistical Methods*, Ames: Iowa State University Press.

Szatrowski, T. H. (1978). Explicit solutions, one iteration convergence and averaging in the multivariate normal estimation problem for patterned means and covariances, *Ann. Inst. Statist. Math.* **30**, 81–88.

Trawinski, I. M., and Bargmann, R. W. (1964). Maximum likelihood with incomplete multivariate data, *Ann. Math. Statist.* **35**, 647–657.

Ware, J. H. (1985). Linear models for the analysis of longitudinal studies, *Am. Statist.* **39**, 95–101.

PROBLEMS

1. Show that the available cases estimates of the means and variances of an incomplete multivariate sample, discussed in Section 3.3, are ML when the data are multivariate normal with unrestricted means and variances and zero correlations, with ignorable nonresponse. Infer situations where the available cases method is appropriate for multivariate data with missing values.
2. Write a computer program for the EM algorithm for bivariate normal data with an arbitrary pattern of missing values.
3. Describe the EM algorithm for bivariate normal data with means (μ_1, μ_2), correlation ρ, and common variance σ^2, and an arbitrary pattern of missing values. If you did Problem 2, modify the program you wrote to handle this model. (*Hint:* Transform to $U_1 = Y_1 + Y_2$, $U_2 = Y_1 - Y_2$.)
4. Derive the expression for the expected information matrix in Section 8.2.2, for the special case of bivariate data.

5. For bivariate data, find the ML estimate of the correlation ρ for (a) a bivariate sample of size m, with means μ_1, μ_2 and variances σ_1^2 and σ_2^2 assumed known, and (b) a bivariate sample of size m, and effectively infinite supplemental samples from the marginal distributions of both variables. Note the rather surprising fact that (a) and (b) yield different answers.
6. Prove the statement before Equation (8.3) that complete-data ML estimates of Σ are obtained from C by simple averaging. (*Hint:* Consider the covariance matrix of the four variables $U_1 = Y_1 + Y_2 + Y_3 + Y_4$, $U_2 = Y_1 - Y_2 + Y_3 - Y_4$, $U_3 = Y_1 - Y_3$, and $U_4 = Y_2 - Y_4$.)
7. Review the discussion in Rubin and Thayer (1978, 1982) and Bentler and Tanaka (1983) on EM for factor analysis.
8. Derive the EM algorithm for the model of Example 8.3 extended with the specification that $\mu \sim N(0, \tau^2)$, where μ is treated as missing data. Then consider the case where $\tau^2 \to \infty$, yielding a flat prior on μ.
9. Examine Beale and Little's (1975) approximate method for estimating the covariance matrix of estimated slopes in Section 8.4.2, for a single predictor X, and data with (a) Y completely observed and X subject to missing values, and (b) X completely observed and Y subject to missing values. Does the method produce the correct asymptotic covariance matrix in either case?
10. Fill in the details leading to the expressions for the mean and variance of y_{j+1} given y_j, y_{j+2} and θ in Example 8.7. Comment on the form of the expected value of y_{j+1} as $\beta \uparrow 1$ and $\beta \downarrow 0$.
11. For Example 8.7, extend the results of Problem 10 to compute the means, variances, and covariance of y_{j+1} and y_{j+2} given y_j and y_{j+3} and θ, for a sequence where y_j and y_{j+3} are observed and y_{j+1} and y_{j+2} are missing.

CHAPTER 9

Models for Partially Classified Contingency Tables, Ignoring the Missing-Data Mechanism

9.1. INTRODUCTION

This chapter concerns the analysis of incomplete data when the variables are categorical. Although interval-scaled variables can be handled by forming categories based on segments of the scale, the ordering between the categories of variables treated in this way, or of other ordinal variables, is not explored here. Recent methods for categorical data that take into account orderings between categories (e.g., Goodman, 1979; McCullagh, 1980) could be extended to handle incomplete data, by applying the likelihood theory of Chapters 6 and 7.

A rectangular ($n \times V$) data matrix consisting of n observations on V categorical variables $Y_1, \ldots, Y_V$ can be rearranged as a V-dimensional contingency table, with cells defined by joint levels of the variables. The entries in the table are counts $\{n_{jkl\cdots t}\}$, where $n_{jkl\cdots t}$ is the number of sampled cases in the cell with $Y_1 = j$, $Y_2 = k$, $Y_3 = l, \ldots, Y_V = t$. If the data matrix has missing items, some of the cases in the preceding contingency table are partially classified. The completely classified cases yield a V-dimensional table of counts $\{m_{jkl\cdots t}\}$ and the incompletely classified cases form supplemental subtables defined by the subset of variables $(Y_1, \ldots, Y_V)$ that are observed. For example, the first eight rows of Table 1.1b represent complete cases in a five-way contingency table with variables Gender, Age Group and Obesity at three time points. The remaining eighteen rows provide data on the six partially classified tables with one or two of the obesity variables missing. We discuss ML estimation for data of this form.

In the next section factorizations of the likelihood analogous to those discussed in Chapter 6 for normal data are applied to special patterns of incomplete categorical data. ML estimation for general patterns via the EM algorithm is discussed in Section 9.3. Section 9.4 considers ML estimation for partially classified data when the classification probabilities are constrained by a loglinear model. Nonignorable nonresponse models for categorical data are deferred until Chapter 11.

A more general type of incomplete data occurs when level j of a particular item, Y_1, say, is not known, but it is known that the case falls into one of a subset S of values of Y_1. If Y_1 is completely missing, then S consists of all the possible values of Y_1. If Y_1 is missing but the value of a less detailed recode Y_1^* of Y_1 is recorded, then S will be a proper subset of the possible values of Y_1. An example of such data subject to coarse and refined classifications is given in Section 9.2.

The missing-data problems considered here should be carefully distinguished from the problem of *structural zeros*, where certain cells contain zero counts because the model assigns them zero probability of containing entries. For example, if Y_1 = year of birth and Y_2 = year of first marriage, and marriages below the age of 10 are ruled out, then cells with $Y_2 \leqslant Y_1 + 9$ are structural zeros in the joint distribution of Y_1 and Y_2. Instances of no data (zero counts) are not considered here as instances of missing data. For discussion of the structural zero problem, see Bishop, Fienberg and Holland (1975, Chapter 5).

9.2. FACTORED LIKELIHOODS FOR MONOTONE MULTINOMIAL DATA

9.2.1. Introduction

In this and the next section we assume that

1. The complete data counts $\mathbf{n} = \{n_{jkl\cdots t}\}$ have a multinomial distribution with index n and probabilities $\theta = \{\pi_{jkl\cdots t}\}$.
2. The missing-data mechanism is ignorable, in the sense discussed in Chapter 5. Thus the likelihood for the probabilities θ is obtained by integrating the complete-data likelihood

$$L(\theta|\mathbf{n}) = \prod_{j,k,l,\ldots,t} \pi_{jkl\cdots t}^{n_{jkl\cdots t}} \tag{9.1}$$

over the missing data. ML estimates of θ are obtained by maximizing the resulting likelihood, subject to the constraint that the cell probabilities sum to 1.

In an alternative model to assumption 1, the cell counts $\{n_{jkl\cdots t}\}$ are independent Poisson random variables with means $\{\mu_{jkl\cdots t}\}$ and cell probabilities $\pi^*_{jkl\cdots t} = \mu_{jkl\cdots t}/\sum_{j,k,l,\ldots,t}\mu_{jkl\cdots t}$. If the nonresponse mechanism is ignorable, likelihood inferences for $\{\pi^*_{jkl\cdots t}\}$ are the same as those for $\{\pi_{jkl\cdots t}\}$ under the multinomial model. This fact follows from arguments analogous to those for the complete data case (Bishop, Fienberg and Holland, 1975). We restrict attention to the multinomial model since it seems more common than the Poisson model in practical situations.

For complete data the likelihood (9.1) yields the ML estimate

$$\hat{\pi}_{jkl\cdots t} = n_{jkl\cdots t}/n$$

with large sample variance

$$\operatorname{Var}(\pi_{jkl\cdots t} - \hat{\pi}_{jkl\cdots t}) = \hat{\pi}_{jkl\cdots t}(1 - \hat{\pi}_{jkl\cdots t})/n.$$

Our objective is to obtain analogous quantities from incomplete data. In this section we discuss special patterns of incomplete data that yield explicit ML estimates.

9.2.2. ML Estimation for Monotone Patterns

We first consider ML estimates for the simple case of a two-way contingency table with one supplemental one-way margin.

EXAMPLE 9.1. *Two-Way Contingency Table with One Supplemental One-Way Margin.* Consider two categorical variables Y_1, with levels $j = 1, \ldots, J$ and Y_2, with levels $k = 1, \ldots, K$. The data consist of m observations $(y_{i1}, y_{i2}, i = 1, \ldots, m)$ with y_{i1} and y_{i2} recorded and $r = n - m$ observations $(y_{i1}, i = m + 1, \ldots, n)$ with y_{i1} recorded and y_{i2} missing. The data pattern is identical to Example 6.1, but the variables are now categorical.

The m completely classified observations can be displayed in a $(J \times K)$ contingency table, with m_{jk} observations in the cell with $y_{i1} = j$, $y_{i2} = k$. The r remaining observations form a supplemental $(J \times 1)$ margin, with r_j units in the cell with $y_{i1} = j$ (see Figure 9.1).

We shall use the standard plus notation for summation over subscripts j and k. For this problem

$$\theta = (\pi_{11}, \pi_{12}, \ldots, \pi_{JK}) \qquad \text{and} \qquad \sum_{j=1}^{J}\sum_{k=1}^{K}\pi_{jk} \equiv \pi_{++} = 1.$$

As in Example 6.1, we adopt the alternative parameter set ϕ corresponding to the marginal distribution of Y_1 and the conditional distribution of Y_2 given Y_1. The loglikelihood of the data can be written

$$l(\phi|\{m_{jk}, r_j\}) = \sum_{j=1}^{J}(m_{j+} + r_j)\ln\pi_{j+} + \sum_{j=1}^{J}\sum_{k=1}^{K}m_{jk}\ln\pi_{k\cdot j}, \tag{9.2}$$

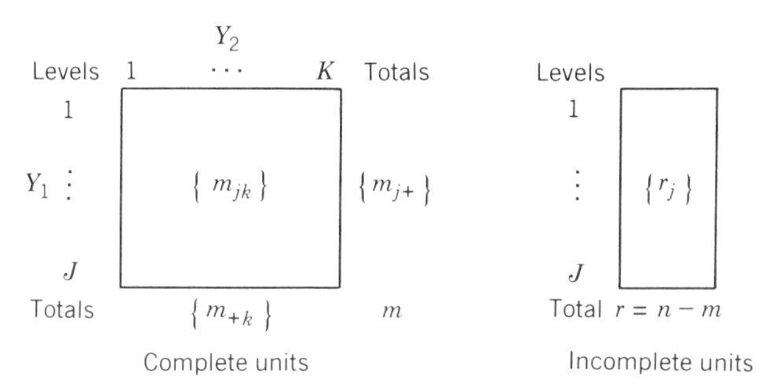

Figure 9.1. The data in Example 9.1.

where the first term is the loglikelihood for the multinomial distribution of the marginal counts $m_{j+} + r_j$, with index n and probabilities π_{j+}, and the second term is the loglikelihood for the conditional multinomial distribution of m_{jk} given m_{j+}, with index m_{j+} and probabilities

$$\pi_{k \cdot j} = \Pr(Y_2 = k | Y_1 = j) = \pi_{jk}/\pi_{j+}.$$

Thus (9.2) is a factorization of the likelihood of the form (6.1), with

$$\phi_1 = \{\pi_{j+}, j = 1, \ldots, J\} \qquad \text{and} \qquad \phi_2 = \{\pi_{k \cdot j}, j = 1, \ldots, J; k = 1, \ldots, K\},$$

and ϕ_1 and ϕ_2 are distinct. Maximizing each component separately, we obtain ML estimates

$$\hat{\pi}_{j+} = \frac{m_{j+} + r_j}{n}, \qquad \hat{\pi}_{k \cdot j} = \frac{m_{jk}}{m_{j+}}$$

and so

$$\hat{\pi}_{jk} = \hat{\pi}_{j+} \hat{\pi}_{k \cdot j} = \frac{[m_{jk} + (m_{jk}/m_{j+}) r_j]}{n}. \tag{9.3}$$

Thus the ML estimate effectively distributes a proportion m_{jk}/m_{j+} of the unclassified observations r_j into the (j, k)th cell.

EXAMPLE 9.2. *Numerical Illustration of Monotone Bivariate Counted Data.* A numerical illustration of the results of Example 9.1 is provided by the data in Table 9.1, where Y_1 is a dichotomous variable and Y_2 is a trichotomous one. The marginal probabilities of Y_1 are estimated from completely and partially classified units:

$$\hat{\pi}_{1+} = 190/410, \qquad \hat{\pi}_{2+} = 220/410.$$

The conditional probabilities of classification for Y_2 given Y_1 are estimated from the completely classified units:

Table 9.1 Numerical Example of the Data Pattern of Figure 9.1

	Complete Units						Incomplete Units		
		Y_2							
		1	2	3	Total				
Y_1	1	20	30	40	90		Y_1	1	100
	2	50	60	20	130			2	90
Total		70	90	60	220		Total		190

$$\hat{\pi}_{1\cdot 1} = 20/90, \qquad \hat{\pi}_{2\cdot 1} = 30/90, \qquad \hat{\pi}_{3\cdot 1} = 40/90,$$

$$\hat{\pi}_{1\cdot 2} = 50/130, \qquad \hat{\pi}_{2\cdot 2} = 60/130, \qquad \hat{\pi}_{3\cdot 2} = 20/130,$$

Combining these estimates gives the probabilities in Eq. (9.3):

$$\hat{\pi}_{11} = (20/90)(190/410), \quad \hat{\pi}_{12} = (30/90)(190/410), \quad \hat{\pi}_{13} = (40/90)(190/410),$$
$$= 0.1030, \qquad = 0.1545, \qquad = 0.2060,$$

$$\hat{\pi}_{21} = (50/130)(220/410), \quad \hat{\pi}_{22} = (60/130)(220/410), \quad \hat{\pi}_{23} = (20/130)(220/410),$$
$$= 0.2064, \qquad = 0.2377, \qquad = 0.0826.$$

In contrast, estimates based on the completely classified cases are as follows:

$$\tilde{\pi}_{11} = 20/220 = 0.0909, \quad \tilde{\pi}_{12} = 30/220 = 0.1364, \quad \tilde{\pi}_{13} = 40/220 = 0.1818,$$

$$\tilde{\pi}_{21} = 50/220 = 0.2273, \quad \tilde{\pi}_{22} = 60/220 = 0.2727, \quad \tilde{\pi}_{23} = 20/220 = 0.0909.$$

The estimates $\{\tilde{\pi}_{jk}\}$ are less efficient than the ML estimates $\{\hat{\pi}_{jk}\}$. However, as in the normal case discussed in Example 6.1, the principal value of ML is its ability to reduce or eliminate bias when the data are not missing completely at random. The estimates $\{\hat{\pi}_{jk}\}$ are ML if the data are MAR, and in particular if the probability that Y_2 is missing depends on Y_1 but not Y_2. The $\{\tilde{\pi}_{jk}\}$ are consistent for $\{\pi_{jk}\}$ in general only if the data are MCAR, that is, missingness is independent of Y_1 and Y_2. Since the marginal distribution of Y_1 appears to be different for the completely and incompletely classified samples (a chi-squared test yields $\chi_1^2 = 5.23$, with associated p value < 0.01), the MCAR assumption is contradicted by these data.

Extensions of this example to other monotone patterns can be developed by analogous factorizations of the likelihood.

EXAMPLE 9.3. *Application to a Six-Way Table.* Fuchs (1982) presents data from the Protective Services Project for Older Persons, a longitudinal study of 165 people designed to assess the effect of enriched social casework

Table 9.2 Partially Classified Contingency Table for Example 9.3

			Male				Female			
			<75		>75		<75		>75	
Mental	Physical	Survival	E[a]	C[a]	E	C	E	C	E	C
		(a) *Fully Categorized*								
Poor	Poor	Deceased	0	2	5	3	0	0	2	1
		Survived	1	0	0	0	0	0	0	1
	Good	Deceased	0	0	2	2	1	1	1	0
		Survived	0	2	2	0	0	0	0	0
Good	Poor	Deceased	0	0	3	1	0	0	1	2
		Survived	3	1	1	2	0	1	1	0
	Good	Deceased	1	1	4	6	2	0	0	2
		Survived	5	10	6	8	3	5	2	4
		(b) *Missing Physical Status*								
Poor	Missing	Deceased	0	0	0	0	0	0	0	0
		Survived	0	0	1	0	0	0	0	0
Good		Deceased	0	0	0	0	0	0	0	0
		Survived	0	0	0	0	0	0	0	0
		(c) *Missing Mental Status*								
Missing	Poor	Deceased	2	0	5	3	1	1	2	0
		Survived	1	1	0	3	0	0	0	1
	Good	Deceased	1	0	0	0	0	0	1	2
		Survived	1	3	2	1	1	1	0	0
		(d) *Missing Both Physical and Mental Status*								
Missing	Missing	Deceased	0	1	2	2	1	0	3	1
		Survived	2	8	1	2	1	1	2	2

[a] E denotes experimental; C denotes control.

services on the well-being of clients (Table 9.2). Investigators collected data on six dichotomized variables, D = survival status (survived, deceased), G = group membership (experimental, control), S = sex (male, female), A = age (less than 75, over 75), P = physical status (poor, good), M = mental status (poor, good). The data on all the variables were available on 101 participants (Table 9.2a). Physical status was missing for one participant (Table 9.2b). Mental status was missing for 33 participants (Table 9.2c). Finally, physical and mental status were both missing on 29 participants (Table 9.2d).

When the information on mental status for the single observation in Table 9.2b is ignored, the data have a monotone pattern, and ML estimates of the cell probabilities can be derived using the factorization

$$\Pr(D, G, A, S, P, M) = \Pr(D, G, A, S)\Pr(P \mid D, G, A, S)\Pr(M \mid D, G, A, S, P).$$

The observed counts for estimating the three distributions on the right-hand side are displayed in Table 9.3a. The resultant expected cell frequencies, which are the estimated cell probabilities multiplied by the total sample size, 164, are displayed in Table 9.3b; for example, the count for the cell with D = survived, G = experimental, A = over 75, S = male, P = good, M = good is

$$164\left(\frac{13}{164}\right)\left(\frac{10}{11}\right)\left(\frac{6}{8}\right) = 8.8636.$$

Changing D = survived to D = deceased yields the expected count

$$164\left(\frac{21}{164}\right)\left(\frac{6}{19}\right)\left(\frac{4}{6}\right) = 4.4211.$$

Hence the estimated conditional probability of survival given G = experimental, A = over 75, S = male, P = good, M = good is (8.8636/(8.8636 + 4.4211)) = 0.6672. This estimate compares with 10/(10 + 6) = 0.6 for the complete cases in Table 9.2.

EXAMPLE 9.4. *Tables with Refined and Coarse Classifications.* The data in Tables 9.4a and b, presented and analyzed by Hocking and Oxspring (1974), illustrate another situation where ML estimates can be found by factoring the likelihood. Table 9.4a gives data on the use of drugs in the treatment of leprosy. The 196 patients are classified according to the degree of infiltration and overall clinical condition after a fixed time over which treatments were administered. The supplemental data in Table 9.4b on 400 different patients are classified coarsely with respect to improvement in health. Such data arise naturally in health surveys where detailed results can be obtained for a small group of subjects, and coarsely classified data can be collected inexpensively for a larger group of individuals.

The likelihood factorizes according to the joint distribution of the combined cell counts from the two tables, classified coarsely as in Table 9.4b, based on all 596 patients, and the conditional distribution of degree of improvement (marked, moderate, or slight) given improvement and degree of infiltration, based on the 196 patients. Resulting ML estimates of the cell probabilities are displayed in Table 9.4c, in a form that illustrates the calculations. The joint probabilities of infiltration and coarsely classified clinical change are obtained by merging the data in (a) and (b), yielding the fractions in the last two columns and the first factors of the first three columns. The latter are multiplied by the conditional probabilities of degree of improvement, calculated from the first three columns of (a). In particular, the top left-hand corner entry is $\hat{\pi}_{11} = (224/596)(11/80) = 0.0517$, compared with $\tilde{\pi}_{11} = 11/196 = 0.0561$ from the finely classified data alone.

Table 9.3 ML Estimation for Monotone Data[a] in Table 9.2a, c, and d, Using Factored Likelihood Method

			Male				Female			
			<75		>75		<75		>75	
Mental	Physical	Survival	Exper.	Control	Exper.	Control	Exper.	Control	Exper.	Control
(a) *Partitioned Table for Monotone Patterns*										
			(i) *Available Information on D, G, A, S*							
		Deceased	4	4	21	17	5	2	10	8
		Survived	13	25	13	16	5	8	5	8
			(ii) *Available Information on D, G, A, S, P*							
	Poor	Deceased	2	2	13	7	1	1	5	3
		Survived	5	2	1	5	0	1	1	2
	Good	Deceased	2	1	6	8	3	1	2	4
		Survived	6	15	10	9	4	6	2	4
		(iii) *Available Information on All Variables (D, G, A, S, P, M) Given in Table 9.2a*								
(b) *Expected Cell Frequencies*										
Poor	Poor	Deceased	1.00	2.67	8.98	5.95	.63	.50	4.76	1.14
		Survived	1.48	.00	.00	.00	.00	.00	.00	2.67
	Good	Deceased	.00	.00	2.21	2.27	1.25	1.00	2.86	.00
		Survived	.00	3.68	2.95	.00	.00	.00	.00	.00
Good	Poor	Deceased	1.00	.00	5.39	1.98	.63	.50	2.38	2.28
		Survived	4.43	2.94	1.18	5.71	.00	1.14	1.67	.00
	Good	Deceased	2.00	1.33	4.42	6.80	2.50	.00	.00	4.57
		Survived	7.09	18.38	8.86	10.29	5.00	6.86	3.33	5.33

[a] The information on the mental status of the individual in Table 9.2b is ignored.
Source: Fuchs (1982), with minor corrections.

Table 9.4 Patients Classified by Degree of Infiltration and Change of Condition

(a) *Finely Classified Data*

Degree of Infiltration	Clinical Change					
	Improvement					
	Marked	Moderate	Slight	Stationary	Worse	Totals
Little	11	27	42	53	11	144
Much	7	15	16	13	1	52
Totals	18	42	58	66	12	196

(b) *Coarsely Classified Data*

Degree of Infiltration	Clinical Change			
	Improvement	Stationary	Worse	Totals
Little	144	120	16	280
Much	92	24	4	120
Totals	236	144	20	400

(c) *ML Estimates of Cell Probabilities from* (*a*) *and* (*b*)

Degree of Infiltration	Clinical Change				
	Improvement				
	Marked	Moderate	Slight	Stationary	Worse
Little	(224/596)(11/80)	(224/596)(27/80)	(224/596)(42/80)	173/596	27/596
Much	(130/596)(7/38)	(130/596)(15/38)	(130/596)(16/38)	37/596	5/596

Source: Hocking and Oxspring (1974).

9.2.3. Estimating the Precision of the ML Estimates

The asymptotic covariance matrix associated with the ML estimates (9.3) can be obtained by calculating the information matrix for the parameters in the factored form of the likelihood, inverting this matrix, and then transforming to the original parameterization using the method outlined in Section 6.1. Alternatively, we can calculate the variances and covariances directly. For example, to calculate the large sample variance of $\hat{\pi}_{jk} = \hat{\pi}_{j+}\hat{\pi}_{k\cdot j}$ in Example 9.1, we write

$$\operatorname{Var}\hat{\pi}_{jk} = E(\operatorname{Var}\hat{\pi}_{jk}|\mathbf{Y}_1) + \operatorname{Var}(E\hat{\pi}_{jk}|\mathbf{Y}_1),$$

where $\mathbf{Y}_1$ is the set of marginal counts of Y_1. Hence

$$\begin{aligned}\operatorname{Var}\hat{\pi}_{jk} &= E[\hat{\pi}_{j+}^2\pi_{k\cdot j}(1 - \pi_{k\cdot j})/m_{j+}] + \operatorname{Var}\{\hat{\pi}_{j+}\pi_{k\cdot j}\} \\ &= \pi_{j+}^2\pi_{k\cdot j}(1 - \pi_{k\cdot j})/m_{j+} + \pi_{k\cdot j}^2\pi_{j+}(1 - \pi_{j+})/n,\end{aligned}$$

asymptotically to order $1/m_{j+}^2$. Some algebra yields

$$\operatorname{Var}\hat{\pi}_{jk} \simeq \frac{\pi_{jk}(1 - \pi_{jk})}{m}\left[1 - \frac{\pi_{k\cdot j} - \pi_{jk}}{1 - \pi_{jk}}\frac{r}{n} + c_j\frac{(1 - \pi_{k\cdot j})}{1 - \pi_{jk}}\right],$$

where $c_j = m\pi_{j+}/m_{j+} - 1$. Substituting estimates of the parameters yields

$$\operatorname{Var}(\pi_{jk} - \hat{\pi}_{jk}) \simeq \frac{\hat{\pi}_{jk}(1 - \hat{\pi}_{jk})}{m}\left[1 - \frac{\hat{\pi}_{k\cdot j} - \hat{\pi}_{jk}}{1 - \hat{\pi}_{jk}}\frac{r}{n} + c_j\frac{(1 - \hat{\pi}_{k\cdot j})}{1 - \hat{\pi}_{jk}}\right]. \tag{9.4}$$

The left side of (9.4) is written in a modified form to indicate that a Bayesian analysis of the asymptotic posterior variance of π_{jk} yields similar results. For the covariances we find

$$\operatorname{Cov}(\pi_{jk} - \hat{\pi}_{jk}, \pi_{jl} - \hat{\pi}_{jl}) \simeq \frac{-\hat{\pi}_{jk}\hat{\pi}_{jl}}{m}\left[1 + \frac{(1 - \hat{\pi}_{j+})}{\hat{\pi}_{j+}}\frac{r}{n} + \frac{c_j}{\hat{\pi}_{j+}}\right], \qquad k \neq l,$$

$$\operatorname{Cov}(\pi_{jk} - \hat{\pi}_{jk}, \pi_{jl} - \hat{\pi}_{jl}) \simeq \frac{-\hat{\pi}_{jk}\hat{\pi}_{jl}}{n}, \qquad (i \neq j).$$

If the data are MCAR, then c_j is small, and (9.4) reduces to an expression analogous to (6.14), namely,

$$\operatorname{Var}(\pi_{jk} - \hat{\pi}_{jk}) \simeq \frac{\hat{\pi}_{jk}(1 - \hat{\pi}_{jk})}{m}\left[1 - \frac{\hat{\pi}_{k\cdot j} - \hat{\pi}_{jk}}{1 - \hat{\pi}_{jk}}\frac{r}{n}\right]. \tag{9.5}$$

In this expression, $\hat{\pi}_{jk}(1 - \hat{\pi}_{jk})/m$ is the estimated variance of the estimate $\tilde{\pi}_{jk} = m_{jk}/m$ obtained by ignoring the supplemental margin, and the remainder on the right side represents the proportionate reduction in variance from using the supplemental margin.

EXAMPLE 9.5. (*Example 9.2 continued*). Let us apply these formulas to the estimates of π_{11} from data in Table 9.1. The estimate $\tilde{\pi}_{11} = 0.0909$ that ignores the supplementary margin has large sample variance $\pi_{11}(1 - \pi_{11})/m$, which on substituting ML estimates yields

$$\mathrm{Var}(\tilde{\pi}_{11}) = \frac{(0.1030)(1 - 0.1030)}{220} = 0.00042.$$

Similarly from (9.5), the ML estimate $\hat{\pi}_{11} = 0.1030$ has estimated large sample variance

$$\begin{aligned}\mathrm{Var}(\hat{\pi}_{11}) &= 0.00042\left[1 - \frac{0.2222 - 0.1030}{0.8970}\right]\frac{190}{410} \\ &= (0.00042)(0.9384) = 0.00039.\end{aligned}$$

Thus the reduction in variance from using the supplemental margin is small. However, as noted in Example 9.2, the data do not appear to be MCAR, so $\tilde{\pi}_{11}$ is biased for π_{11}. Assuming the missing data are MAR, $\hat{\pi}_{11}$ is unbiased for π_{11} and so a rough estimate of the bias of $\tilde{\pi}_1$ as $\tilde{\pi}_{11} - \hat{\pi}_{11} = 0.0121$. Hence a rough estimate of the mean squared error of $\tilde{\pi}_{11}$ is

$$\mathrm{M\hat{s}e}(\tilde{\pi}_{11}) \simeq 0.0121^2 + \mathrm{Var}(\tilde{\pi}_{11}) = 0.00057.$$

Replacing (9.5) by the more appropriate formula (9.4), we obtain for the variance of $\hat{\pi}_{11}$

$$\mathrm{Var}(\hat{\pi}_{11}) \simeq 0.00042(0.9384 + 0.1151) = 0.00044.$$

Thus the ML procedure appears to provide a considerably more precise estimate of π_{11} by accounting for the fact that the data do not appear to be MCAR.

9.3. ML ESTIMATION FOR MULTINOMIAL SAMPLES WITH GENERAL PATTERNS OF MISSING DATA

As with normal data, incomplete multinomial data that do not form a monotone data pattern require an iterative procedure for ML estimation. The EM algorithm is particularly simple, since the loglikelihood is linear in the missing values. For the monotone data in Examples 9.1 and 9.2, ML estimation effectively distributes the partially classified data in the full table, using conditional probabilities calculated from the fully classified data. The E step of the EM algorithm for general patterns has the same function, except that the conditional probabilities are calculated from current estimates of the cell probabilities rather than from the fully classified data. The M step of the EM algorithm calculates new cell probabilities from the completed data. The algorithm first appeared in the statistical literature in a paper by Hartley (1958). We provide a quite general formulation of the algorithm, and then apply the algorithm to a single special case.

Suppose that the hypothetical complete data are a multinomial sample of size n, with J cells $c_1, \ldots, c_J$, n_j observations classified in cell c_j, and parameters $\theta = (\pi_1, \ldots, \pi_J)$, where π_j is the classification probability for cell c_j. The observed data consist of m completely classified observations, with m_j falling in the cell c_j for $j = 1, \ldots, J$ and $m = \sum_{j=1}^{J} m_j$, and $n - m$ partially classified observations, which fall in subsets of the J cells. We partition the partially classified units into K groups, so that within each group, all units have the same set of possible cells. Suppose that r_k partially classified observations fall in the kth group, and let S_k denote the set of cells to which these observations might belong. Furthermore, define the indicator functions $\delta(j \in S_k)$, $j = 1, \ldots, J$, $k = 1, \ldots, K$, where $\delta(j \in S_k) = 1$ if cell c_j belongs to S_k and $\delta(j \in S_k) = 0$ otherwise.

To define the E step of the EM algorithm, let $\{\pi_j^{(t)}, j = 1, \ldots, J\}$ denote the current (tth iterate) estimate of the parameters. The hypothetical complete data belong to the regular exponential family with complete-data sufficient statistics

$$\{n_j : j = 1, \ldots, J\}.$$

Hence the E step consists in calculating

$$n_j^{(t)} = E\{n_j | \text{data}, \pi_1^{(t)}, \ldots, \pi_J^{(t)}\} = m_j + \sum_{k=1}^{K} r_k \psi_{j \cdot S_k}^{(t)},$$

where

$$\psi_{j \cdot S_k}^{(t)} = \pi_j^{(t)} \delta(j \in S_k) \Bigg/ \left(\sum_{l=1}^{J} \pi_l^{(t)} \delta(l \in S_k) \right)$$

is the current estimate of the conditional probability of falling in cell j given that an observation falls in the set of categories S_k. The E step effectively distributes the partially classified observations into the table according to these probabilities.

The M step calculates new parameter estimates as

$$\pi_j^{(t+1)} = n_j^{(t)} / n.$$

This formulation is quite general. The cells may form a multiway table with observations classified by V variables $Y_1, \ldots, Y_V$, where Y_v has j_v categories and $\prod_{v=1}^{V} j_v = J$. The partially classified observations may then arise as supplemental margins where one or more of the classifying variables Y_v are not recorded. Here is a simple numerical example of this incomplete-data pattern.

Table 9.5 A 2 × 2 Table with Supplemental Margins for Both Variables

(1) Classified by Y_1 and Y_2

		Y_2 = 1	2	Total
Y_1	1	100	50	150
	2	75	75	150
Total		175	125	300

(2) Classified by Y_1

Y_1	1	30[a]
	2	60[b]
Total		90

(3) Classified by Y_2

Y_2 = 1	2	Total
28[c]	60[d]	88

Note: The superscripts *a*, *b*, *c*, *d* refer to the partially classified cells and are used in Table 9.6.

EXAMPLE 9.6. *A 2 × 2 Table with Supplemental Data on Both Margins.* ML estimation for two-way tables with supplementary data on both margins was first considered by Chen and Fienberg (1974). Table 9.5 gives data for a (2 × 2) table with supplemental margins for both the classifying variables, analyzed in Little (1982). Table 9.6 shows the first three iterations of the EM algorithm, where initially the cell probabilities are estimated from the completely classified table. These probabilities are then used to allocate the partially classified observations as indicated. For example, the 28 partially classified units with $Y_2 = 1$ have $Y_1 = 1$ with probability $100/(100 + 75)$ and $Y_1 = 2$ with probability $75/(100 + 75)$. Thus of the 28 units, in effect $(28)(100)/175 = 16$ are allocated to $Y_1 = 1$ and $(28)(75)/175 = 12$ are allocated to $Y_1 = 2$. In the next step new probabilities are found from the completed data and the procedure iterates to convergence. Final probabilities of classification after convergence are

$$\hat{\pi}_{11} = 0.28, \quad \hat{\pi}_{12} = 0.17, \quad \hat{\pi}_{21} = 0.24, \quad \hat{\pi}_{22} = 0.31.$$

EXAMPLE 9.7. *Application to PET.* Vardi, Shepp and Kaufman (1985) give an interesting application of the EM algorithm to two-way counted data from positron emission tomography (PET). The description here is from Rubin's (1985) discussion. In PET a "picture" of an organ (say, the brain) is created by collecting counts of emissions in D detectors placed systematically around the organ. The organ is hypothesized to consist of B boxes or pixels, each characterized by a distinct intensity parameter $\lambda(b)$, $b = 1, \ldots, B$ governing the rate of emission. Physical considerations provide a $D \times B$ matrix of known conditional probabilities, $\Pr(\text{detector} = d|\text{pixel} = b)$, for the probability that an emission from pixel b will be recorded in detector d. The objective is to use these known conditional probabilities in conjunction with the observed counts in the D detectors to estimate the intensities (or marginal probabilities of emissions) in the B pixels.

Let $\pi = \{\pi(d, b)\}$ be the $D \times B$ matrix of joint probabilities that an emission

Table 9.6 The EM Algorithm for Data in Table 9.5 Ignoring the Missing-Data Mechanism

		Estimated Probabilities $Y_2 = 1$	Estimated Probabilities $Y_2 = 2$		Fractional Allocation of Units $Y_2 = 1$	Fractional Allocation of Units $Y_2 = 2$	
Step 1							
Y_1	1	100/300	50/300	1	$100 + 20^a + 16^c$	$50 + 10^a + 24^d$	30^a
	2	75/300	75/300	2	$75 + 30^b + 12^c$	$75 + 30^b + 36^d$	60^b
					28^c	60^d	
Step 2							
		136/478	84/478		100 + 18.6 + 15.1	50 + 11.4 + 22.4	
		117/478	141/478		75 + 27.2 + 12.9	75 + 32.8 + 37.6	
Step 3							
		.28	.18		100 + 18.4 + 15.1	50 + 11.6 + 21.9	
		.24	.30		75 + 26.5 + 12.9	75 + 33.5 + 38.1	
Step 4							
		.28	.17				
		.24	.31				

Note: The superscripts in the top right panel indicate the partially classified cells in Table 9.5. For example, of the 28 units with $Y_2 = 1$ (superscript c), 16 are allocated to $Y_1 = 1$ and 12 are allocated to $Y_1 = 2$.

emanates from pixel b and is detected in detector d; π is determined by the $\Pr(d|b)$ and $\lambda(b)$. The hypothetical complete data are n iid observations, $\delta_i = \{\delta_i(d, b)\}$ where $\delta_i(d, b) = 1$ if the ith emission emanated from pixel b and is recorded in detector d and zero otherwise. The observed (or incomplete) data consist of the n row margins of the δ_i, which are $D \times 1$ vectors indicating the detector for the n emissions. The EM algorithm proceeds as follows:

1. Start with some initial guess for λ, say $\lambda^{(0)}$, which implies an initial value for π, $\pi^{(0)}$.
2. At the E step, for $d = 1, \ldots, D$, allocate the observed count in detector d across the B pixels according to the conditional probabilities implied by $\pi^{(0)}$.

3. At the M step use the pixel margin (summed counts across all detectors) to estimate $\lambda^{(1)}$.
4. Repeat the E step with the new estimate of λ, and iterate to convergence.

9.4. LOGLINEAR MODELS FOR PARTIALLY CLASSIFIED CONTINGENCY TABLES

9.4.1. The Complete-Data Case

For a complete V-way contingency table with cell probabilities $\{\pi_{jkl\cdots t}\}$, it is often important to consider models where the cell probabilities have a special structure. For example, independence between the factors corresponds to a model where the probabilities can be expressed in the form

$$\pi_{jkl\cdots t} = \tau\tau_j^{(1)}\tau_k^{(2)}\cdots\tau_t^{(V)} \tag{9.6}$$

for suitable multiplicative factors τ and $\{\tau_j^{(1)}\}, \{\tau_k^{(2)}\}, \ldots, \{\tau_t^{(V)}\}$. It is convenient to express (9.6) as a loglinear model:

$$\ln \pi_{jkl\cdots t} = \alpha + \alpha_j^{(1)} + \alpha_k^{(2)} + \cdots + \alpha_t^{(V)}, \tag{9.7}$$

where $\alpha_j^{(1)} = \ln \tau_j^{(1)}$, and so forth. Different sets of α's on the right side of (9.7) yield the same set of cell probabilities $\{\pi_{jkl\cdots t}\}$, and V constraints are needed to define the α's uniquely. A common choice is to set

$$\sum_{j=1}^{J} \alpha_j^{(1)} = \cdots = \sum_{t=1}^{T} \alpha_t^{(V)} = 0.$$

Equation (9.6) or (9.7) define a loglinear model for the cell probabilities. A more general class of models is obtained by decomposing the logarithm of the cell probabilities into a sum of a constant, main effects as in (9.7), and higher-order associations, and then setting some of the terms in the decomposition to zero. For example, for a $V = 3$-way table, we write

$$\ln \pi_{jkl} = \alpha + \alpha_j^{(1)} + \alpha_k^{(2)} + \alpha_l^{(3)} + \alpha_{jk}^{(12)} + \alpha_{jl}^{(13)} + \alpha_{kl}^{(23)} + \alpha_{jkl}^{(123)}, \tag{9.8}$$

where the α terms are constrained to sum to zero over any of their subscripts. The terms $\{\alpha_j^{(1)}\}$, $\{\alpha_k^{(2)}\}$, $\{\alpha_l^{(3)}\}$ are called the main effects of Y_1, Y_2, and Y_3, respectively. The terms $\{\alpha_{jk}^{(12)}\}$, $\{\alpha_{jl}^{(13)}\}$, $\{\alpha_{kl}^{(23)}\}$ are called two-way associations between Y_1 and Y_2, Y_1 and Y_3, and Y_2 and Y_3, respectively. Finally the terms $\{\alpha_{jkl}^{(123)}\}$ are called three-way associations between Y_1, Y_2, and Y_3. Setting all two- and three-way associations to zero yields the independence model (9.7) for $V = 3$ variables. Other models are obtained by setting other terms to zero in (9.8).

Table 9.7 Hierarchical Loglinear Models for Three-Way Tables

Model	Label	Terms to Set to Zero in (9.8)
(1)	$\{123\}$	None
(2)	$\{12, 23, 31\}$	$\{\alpha_{jkl}^{(123)}\}$
(3)	$\{12, 13\}$	$\{\alpha_{jkl}^{(123)}, \alpha_{kl}^{(23)}\}$
(4)	$\{1, 23\}$	$\{\alpha_{jkl}^{(123)}, \alpha_{jk}^{(12)}, \alpha_{jl}^{(13)}\}$
(5)	$\{23\}$	$\{\alpha_{jkl}^{(123)}, \alpha_{jk}^{(12)}, \alpha_{jl}^{(13)}, \alpha_{j}^{(1)}\}$
(6)	$\{1, 2, 3\}$	$\{\alpha_{jkl}^{(123)}, \alpha_{jk}^{(12)}, \alpha_{jl}^{(13)}, \alpha_{kl}^{(23)}\}$
(7)	$\{2, 3\}$	$\{\alpha_{jkl}^{(123)}, \alpha_{jk}^{(12)}, \alpha_{jl}^{(13)}, \alpha_{kl}^{(23)}, \alpha_{j}^{(1)}\}$
(8)	$\{1\}$	$\{\alpha_{jkl}^{(123)}, \alpha_{jk}^{(12)}, \alpha_{jl}^{(13)}, \alpha_{kl}^{(23)}, \alpha_{k}^{(2)}, \alpha_{l}^{(3)}\}$
(9)	$\{\varnothing\}$	$\{\alpha_{jkl}^{(123)}, \alpha_{jk}^{(12)}, \alpha_{jl}^{(13)}, \alpha_{kl}^{(23)}, \alpha_{j}^{(1)}, \alpha_{k}^{(2)}, \alpha_{l}^{(3)}\}$.

An important class of models obtained in this way are *hierarchical* loglinear models, which have the property that inclusion of a v-way association α_S^* between a set of factors S implies inclusion of all $(v - 1)$-way and lower-order associations and main effects involving subsets of the factors in S. There are 19 hierarchical models for a three-way table. Nine of them are listed in Table 9.7; the remaining 10 can be obtained by permuting the factors in models (3), (4), (5), (7), and (8).

ML estimation for hierarchical models varies in complexity according to the model fitted. In particular, explicit ML estimates can be found for all the models in Table 9.7 except for $\{12, 23, 31\}$, where an iterative fitting procedure is necessary.

Two asymptotically equivalent goodness-of-fit statistics are widely used to compare the fit of loglinear models. The likelihood ratio statistic is

$$G^2 = 2\sum_c n_c \ln \frac{n_c}{\hat{n}_c}, \tag{9.9}$$

where the summation is over all the cells c in the table, n_c is the observed count in cell c, and $\hat{n}_c = n\hat{\pi}_c$ is the expected count in cell c estimated from the model. The Pearson chi-squared statistic is defined as

$$X^2 = \sum_c \frac{(n_c - \hat{n}_c)^2}{\hat{n}_c}. \tag{9.10}$$

If the fitted model is correct, then both G^2 and X^2 are asymptotically chi-squared distributed with degrees of freedom equal to the number of independent restrictions on the cell probabilities. Details on calculating degrees of freedom and more information on loglinear models for complete data are given in Goodman (1970), Haberman (1974), Bishop, Fienberg, and Holland (1975), and Fienberg (1980).

Table 9.8 A 2^3 Contingency Table with Partially Classified Observations

		Survival (S)		
Clinic (C)	Prenatal Care (P)	Died	Survived	
		(a) Completely Classified Cases		
A	Less	3	176	
	More	4	293	
B	Less	17	197	
	More	2	23	$m = 715$ cases
		(b) Partially Classified Cases (Clinic missing)		
	Less	10	150	
	More	5	90	$r = 255$ cases

Source: (a) Bishop, Fienberg and Holland (1975), table 2.4-2. (b) Artificial data.

EXAMPLE 9.8. *A Complete Three-Way Table.* Table 9.8a presents a 2^3 contingency table on survival of infants, previously analyzed in Bishop, Fienberg and Holland (1975, table 1.4-2). Table 9.9 shows estimated cell probabilities and goodness-of-fit statistics for selected loglinear models fitted to these data.

The model {SPC} in Table 9.9a places no constraints on the cell probabilities and fits the observed cell proportions perfectly. Hence the goodness-of-fit statistics are both zero with zero degrees of freedom. Two unsaturated models in Table 9.9b and c have very low values of G^2 and X^2, indicating good fits, namely, the model {SC, PC}, which indicates that survival is related to clinic, but survival and prenatal care are not associated conditional on clinic, and the model {SC, PC, SP}, which adds the association SP to the previous model. Since the difference in fits is negligible and the former model is more parsimonious, it will usually be preferred. The model {SP, SC} fits the data poorly and is included for illustrative purposes.

9.4.2. Loglinear Models for Partially Classified Tables

As with the saturated models in Sections 9.2 and 9.3, ML estimation of loglinear models involves distributing the partially classified counts into the full table using estimated conditional probabilities and then estimating the classification probabilities from the filled-in table. The only difference is that all probabilities are estimated subject to the constraints imposed by the loglinear model. These constraints can increase the computation required to compute ML estimates for two reasons. First, factorizations of the likelihood

Table 9.9 Estimated Cell Probabilities $\{\hat{\pi}_{jkl}\} \times 100$ from Saturated Model {SPC} and Three Loglinear Models, Fitted to Data in Table 9.8a

Clinic	Prenatal Care (P)	Survival (S): Died	Survival (S): Survived	Goodness-of-Fit
		(a) *Model*: {SPC}		
A	Less	0.42	24.62	
	More	0.56	40.98	
				df = 0, $G^2 = 0$, $X^2 = 0$
B	Less	2.38	27.55	
	More	0.28	3.22	
		(b) *Model*: {SP, SC, PC}		
A	Less	0.39	24.64	
	More	0.59	40.95	
				df = 1, $G^2 = 0.04$, $X^2 = 0.04$
B	Less	2.41	27.52	
	More	0.25	3.24	
		(c) *Model*: {SC, PC}		
A	Less	0.36	24.67	
	More	0.62	40.92	
				df = 2, $G^2 = 0.08$, $X^2 = 0.08$
B	Less	2.38	27.55	
	More	0.28	3.22	
		(d) *Model*: {SP, SC}		
A	Less	0.76	35.51	
	More	0.22	30.08	
				df = 2, $G^2 = 188.1$, $X^2 = 169.5$
B	Less	2.04	16.66	
	More	0.62	14.11	

for monotone patterns do not necessarily lead to explicit ML estimates, since the parameters in the factors are not necessarily distinct. Second, the M step of the EM algorithm for nonmonotone patterns may itself involve iteration. We provide one example for each of these situations.

EXAMPLE 9.9. *ML Estimates for an Incomplete Three-Way Table (Example 9.8 continued).* Suppose that the supplemental data in Table 9.8b are added to the data in Table 9.8a analyzed in Example 9.8. Survival (S) and Prenatal Care (P) are recorded in the supplemental data, but Clinic (C) is not recorded. The resulting incomplete data form a monotone pattern with P and S more observed than C.

The likelihood for the combined data in Tables 9.8a and b factors into a term for the distribution of SP, involving all $r + m = 970$ cases, and a term for the distribution of C given SP, involving the $m = 715$ completely classified cases. These two distributions involve distinct parameters for the models {SPC}, {SP, SC, PC}, and {SP, SC}. Hence ML estimates can be derived for these models by the factored likelihood method of Chapter 6. Table 9.10a shows ML estimates of $100\pi_{jkl}$ for the saturated model {SPC}, calculated by the methods of Section 9.2. Tables 9.10b and c show ML estimates for {SP, SC, PC} and {SP, SC}. Since the {SP} margin is fitted in these models, estimates of the probabilities in this margin are the same as those for {SPC}. The conditional probabilities that $C = 1$ given SP are obtained from the appropriate model applied to the 715 complete cases. For {SP, SC} this calculation is noniterative, but for {SP, SC, PC} it is iterative. The two sets of ML parameter estimates are combined as for the saturated models to obtain ML estimates of the joint probabilities π_{jkl}, by Property 5.1 of ML estimates.

Parameters of the distributions of SP and C given SP are not distinct for the model {SC, PC}, so the factored likelihood method cannot be applied. Table 9.11 shows four iterations of the EM algorithms for this model; estimates of $100\pi_{jkl}$ are unchanged between iterations 4 and 5, to two decimal places. Calculations in the E step were carried out by hand calculator, and the M step was carried out by fitting {SC, PC} to the filled-in table, using a standard loglinear model program (BMDP4F in the BMDP Statistical Software Package; see Dixon, 1983). Filled-in counts from the E step were converted to integers by multiplying by 100, since the loglinear model program truncates real number inputs to integers. For larger problems the E step can be programmed to reduce computational burden. Note that the M step here is noniterative. The same procedure works when the M step is itself iterative, but then the EM algorithm involves a double iteration.

In the preceding example starting values for the EM algorithm were based on analysis of the completely classified table. With sparse tables containing zero cells, this procedure can yield unsatisfactory starting values, as discussed in Fuchs (1982). In particular, suppose a marginal table corresponding to a term in the model has an empty cell in the fully categorized table, and the same cell has a positive count in the supplemental table. If starting values are based on the fully categorized table, then the EM algorithm never allows the zero cell to attain a nonzero probability, thus contradicting the supplemental information. This problem can be avoided by forming starting values after adding positive values to the cells of the completely classified table, so that initial estimates are in the interior of the parameter space. In subsequent iterations these added values can be discarded.

Table 9.10 ML Estimates for Models {SPC}, {SP, SC, PC}, and {SP, SC} Fitted to Data in Tables 9.8a and b

Clinic	Prenatal Care	Survival: Died	Survival: Survived
		(a) *Model*: {SPC}	
A	Less	100(3/20)(30/970) = 0.46	100(176/373)(523/970) = 25.44
	More	100(4/6)(11/970) = 0.76	100(293/316)(406/970) = 38.81
B	Less	100(17/20)(30/970) = 2.63	100(197/373)(523/970) = 28.48
	More	100(2/6)(11/970) = 0.38	100(23/316)(406/970) = 3.05
			Total = 100.0
		(b) *Model*: {SP, SC, PC}	
A	Less	100(2.8/20)(30/970) = 0.43	100(176.2/373)(523/970) = 25.47
	More	100(4.2/6)(11/970) = 0.79	100(292.8/316)(406/970) = 38.78
B	Less	100(17.2/20)(30/970) = 2.66	100(196.8/373)(523/970) = 28.45
	More	100(1.8/6)(11/970) = 0.34	100(23.2/316)(406/970) = 3.07
			Total = 100.0
		(c) *Model*: {SP, SC}	
A	Less	100(5.4/20)(30/970) = 0.84	100(253.9/373)(523/970) = 36.70
	More	100(1.6/6)(11/970) = 0.30	100(215.1/316)(406/970) = 28.49
B	Less	100(14.6/20)(30/970) = 2.26	100(119.1/373)(523/970) = 17.22
	More	100(4.4/6)(11/970) = 0.83	100(100.9/316)(406/970) = 13.26
			Total = 100.0

Table 9.11 ML Estimates for Model {SC, PC} Fitted to Data in Tables 9.8a and b, via the EM Algorithm

			M Step: Estimated Cell Probabilities × 100 Survival		E Step: Filled-In Cell Counts Survival	
Iteration	Clinic	Prenatal	Died	Survived	Died	Survived
1	A	Less	.36	24.67	3 + (10)(.36)/2.74 = 4.33	176 + 150(24.62)/52.22 = 246.86
		More	.62	40.92	4 + 5(.62)/.90 = 7.44	293 + 90(40.92)/44.14 = 376.44
	B	Less	2.38	27.56	17 + 10(2.38)/2.74 = 25.67	197 + 150(27.56)/52.22 = 276.14
		More	.28	3.22	2 + 5(.28)/.90 = 3.56	23 + 90(3.22)/44.14 = 29.56
2	A	Less	.48	25.42	4.50	246.84
		More	.73	38.84	7.56	376.32
	B	Less	2.72	28.40	25.50	276.16
		More	.30	3.12	3.44	29.68
3	A	Less	.49	25.42	4.55	246.83
		More	.75	38.82	7.59	376.31
	B	Less	2.69	28.41	25.45	276.17
		More	.30	3.12	3.41	29.69
4	A	Less	.50	25.42	4.56	246.83
		More	.76	38.82	7.60	376.31
	B	Less	2.68	28.41	25.44	276.17
		More	.29	3.12	3.40	29.69

9.4.3. Goodness-of-Fit Tests for Partially Classified Data

Chi-squared goodness-of-fit statistics analogous to (9.9) and (9.10) can be calculated for partially classified tables by summing over the cells in the complete *and* partially classified supplemental tables. Note that unlike the complete-data case, nonzero values of X^2 and G^2 are obtained for the saturated model ({SPC} in Example 9.9); the values of X^2 and G^2 for the saturated model provide tests for whether the data are MCAR.

Chi-squared statistics for restricted models can be obtained by calculating G^2 (or X^2) for the restricted model and the saturated model and then subtracting the two quantities (Fuchs, 1982). The resulting difference has the same number of degrees of freedom as the chi-squared test for the restricted model complete data.

This procedure appears to assume that data are MCAR, but in fact the tests remain valid provided the missing data are MAR. Under the latter assumption the component of the loglikelihood for the missing-data mechanism cancels out when the values of G^2 (or X^2) for the two models are subtracted.

EXAMPLE 9.10. *(Example 9.9 continued) Goodness-of-Fit Statistics for an Incomplete Three-Way Table.* Goodness-of-fit statistics for the saturated model {SPC} in Example (9.9) are

$$X^2(\mathrm{SPC}) = 7.96, \quad G^2(\mathrm{SPC}) = 7.80, \quad \mathrm{df} = 3.$$

To calculate degrees of freedom (df), note that there are $8 + 4 = 12$ cells of data, yielding 11 degrees of freedom for estimating 7 cell probabilities and 1 response probability, or 8 parameters. Hence $\mathrm{df} = 11 - 7 - 1 = 3$. Since the 95th percentile of the chi-squared distribution with 3 df is 7.815, the null hypothesis that the data are MCAR yields a p-value of less than 0.05 using X^2, and approximately 0.05 using G^2. The unsaturated models yield

$$X^2(\mathrm{SP,SC,PC}) = 7.99, \quad G^2(\mathrm{SP,SC,PC}) = 7.84, \quad \mathrm{df} = 11 - 6 - 1 = 4,$$

$$X^2(\mathrm{SP,PC}) = 8.29, \quad G^2(\mathrm{SP,SC}) = 8.00, \quad \mathrm{df} = 11 - 5 - 1 = 5,$$

$$X^2(\mathrm{SP,SC}) = 178.55, \quad G^2(\mathrm{SP,SC}) = 195.92, \quad \mathrm{df} = 11 - 5 - 1 = 5.$$

Subtracting the chi-squared values for the saturated model yields

$$\Delta X^2(\mathrm{SP,SC,PC}) = 0.03, \quad \Delta G^2(\mathrm{SP,SC,PC}) = 0.04, \quad \mathrm{df} = 8 - 6 - 1 = 1,$$

$$\Delta X^2(\mathrm{SC,PC}) = 0.33, \quad \Delta G^2(\mathrm{SP,PC}) = 0.20, \quad \mathrm{df} = 8 - 5 - 1 = 2,$$

$$\Delta X^2(\mathrm{SP,SC}) = 170.59, \quad \Delta G^2(\mathrm{SP,SC}) = 188.12, \quad \mathrm{df} = 8 - 5 - 1 = 2,$$

which can be compared with the goodness-of-fit statistics based on the completely classified cases in Table 9.9. We conclude as before that {SP, PC} is the preferred model.

REFERENCES

Bishop, Y. M. M., Fienberg, S. E., and Holland, P. W. (1975). *Discrete Multivariate Analysis: Theory and Practice*. Cambridge, MA: MIT Press.

Chen, T., and Fienberg, S. E. (1974). Two-dimensional contingency tables with both completely and partially classified data, *Biometrics* **30**, 629–642.

Dixon, W. J. (Ed.) (1983). *BMDP Statistical Software, 1983 revised printing*, Berkeley CA: University of California Press.

Fienberg, S. E. (1980). *The Analysis of Crossclassified Data*, 2nd ed. Cambridge, MA: MIT Press.

Fuchs, C. (1982). Maximum likelihood estimation and model selection in contingency tables with missing data, *J. Am. Statist. Assoc.* **77**, 270–278.

Goodman, L. A. (1970). The multivariate analysis of qualitative data: Interaction among multiple classifications, *J. Am. Statist. Assoc.* **65**, 225–256.

Goodman, L. A. (1979). Simple Models for the Analysis of Association in Crossclassifications Having Ordered Categories, *J. Am. Statist. Assoc.* **74**, 537–552.

Haberman, S. J. (1974). *The Analysis of Frequency Data.* Chicago: University of Chicago Press.

Hartley, H. O. (1958). Maximum likelihood estimation from incomplete data, *Biometrics* **14**, 174–194.

Hocking, R. R., and Oxspring, H. H. (1974). The analysis of partially categorized contingency data, *Biometrics* **30**, 469–483.

Little, R. J. A. (1982). Models for nonresponse in sample surveys, *J. Am. Statist. Assoc.* **77**, 237–250.

McCullagh, P. (1980). Regression models for ordinal data, *J. Roy. Statist. Soci.* **B42**, 109–142.

Rubin, D. B. (1985), Comment on "A statistical model for positron emission tomography," *J. Am. Statist. Assoc.* **80**, 31–32.

Vardi, Y., Shepp, L. A., and Kaufman, L. (1985). A statistical model for positron emission tomography, *J. Am. Statist. Assoc.* **80**, 8–37.

PROBLEMS

1. Show that for complete data the Poisson and multinomial models for multiway counted data yield the same likelihood-based inferences for the cell probabilities. Show that the result continues to hold when data are MAR.
2. Derive ML estimates and associated variances for the likelihood (9.1). (*Hint:* Remember the constraint that the cell probabilities sum to 1.)
3. Verify the results of the chi-squared test for the MCAR assumption in Example 9.2.

4. Compute the proportion of missing information in Example 9.2, using the methods of Sections 7.5 and 7.6.
5. Calculate the expected cell frequencies in the first column of data in Table 9.3b, and compare the answers with those obtained from complete cases.
6. Suppose that in Example 9.3 there are no cases with pattern *d*. Which parameters are inestimable, in that they do not appear in the likelihood? Estimate the cell probabilities, assuming specific values for the inestimable parameters. (See Section 6.6.)
7. State in words the assumption about the missing-data mechanism under which the estimates in Table 9.4c are ML for Example 9.4.
8. Fill in the details in the derivation of Eq. (9.4).
9. Replicate the calculations of Example 9.4 for estimates of π_{12}.
10. Redo Example 9.4 assuming that the coarsely classified data in Table 9.4 were summarized as “Improvement” or “No Improvement” (stationary or worse).
11. Compute the EM algorithm for the data in Table 9.5 with values superscripted *a*, *b* and *c*, *d* in the supplemental margins interchanged. Compare the ML estimate of the odds ratio $\pi_{11}\pi_{22}\pi_{12}^{-1}\pi_{21}^{-1}$ with the estimate from complete cases. Are they identically equal?
12. Show that in Example 9.9 the factors in the factored likelihood are distinct for models {SP, SC, PC} and {SP, SC}, but are not distinct for {SC, PC}.
13. Display explicit ML estimates for all the models in Table 9.7 except for {12, 23, 31}.
14. Using results from Problem 12, derive the estimates in Table 9.9 for the models {SPC}, {SC, PC}, and {SP, SC}.
15. Compute ML estimates for the model {SP, SC} for the full data in Table 9.8, with the counts in the supplemental Table 9.8b increased by a factor of 10.
16. Why can starting values including zero probabilities disrupt proper performance of EM? (Hint: Consider the loglikelihood.)

CHAPTER 10

Mixed Normal and Nonnormal Data with Missing Values, Ignoring the Missing-Data Mechanism

10.1. INTRODUCTION

In Chapter 8 we considered a variety of missing-data models for continuous variables, based on the multivariate normal distribution. The role of categorical variables was confined to that of fully observed covariates in regression models. In Chapter 9 we discussed models for categorical variables with missing values. In this chapter we consider missing-data methods for mixtures of normal and nonnormal variables. Work on this topic has been rather limited and confined to models that assume the missing-data mechanism is ignorable.

Little and Schluchter (1985) discuss a model for missing data with mixed normal and categorical variables and provide relatively simple and computationally feasible EM algorithms with missing data. The basic version of this model is presented in Section 10.2 and important extensions are outlined in Section 10.3. Relationships with previously considered algorithms are examined in Section 10.4. Finally, in Section 10.5 we consider models for mixtures of normal distributions that yield robust ML estimates of the means and covariance matrix and derivative parameters, both for complete data and for data with missing values.

10.2. THE GENERAL LOCATION MODEL

10.2.1. The Complete-Data Model and Parameter Estimates

Suppose that the hypothetical complete data consist of a random sample of size n on K continuous variables (X) and V categorical variables (Y). Categorical variable j has I_j levels, so that the categorical variables define a V-way contingency table with $C = \prod_{j=1}^{V} I_j$ cells. For subject i, let x_i be the $(1 \times K)$ vector of continuous variables and y_i the $(1 \times V)$ vector of categorical variables. Also construct from y_i the $(1 \times C)$ vector w_i, which equals E_m if case i belongs to cell m of the contingency table, where E_m is a $(1 \times C)$ vector with 1 as the mth entry and 0's elsewhere.

Olkin and Tate (1961) define the general location model for the distribution of (x_i, w_i) in terms of the marginal distribution of w_i and the conditional distribution of x_i given w_i:

1. The w_i are iid multinomial random variables with cell probabilities

$$\Pr(w_i = E_m) = \pi_m, \qquad m = 1, \ldots, C; \sum \pi_m = 1.$$

2. Given that $w_i = E_m$,

$$(x_i | w_i = E_m) \overset{\text{iid}}{\sim} N_K(\mu_m, \Omega).$$

the K-variate normal distribution with mean $\mu_m = (\mu_{m1}, \ldots, \mu_{mK})$ and covariance matrix Ω. We write $\Pi = (\pi_1, \ldots, \pi_C)$ for the $(1 \times C)$ vector of cell probabilities and $\Gamma = \{\mu_{mk}\}$ for the $(C \times K)$ matrix of cell means. There are $C - 1 + KC + \frac{1}{2}K(K + 1)$ parameters $\theta = (\Pi, \Gamma, \Omega)$ in the model.

The following properties of this model are worthy of attention: (i) The within cell covariance matrix Ω is assumed the same across all the cells of the contingency table. (ii) If a particular binary variable (say Y_1), with values 1 and 0, is chosen as a dependent variable, then a logistic regression model results. That is, the conditional distribution of Y_1 given the other variables is Bernoulli with $\Pr(Y_1 = 1) = e^L/(1 + e^L)$, where L is linear in the other variables. (iii) If a particular continuous variable (say X_1) is chosen as a dependent variable, then a normal linear regression model results. That is, the conditional distribution of X_1 given the other variables is normal with mean given by a linear combination of the other variables, and constant variance.

Properties (ii) and (iii) imply that ML estimates for certain logistic regression models with missing values, and for certain linear regression models with missing continuous and categorical predictors, can be found by finding ML estimates of $\theta = \{\Pi, \Gamma, \Omega\}$ and then transforming them to yield parameters of the appropriate conditional distribution. The transformations are easily computed via the sweep operator, as discussed in Section 10.2.3.

The complete-data loglikelihood for this model is

$$
\begin{aligned}
l(\Gamma, \Omega, \Pi) &= \sum_{i=1}^{n} \ln f(x_i | w_i, \Gamma, \Omega) + \sum_{i=1}^{n} \ln f(w_i | \Pi) \\
&= h(\Omega) - \tfrac{1}{2} \operatorname{tr}\left(\Omega^{-1} \sum_{i=1}^{n} x_i^{\mathrm{T}} x_i \right) + \operatorname{tr} \Omega^{-1} \Gamma \left(\sum_{i=1}^{n} w_i^{\mathrm{T}} x_i \right) \\
&\quad + \sum_{m=1}^{C} \left[\left(\sum_{i=1}^{n} w_{im} \right) (\ln \pi_m - \tfrac{1}{2} \mu_m \Omega^{-1} \mu_m^{\mathrm{T}}) \right],
\end{aligned}
\tag{10.1}
$$

where w_{im} is the mth component of w_i, tr means "trace of the matrix," and $h(\Omega) = -\frac{1}{2}n\{K \ln(2\pi) + \ln(|\Omega|)\}$. Maximizing (10.1) yields complete-data ML estimates

$$
\begin{aligned}
\hat{\Pi} &= n^{-1} \sum w_i, \\
\hat{\Gamma} &= (\sum x_i^{\mathrm{T}} w_i)(\sum w_i^{\mathrm{T}} w_i)^{-1}, \\
\hat{\Omega} &= n^{-1} \sum (x_i - w_i \hat{\Gamma})^{\mathrm{T}} (x_i - w_i \hat{\Gamma}),
\end{aligned}
\tag{10.2}
$$

which are simply the observed cell proportions, the observed cell means, and the pooled within-cell covariance matrix of X, respectively.

10.2.2. ML Estimation with Missing Values

Now suppose some of the X's and W's are missing. For subject i, let $x_{\text{obs},i}$ denote the vector of observed continuous variables, $x_{\text{mis},i}$ denote the vector of missing continuous variables, and S_i denote the set of cells in the contingency table where subject i could lie, given the observed categorical variables. We now consider the EM algorithm for ML estimation of θ given data $\{x_{\text{obs},i}, S_i \colon i = 1, \ldots, n\}$.

The density (10.1) belongs to the regular exponential family with complete-data sufficient statistics $\sum x_i^{\mathrm{T}} x_i$, $\sum w_i^{\mathrm{T}} x_i$, and $\sum w_i$, which are, respectively, the raw sum of squares and cross products of the X's, the cell totals of the X's, and the cell frequencies. Hence we can apply the simplified form of the EM algorithm of Section 7.6. At iteration t the E step computes the expected values of the complete-data sufficient statistics given data $\{x_{\text{obs},i}, S_i \colon i = 1, \ldots, n\}$ and current parameter estimates $\theta^{(t)} = (\Gamma^{(t)}, \Omega^{(t)}, \Pi^{(t)})$. The contributions from case i are

E Step

$$T_{1i} = E(x_i^{\mathrm{T}} x_i | x_{\text{obs},i}, S_i, \theta^{(t)}), \tag{10.3}$$

$$T_{2i} = E(w_i^{\mathrm{T}} x_i | x_{\text{obs},i}, S_i, \theta^{(t)}), \tag{10.4}$$

$$T_{3i} = E(w_i | x_{\text{obs},i}, S_i, \theta^{(t)}). \tag{10.5}$$

Details of the E step computations are given in Section 10.2.3. The M step computes the complete-data ML estimates (10.2) with complete-data sufficient statistics replaced by their estimates from the E step:

M Step

$$\Pi^{(t+1)} = n^{-1} \sum_{i=1}^{n} T_{3i}$$

$$\Gamma^{(t+1)} = D^{-1}\left(\sum_{i=1}^{n} T_{2i}\right) \tag{10.6}$$

$$\Omega^{(t+1)} = n^{-1}\left[\sum_{i=1}^{n} T_{1i} - \left(\sum_{i=1}^{n} T_{2i}\right)^{\mathrm{T}} D^{-1}\left(\sum_{i=1}^{n} T_{2i}\right)\right]$$

where D is a matrix with elements of $\sum T_{3i}$ along the main diagonal and 0's elsewhere. The algorithm then returns to the E step to recompute (10.3)–(10.5) with the new parameter estimates, and cycles back and forth between E and M steps until convergence.

EXAMPLE 10.1. *St. Louis Risk Research Data.* Little and Schluchter (1985) analyze the data in Table 10.1 from the St. Louis Risk Research Project. One objective of the study was to evaluate the effects of parental psychological disorders on various aspects of the development of their children. In a preliminary study, data on $n = 69$ families with two children were collected. Families were classified according to risk group of the parent (G), a trichotomy defined as follows:

1. ($G = 1$), a normal group of control families from the local community.
2. ($G = 2$), a moderate-risk group where one parent was diagnosed as having secondary schizo-affective or other psychiatric illness or where one parent had a chronic physical illness.
3. ($G = 3$), a high-risk group where one parent had been diagnosed as having schizophrenia or an affective mental disorder.

Two other categorical outcome variables are displayed, D_1 = number of symptoms for first child (1 = low, 2 = high) and D_2 = number of symptoms for second child (1 = low, 2 = high). Thus there are $V = 3$ categorical variables that form a $3 \times 2 \times 2$ contingency table with $C = 12$ cells. There are also $K = 4$ continuous variables R_1, V_1, R_2, V_2, where R_c and V_c are standardized reading and verbal comprehension scores for the cth child in a family, $c = 1, 2$. The variable G is always observed, but the other variables are missing in a variety of different combinations.

Table 10.1 St. Louis Risk Research Data for Example 10.1

Low Risk ($G = 1$)						Moderate Risk ($G = 2$)						High Risk ($G = 3$)					
First Child			*Second Child*			*First Child*			*Second Child*			*First Child*			*Second Child*		
R_1	V_1	D_1	R_2	V_2	D_2	R_1	V_1	D_1	R_2	V_2	D_2	R_1	V_1	D_1	R_2	V_2	D_2
110	—	—	—	150	1	88	85	2	76	78	—	98	110	—	112	103	2
118	165	1	—	130	2	—	98	—	114	133	—	127	138	1	92	118	1
116	145	2	114	125	—	108	103	2	90	100	2	113	—	—	—	—	—
—	—	—	126	—	—	113	—	2	95	115	2	107	93	—	92	75	—
118	140	1	118	123	—	—	65	—	97	68	2	—	—	1	101	—	2
—	120	—	105	128	—	118	—	2	—	—	2	—	—	—	87	98	2
—	—	—	96	113	—	92	—	2	—	—	—	114	—	2	—	—	2
138	163	1	130	140	—	90	—	1	110	—	2	56	58	2	88	105	1
115	153	1	—	—	—	98	123	—	96	88	—	96	95	1	87	100	2
—	145	2	139	185	2	113	110	—	112	115	—	126	135	2	118	133	—
126	138	1	105	133	1	102	130	—	114	120	—	—	—	—	130	195	—
120	160	—	109	150	—	89	113	2	130	135	—	—	—	—	116	—	2
—	133	—	98	108	—	90	80	2	91	75	2	64	45	2	82	53	2
—	—	—	115	140	2	—	—	—	109	88	2	128	—	2	121	—	2
115	158	2	—	135	1	75	63	1	88	73	1	—	120	1	108	118	—
112	115	2	93	140	—	93	—	1	—	—	—	—	—	—	100	140	2
133	168	1	126	158	2	—	—	—	115	—	2	105	138	1	74	75	1
118	180	1	116	148	—	123	170	1	115	138	2	88	118	—	84	103	—
123	—	1	110	155	1	114	130	2	104	123	2						
100	—	1	101	120	1	—	—	2	113	123	2						
118	138	1	—	110	1	113	—	2	—	—	2						
103	108	—	—	—	—	117	—	—	82	103	2						
121	155	1	—	100	2	122	—	1	114	—	2						
—	—	—	—	—	2	105	—	2	—	—	1						
—	—	—	104	118	1												
—	—	—	87	85	1												
—	—	—	—	63	—												

An analysis of the missing-data pattern suggests that all the parameters of the general location model are estimable, despite the sparseness of the data matrix. For example, even though R_2 is not observed in the fully classified table when $G = 1$, $D_1 = 2$, $D_2 = 1$, there are five other families with R_2 measured that could possibly be in that cell and these observations provide the information to estimate the mean of R_2 in that cell.

ML estimates computed using the EM algorithm under the unrestricted model are displayed in Table 10.2 (Model A). The maximized loglikelihood under the unrestricted model was -872.73. Perhaps because of the relatively high degree of missingness for the categorical variables D_1 and D_2, several local maxima of the loglikelihood were found, and up to 50 iterations were required for convergence of the loglikelihood to two decimal places, depending on the initial estimates used to start the algorithm. Substantial differences were found between a few of the estimated cell means corresponding to different maxima of the loglikelihood. See Little and Schluchter (1985) for details. Such occurences suggest that drawing inferences requires care, since the data set is not large enough to support conclusions based on assumptions of asymptotic normality.

10.2.3. Details of the E-Step Calculations

We now describe in more detail how the quantities $\{T_{1i}, T_{2i}, T_{3i}, i = 1, \ldots, n\}$ are computed in Eqs. (10.3)–(10.5). All parameters in the expressions that follow are equal to the current parameter estimates in $\theta^{(t)}$. Calculation of T_{3i} involves finding $E(w_i | x_{\text{obs},i}, S_i, \theta^{(t)})$ for each subject $i = 1, \ldots, n$. The mth component of this vector will be denoted $\omega_{im} = \Pr(w_i = E_m | x_{\text{obs},i}, S_i, \theta^{(t)})$. That is, for $m = 1, \ldots, C$, ω_{im} is the conditional posterior probability that subject i belongs in cell m, given the observed continuous variables $x_{\text{obs},i}$, the knowledge that subject i is restricted to be in one of the cells in S_i, and $\theta = \theta^{(t)}$. This is positive when $m \in S_i$, where it takes the form

$$\omega_{im} = \exp(\delta_m) \Big/ \sum_{m \in S_i} \exp(\delta_m), \tag{10.7}$$

where

$$\delta_m = x_{\text{obs},i}\, \Omega_{\text{obs},i}^{-1}\, \mu_{\text{obs},i,m}^{\text{T}} - \tfrac{1}{2}\mu_{\text{obs},i,m}\, \Omega_{\text{obs},i}^{-1}\, \mu_{\text{obs},i,m}^{\text{T}} + \ln(\pi_m) \tag{10.8}$$

and $\mu_{\text{obs},i,m}$ and $\Omega_{\text{obs},i}$ are the mean and covariance matrix in cell m of the continuous variables $x_{\text{obs},i}$ present for subject i.

To calculate T_{1i} and T_{2i}, write the continuous variables for subject i as $\{x_{ij}, j = 1, \ldots, K\}$. If x_{ij} is missing, define $\hat{x}_{ij}^{(m)} = E(x_{ij} | x_{\text{obs},i}, w_i = E_m, \theta^{(t)})$, the predicted value of x_{ij} from the regression in cell m of X_j on $x_{\text{obs},i}$, evaluated at $\theta = \theta^{(t)}$. The element in the mth row and jth column of T_{2i}, for $m = 1, \ldots, C$

Table 10.2 Maximum Likelihood Estimates for Data in Table 10.1[a]

(a) *Expected Frequencies and Cell Means*

Cell			Expected Frequencies		Cell Means							
					R_1		R_2		V_1		V_2	
G	D_1	D_2	A	B	A	B	A	B	A	B	A	B
1	1	1	10.2	4.8	110.2	113.6	99.8	103.0	133.7	140.9	119.4	129.5
1	1	2	9.0	8.8	123.4	122.8	116.0	115.4	161.1	160.1	132.1	131.0
1	2	1	3.6	3.7	111.2	105.3	110.0	101.7	147.7	136.9	126.9	111.6
1	2	2	4.2	9.7	118.0	114.5	111.9	111.1	123.9	120.8	151.4	148.0
2	1	1	2.2	4.3	87.6	88.4	101.1	101.5	81.1	81.7	103.3	104.2
2	1	2	7.2	7.8	104.3	104.4	109.4	109.6	134.6	134.8	109.6	109.9
2	2	1	2.3	3.3	96.4	96.1	134.5	134.3	122.6	122.0	146.1	145.3
2	2	2	12.3	8.6	106.7	106.6	97.0	96.8	104.3	104.5	102.4	102.3
3	1	1	2.1	3.2	115.8	115.7	82.9	82.8	137.7	137.5	96.3	96.0
3	1	2	7.8	5.9	105.7	100.7	100.8	96.1	127.9	119.4	128.3	117.1
3	2	1	1.0	2.5	56.2	76.2	88.2	108.3	58.3	90.4	105.4	148.6
3	2	2	7.1	6.4	107.3	107.4	107.0	107.3	107.2	107.2	104.8	104.8

(b) *Standard Deviations and Correlations*

	Standard Deviations				Correlations					
Model	R_1	R_2	V_1	V_2	(R_1, R_2)	(R_1, V_1)	(R_1, V_2)	(R_2, V_1)	(R_2, V_2)	(V_1, V_2)
A	13.2	11.9	20.7	24.1	.701	.832	.825	.663	.835	.885
B	13.1	11.9	20.1	23.3	.685	.832	.822	.654	.836	.881

[a] A = model with no restrictions on means or cell probabilities. B = model with no restrictions on means, cell probabilities restricted so that D_1 and D_2 are independent of G.

and $j = 1, \ldots, K$, is obtained by multiplying x_{ij} or its estimate by the conditional posterior probability that subject i falls in cell m:

$$E(w_{im}x_{ij}|x_{\text{obs},i}, S_i, \theta^{(t)}) = \begin{cases} \omega_{im}\hat{x}_{ij}^{(m)} & \text{if } x_{ij} \text{ is missing,} \\ \omega_{im}x_{ij} & \text{if } x_{ij} \text{ is present.} \end{cases}$$

When both x_{ij} and x_{ik} are missing, let $\sigma_{jk,\text{obs},i}$ denote the conditional covariance of x_{ij} and x_{ik} given $x_{\text{obs},i}$ and given that $w_i = E_m$. Then the jk element of T_{1i}, for $j, k = 1, \ldots, K$, is

$$\begin{aligned} &E(x_{ij}x_{ik}|x_{\text{obs},i}, S_i, \theta^{(t)}) \\ &= \sum_{m \in S_i} \omega_{im} E(x_{ij}x_{ik}|x_{\text{obs},i}, w_i = E_m, \theta^{(t)}) \\ &= \begin{cases} x_{ij}x_{ik}, & x_{ij}, x_{ik} \text{ both present;} \\ x_{ik}\sum_{m \in S_i} \omega_{im}\hat{x}_{ij}^{(m)}, & x_{ij} \text{ missing, } x_{ik} \text{ present;} \\ x_{ij}\sum_{m \in S_i} \omega_{im}\hat{x}_{ik}^{(m)}, & x_{ij} \text{ present, } x_{ik} \text{ missing;} \\ \sigma_{jk\cdot\text{obs},i} + \sum_{m \in S_i} \omega_{im}\hat{x}_{ij}^{(m)}\hat{x}_{ik}^{(m)}, & x_{ij}, x_{ik} \text{ both missing.} \end{cases} \end{aligned}$$

The computations are easily performed by sweep operations, as introduced in Section 6.5. Consider the matrix

$$M = \begin{bmatrix} \hat{\Omega}_{\text{obs},i} & \hat{\Omega}_{\text{cov},i}^{\text{T}} & \hat{\Gamma}_{\text{obs},i}^{\text{T}} \\ \hat{\Omega}_{\text{cov},i} & \hat{\Omega}_{\text{mis},i} & \hat{\Gamma}_{\text{mis},i}^{\text{T}} \\ \hat{\Gamma}_{\text{obs},i} & \hat{\Gamma}_{\text{mis},i} & P \end{bmatrix},$$

where P is a $C \times C$ diagonal matrix, having mth diagonal element equal to $2 \ln \pi_m$, for $m = 1, \ldots, C$, and

$$\hat{\Omega} = \begin{bmatrix} \hat{\Omega}_{\text{obs},i}^{\text{T}} & \hat{\Omega}_{\text{cov},i}^{\text{T}} \\ \hat{\Omega}_{\text{cov},i} & \hat{\Omega}_{\text{mis},i} \end{bmatrix} \quad \text{and} \quad \hat{\Gamma} = [\hat{\Gamma}_{\text{obs},i}, \hat{\Gamma}_{\text{mis},i}]$$

are current estimates of Ω and Γ, partitioned according to the observed and missing X variables in case i. Sweeping on the elements of M corresponding to present X's yields

$$\text{SWP}[x_{\text{obs},i}]M = \begin{bmatrix} G_{11} & G_{12}^{\text{T}} & G_{13}^{\text{T}} \\ G_{12} & G_{22} & G_{23}^{\text{T}} \\ G_{13} & G_{23} & G_{33} \end{bmatrix}$$

where $G_{11} = -\hat{\Omega}_{\text{obs},i}^{-1}$; $G_{12} = \hat{\Omega}_{\text{obs},i}^{-1}\hat{\Omega}_{\text{cov},i}$ are regression coefficients of the missing X's on $x_{\text{obs},i}$; $G_{22} = \hat{\Omega}_{\text{mis},i} - \hat{\Omega}_{\text{cov},i}^{\text{T}}\hat{\Omega}_{\text{obs},i}^{-1}\hat{\Omega}_{\text{cov},i}$ contains the residual variances and covariances $\sigma_{jk\cdot\text{obs},i}$ for $x_{ij}, x_{ik} \in x_{\text{obs},i}$; $G_{13} = \hat{\Omega}_{\text{obs},i}^{-1}\hat{\Gamma}_{\text{obs},i}$ yields the coefficients of $x_{\text{obs},i}$ in the linear discriminant function (10.8); and the mth diagonal element of $\frac{1}{2}G_{33} = \frac{1}{2}P - \frac{1}{2}\Gamma_{\text{obs},i}\Omega_{\text{obs},i}^{-1}\Gamma_{\text{obs},i}^{\text{T}}$ equals the sum of the second and third terms on the right side of (10.8). Thus G_{13} and G_{33}, together with π_m, yield the linear discriminant function δ_m and hence ω_{im} as in (10.7). Considerable savings in computation may be obtained if subjects with the same pattern of missing X's are grouped together to avoid unnecessary sweep operations.

10.3. EXTENSIONS OF THE GENERAL LOCATION MODEL THAT PLACE CONSTRAINTS ON THE PARAMETERS

10.3.1. Introduction

The model of Section 10.2 specifies a distinct mean vector μ_m for each cell m of the table and makes no restrictions on the cell probabilities, other than the obvious restriction that $\sum \pi_m = 1$. In this section we describe a more general model that allows ANOVA-like restrictions on the μ_m and models the π_m by a restricted loglinear model. This more general model is considered for complete-data discriminant analysis by Krzanowski (1980, 1982).

10.3.2. Restricted Models for the Cell Means

For $r \leqslant C$, let z_i be a $1 \times r$ vector of design variables for case i, which can be obtained from the cell indicator vector w_i as $z_i = w_i A$, where A is a known $C \times r$ matrix that represents the chosen design. The more general model specifies that the conditional distribution of x_i given w_i depends on w_i only through z_i, in that $f(x_i|w_i) \sim N_k(z_i B; \Omega)$, where B is a $(r \times k)$ matrix of unknown parameters. Note that $E(x_i|w_i) = w_i AB$, so that $\Gamma = AB$. In the model of Section 10.2, A is the $C \times C$ identity matrix.

10.3.3. Loglinear Models for the Cell Probabilities

Another way of reducing the dimensionality of the model is to constrain the cell probabilities Π by a loglinear model, as discussed in Section 9.4. For example, suppose the cells are formed by a joint classification of three categorical variables Y_1, Y_2, and Y_3 with, respectively, I_1, I_2, and I_3 levels, and $C = I_1 \times I_2 \times I_3$. Let us revise the notation so that π_{jkl} is the probability that $y_1 = j$, $y_2 = k$, $y_3 = l$, for $j = 1, \ldots, I_1$, $k = 1, \ldots, I_2$, and $l = 1, \ldots, I_3$. The loglinear models are obtained by writing

$$\ln \pi_{jkl} = \alpha + \alpha_j^{(1)} + \alpha_k^{(2)} + \alpha_l^{(3)} + \alpha_{jk}^{(12)} + \alpha_{jl}^{(13)} + \alpha_{kl}^{(23)} + \alpha_{jkl}^{(123)},$$

and setting subsets of the α terms equal to zero. See Section 9.4 for more details.

10.3.4. Modifications to the Algorithm of Section 10.2.2

For a general V-way table with $C = \prod_{j=1}^{V} I_j$ cells, let α denote the nonzero α terms in the loglinear model, and write $\pi_m(\alpha)$ for the constrained probability of falling in cell m, $m = 1, \dots, C$. We now sketch modifications of the algorithm in Section 10.2.2 when the reduced models in Section 10.3.2 and 10.3.3 are fitted to incomplete data.

For a particular choice of the models in Sections 10.3.2 and 10.3.3, let $\alpha^{(0)}$, $\Omega^{(0)}$, and $B^{(0)}$ be initial estimates of the parameters, perhaps calculated from complete cases. Also let $\Gamma^{(0)} = AB^{(0)}$, where A is a known design matrix, and $\pi_m^{(0)} = \pi_m(\alpha^{(0)})$, $m = 1, \dots, C$. The restricted models of Sections 10.3.2 and 10.3.3 lie in the regular exponential family, with complete-data minimal sufficient statistics $\sum x_i^{\mathrm{T}} x_i$, $\sum w_i^{\mathrm{T}} w_i A$ and linear combinations of the counts $\sum w_i$ determined by the margins fitted in the loglinear model. Since these quantities are linear functions of the complete-data sufficient statistics for the model in Section 10.2, the E step consists in calculating $\sum T_{1i}$, $\sum T_{2i}$, and $\sum T_{3i}$ via Eqs. (10.3)–(10.5), and then forming the linear combinations of these functions that yield the complete-data minimal sufficient statistics for the reduced model.

The M-step calculations differ from those for the unrestricted model, yielding estimates of Γ, Ω, and Π that satisfy the model restrictions. For estimates of Π, first form the multiway table with cell frequencies given in the vector $\sum T_{3i}$ [Eq. (10.5)]. This table contains fractional entries from the partially classified counts distributed into the table in the E step. The updated cell frequencies are obtained by fitting the assumed loglinear model to the frequencies in $\sum T_{3i}$ by a complete-data method, which may itself be iterative if explicit estimates are not available. The probabilities in the fitted table are the new estimates of $\{\pi_m(\alpha)\}$, used for the next M step. Actually, when explicit noniterative estimates are not available in the M step, it may be sufficient in early iterations to update the estimates of the cell probabilities in the M step by going through a single cycle of the iterative proportional fitting algorithm, thus avoiding the need for two levels of iterations. The resulting algorithm is a generalized EM algorithm as defined in Section 7.3, because the iterative proportional fitting algorithm has the general property that it increases the likelihood of the data being fit after each cycle, until convergence occurs (Brown, 1959).

With complete data, the ML estimates of B and Ω (e.g., Anderson, 1958, Ch. 8) are $\hat{B}(\sum z_i^{\mathrm{T}} z_i)^{-1} \sum z_i^{\mathrm{T}} x_i$ and $\hat{\Omega} = n^{-1} \sum (x_i - z_i \hat{B})^{\mathrm{T}} (x_i - z_i \hat{B})$. The M-step estimates of B and Ω are obtained by writing $z_i = w_i A$ in the preceding equations for $\hat{B}$ and $\hat{\Omega}$, and then replacing $\sum x_i^{\mathrm{T}} x_i$, $\sum w_i^{\mathrm{T}} x_i$, and $\sum w_i^{\mathrm{T}} w_i$ by $\sum T_{1i}$, $\sum T_{2i}$, and D, respectively, where D is a matrix with elements of $\sum T_{3i}$ on the diagonal and zeros elsewhere. The updated estimates of B, Γ, and Ω in the M step of iteration t are then

$$B^{(t+1)} = (A^{\mathrm{T}}DA)^{-1}A^{\mathrm{T}}(\textstyle\sum T_{2i}) \tag{10.9}$$

$$\Gamma^{(t+1)} = AB^{(t+1)}, \tag{10.10}$$

and

$$\Omega^{(t+1)} = n^{-1}\left[\sum_{i=1}^{n} T_{1i} - \left(\sum_{i=1}^{n} T_{2i}\right)^{\mathrm{T}} A(A^{\mathrm{T}}DA)^{-1}A^{\mathrm{T}}\left(\sum_{i=1}^{n} T_{2i}\right)\right]. \tag{10.11}$$

When no restrictions are placed on the means, A is the $C \times C$ identity matrix and the equations for $\Omega^{(t+1)}$ and $\Gamma^{(t+1)}$ in (10.9)–(10.11) are equivalent to their counterparts in Eq. (10.6). The new estimates $\Pi^{(t+1)}$, $\Gamma^{(t+1)}$, and $\Omega^{(t+1)}$ are then input to the next E step, given by Eqs. (10.3), (10.4), and (10.5).

10.3.5. Restricted Models for the St. Louis Data

EXAMPLE 10.2. (*Example 10.1 continued*). In Section 10.2.2 the unrestricted location model was fitted to the data in Table 10.1. The model is somewhat overparameterized, with 69 parameters for 69 incomplete cases. In this section we fit and test models with fewer parameters that correspond to hypotheses of substantive interest. In particular, suppose we wish to test the hypothesis that the occurrence of adverse psychiatric symptoms in children is unrelated to the risk group of the parent. This hypothesis implies that

$$\pi_{jkl} = \pi_{j++}\pi_{+kl}, \qquad j = 1, 2, 3;\ k, l = 1, 2,$$

where π_{jkl} is the probability associated with level j of G, and levels k and l, respectively, of D_1 and D_2. No restrictions are placed on the cell means of the continuous variables. Little and Schluchter (1985) fit this constrained model to the data, using the method of Section 10.3.4.

ML estimates for the restricted model are shown in Table 10.2 (Model B). The maximized loglikelihood was -877.64. Recall that the loglikelihood for the full model fitted in Section 10.2.2 was -872.73. The likelihood ratio chi-square statistic for testing whether D_1 and D_2 are independent of G was thus $2(-872.73 + 877.64) = 9.82$ with six degrees of freedom, suggesting no evidence of lack of fit. Another local maximum (loglikelihood $= -877.72$) was found for this model.

In search for simpler models, Little and Schluchter next fit the model where the $G \times D_1$, $G \times D_2$, and $G \times D_1 \times D_2$ interaction effects on the means of the continuous variables are set to zero, with the same restricted model for the cell probabilities. The restrictions on the means of continuous variables can be written as $E(x_i|z_i) = z_iB$, where B is a 6×4 matrix of parameters, and $z_i = w_iA$, where

$$A^{\mathrm{T}} = \begin{bmatrix} 1 & 1 & 1 & 1 & 1 & 1 & 1 & 1 & 1 & 1 & 1 & 1 \\ 1 & 1 & 1 & 1 & 0 & 0 & 0 & 0 & -1 & -1 & -1 & -1 \\ 0 & 0 & 0 & 0 & 1 & 1 & 1 & 1 & -1 & -1 & -1 & -1 \\ 1 & 0 & 0 & -1 & 1 & 0 & 0 & -1 & 1 & 0 & 0 & -1 \\ 0 & 1 & 0 & -1 & 0 & 1 & 0 & -1 & 0 & 1 & 0 & -1 \\ 0 & 0 & 1 & -1 & 0 & 0 & 1 & -1 & 0 & 0 & 1 & -1 \end{bmatrix}$$

and the 12 cells in the vector w_i are arranged such that the index of D_2 changes fastest and the index of G changes the most slowly. This model reduces the number of parameters needed to describe the means from 48 to 24.

Again, multiple local maxima of the likelihood function were found. The global maximized loglikelihood for this model was -910.46, so that the likelihood ratio chi-square testing the fit of this model versus the full model was $\chi^2 = 75.46$ ($df = 30$), suggesting that the reduced model does not fit the data. The authors also fit the model where only the three-way $G \times D_1 \times D_2$ interaction effect was set to zero, with the same restriction on cell probabilities. This model showed evidence of lack of fit when compared to the full model ($\chi^2 = 59.39, df = 14$). P values were not reported because they are inappropriate given the multiple maxima of the likelihood. Nonetheless, these results suggest that the degree to which parental mental health affects reading and verbal comprehension performance in the child depends on the psychological state of the child, as one would expect.

10.4. RELATIONSHIPS WITH OTHER EM ALGORITHMS FOR SPECIAL MISSING-DATA PATTERNS

In the absence of categorical variables Y, the algorithm in Section 10.2 reduces to the multivariate normal EM algorithm discussed in Section 8.2.1. If categorical variables are incomplete and no continuous variables are present, then the data can be arranged as a multiway contingency table with partially classified supplemental margins. The algorithm then corresponds to ML estimation for partially classified contingency tables, as discussed in Section 9.4.

More generally, the algorithms of Sections 8.2.1 and 9.4 can also be applied to the data pattern of Figure 10.1. In these data the V categorical variables are more observed than the K continuous variables in that all the categorical variables are present in cases where one or more of the continuous variables are present. Following the factored likelihood theory of Chapter 6, ML estimates for the model of Section 10.3 can be obtained as follows:

	Variables	
Cases	$Y_1 \cdots Y_V$	$X_1 \cdots X_K$
1	$1 \cdots 1$	$\times \cdots \times$
$\vdots$	$\vdots \quad \vdots$	$\vdots \quad \vdots$
m	$1 \cdots 1$	$\times \cdots \times$
$m+1$	$\times \cdots \times$	$0 \cdots 0$
$\vdots$	$\vdots \quad \vdots$	$\vdots \quad \vdots$
n	$\times \cdots \times$	$0 \cdots 0$

Figure 10.1. Pattern of missing data leading to simpler ML estimates. 1, Observed; 0, missing; $\times$, observed or missing. Source: Little and Schluchter (1985).

1. Estimate the parameters of the joint distribution of Y from the first V columns of Figure 10.1. Since these data are entirely categorical, ML algorithms for partially classified contingency tables apply here.
2. Estimate the parameters of the conditional distribution of X given Y from the first m rows of Figure 10.1. The multivariate normal EM algorithm can be used here, even though categorical variables are present. Dummy variables representing the effects z_i in the ANOVA design are included in the multivariate normal EM algorithm, treating them as if they were continuous variables. Elements corresponding to these variables are then swept in the final estimated covariance matrix of all the variables from the algorithm, yielding estimates $\hat{B}$, $\hat{\Omega}$ of the parameters of the conditional distribution of X given Y. These are true ML, by an application of the theory of Chapter 6 for factored likelihoods.

The algorithm of Section 10.2.2, with the modifications for the restricted model in Section 10.3.4, also provides ML estimates of B and Ω (and Π) when the categorical variables are completely observed, but the algorithm differs from the normal EM algorithm, since sweep operations are confined to the covariance matrix of the continuous variables. (See Section 10.2.3 for details.) The main advantage of the algorithm occurs when the data do not conform to the pattern in Figure 10.1, since then the methods of Sections 8.2.1 and 9.4 cannot be applied to produce ML estimates.

When the continuous variables are completely observed and Y consists of a k-category variable that is completely missing, then the algorithm of Section 10.2.2 reduces to that of Day (1969) for k multivariate normal mixtures. Since the algorithm still works with incompletely recorded continuous variables, it provides an extension of Day's algorithm to incomplete data. As with many mixture models, multiple maxima of the likelihood are a definite possibility

Table 10.3 Results of the EM for Mixtures Applied to Darwin's Data

(a) *Darwin's Data on Differences in Heights of Self-fertilized and Cross-fertilized Plants*:

1	2	3	4	5	6	7	8	9	10	11	12	13	14	15
−67	−48	6	8	14	16	23	24	28	29	41	49	56	60	75

(b) *Results Assuming One Normal Population*:

−2 loglikelihood	$\hat{\mu}$	$\hat{\sigma}^2$
122.9	20.93	1329.7

(c) *Results of EM Iterations for Two-Component Normal*

Starting Specifications (Observations in Component 1)	−2 loglikelihood	$\hat{\mu}_1$	$\hat{\mu}_2$	$\hat{\sigma}^2$	$15\hat{p}$
1	116.0	−57.4	33.0	385.4	2.00
15	122.9	21.62	20.91	1330.0	0.957
Any subset of {1...9}	116.0	−57.4	33.0	385.4	2.00
Any subset of {10...15}	122.9	(estimates vary, depending on exact starting value)			

(Aitkin and Rubin, 1985), so it is advisable to run the algorithm for a variety of choices of starting values for the parameters.

EXAMPLE 10.3. *A Univariate Mixture Model For Biological Data.* Aitkin and Wilson (1980) examined the behavior of the EM algorithm for mixture models on several small data sets. One example was Darwin's data on differences in heights of pairs of self-fertilized and cross-fertilized plants, displayed in Table 10.3a. The standard ML estimates, assuming a single normal sample with mean μ and variance σ^2, are displayed in Table 10.3b along with −2 loglikelihood (omitting the constant $n \ln 2\pi$). A two-component normal mixture with means μ_1 and μ_2, common variance σ^2, and mixing proportion p was fit to these data, using EM starting from a variety of initial values. All starting values were obtained by specifying an initial guess as to which observations belonged to component 1 and component 2 (i.e., initial posterior probabilities of component membership were all zero or one), and then applying the M step to obtain initial parameter estimates. The results of these iterations are displayed in Table 10.3c and demonstrate the sensitivity of the final estimate to starting values. The likelihood appears to be bimodal, with a high peaked mode at the estimates obtained starting from the first or third starting values and a low broad mode at the estimates obtained from the second starting value.

10.5. ML ALGORITHMS FOR ROBUST ESTIMATION

10.5.1. Introduction

Example 10.3 modeled data as a mixture of normal distributions with different means and the same variance. Another model, particularly useful for robust estimation, describes data as a mixture of normal distributions with the same mean and different variances. Consider a random sample $\{x_i: i = 1, \dots, n\}$ of values subject to contamination, and let y_i be an unobserved Bernoulli variable taking the value 1 if x_i is contaminated and 2 if x_i is not contaminated. We suppose that x_i given $y_i = 1$ is normal with mean μ and variance σ^2, and x_i given $y_i = 2$ is normal with mean μ and variable σ^2/λ, where λ is assumed known. For example, $\lambda = 0.1$ if the contamination is regarded as inflating the variance by a factor of 10. The result is a *contaminated normal model*. The ML estimate of μ downweights outlying values, as we shall in the next section.

Models of this type can be fitted by the EM algorithm, treating the values of y_i as missing data. We present a general mixture model for robust estimation that includes the contaminated normal as a special case, but also yields models where the marginal distribution of x_i is a t distribution. Section 10.5.2 considers the case of univariate x_i, presented in Dempster, Laird, and Rubin (1977, 1980). The case of multivariate x_i, described in Rubin (1983), is considered in Section 10.5.3, and Section 10.5.4 extends the analysis to handle multivariate x_i when some values are missing (Little, 1986).

10.5.2. Robust Estimation from a Univariate Sample

Let $X = (x_1, \dots, x_n)^{\mathrm{T}}$ be a random sample from a population such that

$$x_i | \theta, q_i \overset{\text{ind}}{\sim} N(\mu, \sigma^2/q_i),$$

where the q_i are unobserved iid positive scalar random variables with known density $h(q_i)$. The object is to derive ML estimates $\hat{\theta} = (\hat{\mu}, \hat{\sigma}^2)^{\mathrm{T}}$ of $\theta = (\mu, \sigma^2)^{\mathrm{T}}$ by the EM algorithm treating the value of $Q = (q_1, \dots, q_n)^{\mathrm{T}}$ as missing data.

If X and Q were observed, then ML estimates of (μ, σ^2) would be found by weighted least squares:

$$\hat{\mu} = \sum_{i=1}^{n} q_i x_i \bigg/ \sum_{i=1}^{n} q_i = s_1/s_0, \tag{10.12}$$

$$\hat{\sigma}^2 = \sum_{i=1}^{n} \frac{q_i(x_i - \hat{\mu})^2}{n} = \frac{s_2 - s_1^2/s_0}{n}, \tag{10.13}$$

where $s_0 = \sum_{i=1}^{n} q_i$, $s_1 = \sum_{i=1}^{n} q_i x_i$, and $s_2 = \sum_{i=1}^{n} q_i x_i^2$ are the complete-data sufficient statistics as defined in Section 7.6. Hence, when Q is not observed, the $(t + 1)$th iteration of EM is as follows:

E Step

Estimate s_0, s_1, and s_2 by their conditional expectations, given X and current estimates of θ $(\mu^{(t)}, \sigma^{(t)2})$. Since s_0, s_1, and s_2 are linear in the q_i's, the E step reduces to finding estimated weights

$$w_i^{(t)} = E(q_i | x_i, \hat{\mu}^{(t)}, \hat{\sigma}^{(t)2}). \tag{10.14}$$

M Step

Compute new estimates $(\mu^{(t+1)}, \sigma^{(t+1)2})$ from (10.12) and (10.13), with (s_0, s_1, s_2) replaced by their estimates from the E step, that is, with q_i replaced by $w_i^{(t)}$ from (10.14).

Hence the EM algorithm is a form of iteratively reweighted least squares. The exact form of the estimated weights (10.14) depends on the assumed distribution for q_i. Dempster, Laird, and Rubin (1977, 1980) discuss two useful choices that yield simple weights.

EXAMPLE 10.4. *The Univariate Contaminated Normal Model.* Suppose that $h(q_i)$ is concentrated at two values of q_i, 1 and λ, such that

$$h(q_i) = \begin{cases} 1 - \pi & \text{if } q_i = 1, \\ \pi & \text{if } q_i = \lambda, \text{ known} \\ 0 & \text{otherwise,} \end{cases} \tag{10.15}$$

where $0 < \pi < 1$. Then the marginal distribution of x_i is a mixture of $N(\mu, \sigma^2)$ and $N(\mu, \sigma^2/\lambda)$, that is, we have the contaminated normal model described in Section 10.5.1, with probability of contamination π. A simple application of Bayes's theorem yields

$$\begin{aligned} w_i &\equiv E(q_i | x_i, \mu, \sigma^2) \\ &= \frac{1 - \pi + \pi\lambda^{3/2} \exp\{(1 - \lambda)d_i^2/2\}}{1 - \pi + \pi\lambda^{1/2} \exp\{(1 - \lambda)d_i^2/2\}}, \end{aligned} \tag{10.16}$$

where

$$d_i^2 = (x_i - \mu)^2/\sigma^2. \tag{10.17}$$

The weights $w_i^{(t)}$ for EM are obtained by substituting current estimates $\mu^{(t)}$ and $\sigma^{(t)}$ in (10.17) for d_i^2 and then using (10.16). Observe that values x_i far from the mean have large values of d_i^2 and (for $\lambda < 1$) reduced weights in the M step. Thus the algorithm leads to a robust estimate of μ that downweights outlying cases.

EXAMPLE 10.5. *ML Estimation for a Sample from the t Distribution.* The second model considered by Dempster, Laird, and Rubin (1977, 1980) sets $q_i = k_i/\alpha$, where k_i has the gamma distribution:

$$h(k_i) = k_i^{\alpha - 1} \exp(-k_i)/\Gamma(\alpha), \tag{10.18}$$

$\Gamma(\cdot)$ here denoting the gamma function. The marginal distribution of x_i is then Student's t with mean μ, scale σ^2, and degrees of freedom $v = 2\alpha$. The model thus provides ML estimates for a t sample with known degrees of freedom v.

Define $\beta_i^* = 1 + 0.5d_i^2/(\alpha - 1)$, where d_i^2 is given by (10.17). Then it is readily shown that given x_i, $k_i\beta_i^*$ has the gamma distribution (10.18) with α replaced by $\alpha + 1/2$. Hence

$$w_i = E(k_i|x_i, \mu, \sigma^2)/\alpha = (v + 1)/(v + d_i^2) \tag{10.19}$$

As in the previous example, weights $w_i^{(t)}$ are found by substituting current estimates of the parameters μ and σ^2 in (10.17) for d_i^2 and then using (10.19). The effect of (10.19) is to downweight outliers, with a degree of downweighting inversely related to the value of v. The latter parameter may be fixed at a plausible value (such as 4), or the algorithm repeated for a grid of values of v, and the one chosen that produces the largest maximized loglikelihood. Joint ML estimation of v, μ, and σ^2 is also possible at the cost of a somewhat more complex M step.

A straightforward and important practical extension of the models in Examples 10.4 and 10.5 is to model the mean as a linear combination of predictors X, yielding a weighted least squares algorithm for linear regression with contaminated normal or t errors (Rubin, 1983). Pettitt (1985) describes ML estimation for the contaminated normal and t models when the values of X are grouped and rounded.

10.5.3. Robust Estimation of the Means and Covariance Matrix: Complete Data

Rubin (1983) generalizes the model of Section 10.5.2 to multivariate data, and applies it to derive ML estimates for contaminated multivariate normal and multivariate t samples. Let x_i be a $(1 \times K)$ vector of values of variables X_1, ..., X_K. We suppose that x_i has the K-variate normal distribution

$$(x_i|\theta, q_i) \overset{\text{ind}}{\sim} N_K(\mu, \Psi/q_i) \tag{10.20}$$

where the q_i are unobserved iid positive scalar random variables with known density $h(q_i)$. ML estimates of μ and Ψ can be found by applying the EM algorithm, treating $Q = (q_1, \ldots, q_n)^{\mathrm{T}}$ as missing data.

If Q were observed, ML estimates of μ and Ψ would be the multivariate analogs of (10.12) and (10.13):

$$\hat{\mu} = \sum_{i=1}^{n} q_i x_i \bigg/ \sum_{i=1}^{n} q_i = s_1/s_0, \tag{10.21}$$

$$\hat{\Psi} = \sum_{i=1}^{n} \frac{q_i(x_i - \hat{\mu})^{\mathrm{T}}(x_i - \hat{\mu})}{n} = \frac{s_2 - s_1^T s_1/s_0}{n}, \tag{10.22}$$

where $s_0 = \sum_{i=1}^n q_i$, $s_1 = \sum_{i=1}^n q_i x_i$, and $s_2 = \sum_{i=1}^n q_i x_i^T x_i$ are complete data sufficient statistics. Hence when Q is not observed, the $(t + 1)$st iteration of EM is as follows:

E Step

Estimate s_0, s_1, and s_2 by their conditional expectations given X and current estimates $(\mu^{(t)}, \Psi^{(t)})$ of the parameters. Since s_0, s_1, and s_2 are linear in the q_i's, the E step again reduces to finding estimated weights

$$w_i^{(t)} = E(q_i | x_i, \mu^{(t)}, \Psi^{(t)}).$$

M Step

Compute new estimates $(\mu^{(t+1)}, \Psi^{(t+1)})$ from (10.21) and (10.22), with s_0, s_1, and s_2 replaced by their estimates from the E step.

If the q_i are distributed as (10.15), the marginal distribution of x_i is a mixture of $N(\mu, \Psi)$ and $N(\mu, \Psi/\lambda)$, that is, we obtain a contaminated K-variate normal model. The weights are then given by the following generalizations of (10.16) and (10.17):

$$w_i \equiv E(q_i | x_i, \mu, \Psi) = \frac{1 - \pi + \pi\lambda^{K/2+1} \exp\{(1 - \lambda)d_i^2/2\}}{1 - \pi + \pi\lambda^{K/2} \exp\{(1 - \lambda)d_i^2/2\}} \tag{10.23}$$

where d_i^2 is now the squared distance for case i:

$$d_i^2 = (x_i - \mu)\Psi^{-1}(x_i - \mu)^{\mathrm{T}}. \tag{10.24}$$

The model downweights cases with large values of d_i^2.

If, on the other hand, the distribution of $k_i = \alpha q_i$ is gamma as in (10.18), the marginal distribution of x_i is multivariate t with mean μ, scale Ψ, and degrees of freedom $v = 2\alpha$. The weights are then given by the following generalization of (10.19):

$$w_i = E(q_i | x_i, \mu, \Psi)/\alpha = (v + K)/(v + d_i^2), \tag{10.25}$$

where d_i^2 is again given by (10.24).

Rubin (1983) also considers extensions of these models to multivariate regression.

10.5.4. Robust Estimation of the Mean and Covariance Matrix from Data with Missing Values

Little (1986) extends these algorithms to situations where some values of X are missing. Let $x_{\mathrm{obs},i}$ denote the set of variables observed in case i, let $x_{\mathrm{mis},i}$

denote the missing variables, and write $X_{\text{obs}} = \{x_{\text{obs},i}: i = 1, \ldots, n\}$ and $X_{\text{mis}} = \{x_{\text{mis},i}: i = 1, \ldots, n\}$. We assume that (1) $\{x_i\}$ has the distribution given by (10.20) and (2) the missing data are MAR. ML estimates of μ and Ψ are found by applying the EM algorithm, treating the values of X_{mis} and Q as missing data.

The M step is identical to the M step when X is completely observed, described in the previous section. The E step estimates the complete-data sufficient statistics s_0, s_1, and s_2 by their conditional expectations, given X_{obs} and current estimate $\theta^{(t)} = (\mu^{(t)}, \Psi^{(t)})$ of the parameter θ. We find

$$E(s_0 | \theta^{(t)}, X_{\text{obs}}) = E\left(\sum_{i=1}^{n} q_i | \theta^{(t)}, x_{\text{obs},i}\right) = \sum_{i=1}^{n} w_i^{(t)},$$

where $w_i^{(t)} = E(q_i | \theta^{(t)}, x_{\text{obs},i})$; the jth component of $E(s_1 | \theta^{(t)}, X_{\text{obs}})$ is

$$E\left(\sum_{i=1}^{n} q_i x_{ij} | \theta^{(t)}, X_{\text{obs}}\right) = \sum_{i=1}^{n} E\{q_i E(x_{ij} | \theta^{(t)}, x_{\text{obs},i}, q_i) | \theta^{(t)}, x_{\text{obs},i}\}$$

$$= \sum_{i=1}^{n} w_i^{(t)} \hat{x}_{ij}^{(t)},$$

where $\hat{x}_{ij}^{(t)} = E(x_{ij} | \theta^{(t)}, x_{\text{obs},i})$, since the conditional mean of x_{ij} given $\theta^{(t)}$, $x_{\text{obs},i}$ and q_i does not depend on q_i. Finally the (j, k)th element of $E(s_2 | \theta^{(t)}, X_{\text{obs}})$ is

$$E\left(\sum_{i=1}^{n} q_i x_{ij} x_{ik} | \theta^{(t)}, X_{\text{obs}}\right) = \sum_{i=1}^{n} E\{q_i E(x_{ij} x_{ik} | \theta^{(t)}, x_{\text{obs},i}, q_i) | \theta^{(t)}, x_{\text{obs},i}\}$$

$$= \sum_{i=1}^{n} (w_i^{(t)} \hat{x}_{ij}^{(t)} \hat{x}_{ik}^{(t)} + \psi_{jk\cdot\text{obs},i}^{(t)}),$$

where the adjustment $\psi_{jk\cdot\text{obs},i}^{(t)}$ is zero if x_{ij} or x_{ik} are observed, and q_i times the residual covariance of x_{ij} and x_{ik} given $x_{\text{obs},i}$, if x_{ij} and x_{ik} are both missing. The quantities $\hat{x}_{ij}^{(t)}$ and $\psi_{jk\cdot\text{obs},i}^{(t)}$ are found by sweeping on the current estimate $\Psi^{(t)}$ of Ψ to make $x_{\text{obs},i}$ predictor variables, computations identical to those of the normal EM algorithm (Section 8.2.1). The only modification needed to the latter algorithm is to weight by $w_i^{(t)}$ the sums and sums of squares and cross products passed to the M step.

The weights $w_i^{(t)}$ for the contaminated normal and t models are simple modifications of the weights when data are complete: they are given by Eqs. (10.23) and (10.25), respectively, with (i) K replaced by K_i, the number of observed variables in case i, and (ii) the squared distance (10.24) computed using only the observed variables in case i.

Both the multivariate t and contaminated normal models downweight cases with large squared distances, d_i^2. The distribution of the weights, however, is quite different for the two models, as the following example illustrates.

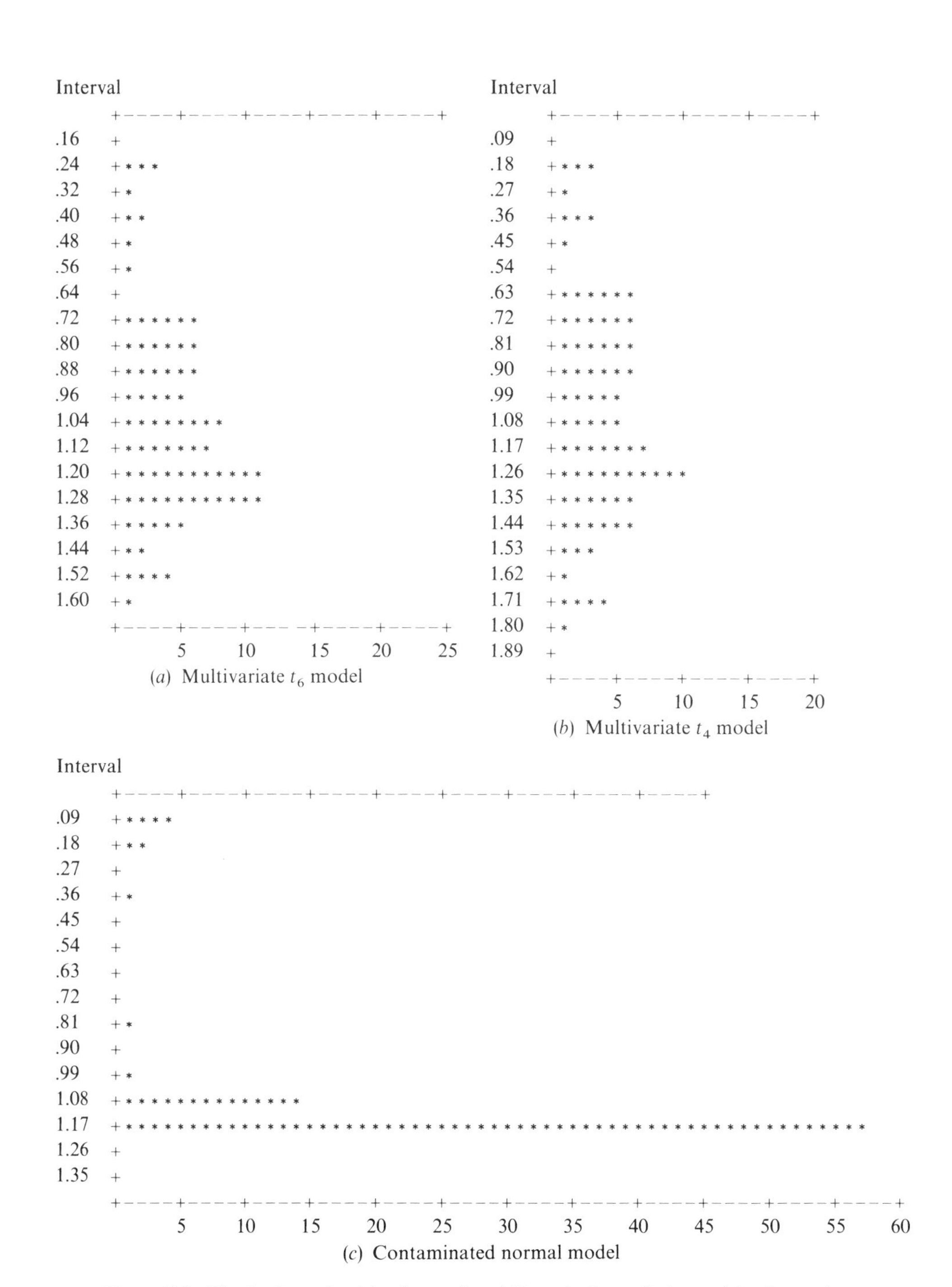

Figure 10.2. Distributions of weights from robust ML methods, applied to multivariate t_4 data. *Weights are scaled to average to one.

EXAMPLE 10.6. *Distribution of Weights from Multivariate t and Contaminated Multivariate Normal Models.* As an illustration, Figure 10.2 shows the distribution of weights for (*a*) the multivariate t with $v = 6.0$, (*b*) the multivariate t with $v = 4.0$, and (*c*) the contaminated normal model with $\pi = 0.1$, $\lambda = 0.077$, for an artificially generated multivariate t_4 data set with $K = 4$ variables, $n = 80$ cases, and 72 of the 320 values randomly deleted. Observe that the t weights are more dispersed for $v = 4$ than for $v = 6$, and the downweighting for the contaminated normal model tends to be concentrated in a few outlying cases.

Little (1986) shows by a simulation study that estimates from these models can produce estimates of means, slopes, and correlations that are protected against outliers when the data are nonnormal, with minor sacrifices of efficiency when the data are, in fact, normal. A graphical procedure for assessing normality is also presented.

10.5.5. Extensions of the Model

A limiting feature of the models described here is that the scaling quantity q_i applied to model longer-than-normal tails is the same for all the variables in the data set. It may be desirable to allow different scaling factors for different variables, reflecting, for example, different degrees of contamination. In particular, for robust regression with missing predictors, it may be more appropriate to confine the scaling quantity to the outcome variable.

Unfortunately, if the models are extended to allow different scaling factors for different variables, the simplicity of the E step of the EM algorithm is lost for general patterns of missing data. Some exceptions worth mentioning are based on the fact that the models can be readily extended to handle a set of fully observed covariates Z, such that y_i has a multivariate normal linear regression on z_i with mean $\sum \beta_j z_{ij}$ and covariance matrix Ψ/q_i, conditional on the unknown scaling quantity q_i defined as before. Thus, suppose the data can be arranged in a *monotone* missing data pattern, with the variables arranged in blocks $X_1, \ldots, X_K$ such that for $j = 1, \ldots, K-1$, X_j is observed for all cases where X_{j+1} is observed. Then the joint distribution of $X_1, \ldots, X_K$ can be expressed as the product of distributions

$$f(X_1, \ldots, X_K, \phi) = f(X_1, \phi) f(X_2 | X_1, \phi) \cdots f(X_K | X_1, \ldots, X_{K-1}, \phi),$$

as discussed in Chapter 6. The conditional distribution $f(X_j | X_1, \ldots, X_{j-1})$ in this factorization can then be modeled as multivariate normal with mean $\sum_{u=1}^{j-1} \beta_u X_u$ and covariance matrix $\Psi_{j \cdot 1 \ldots j-1}/q_{ij}$, where now the scaling factor q_{ij} can be allowed to vary for different values of j. The parameters of each component of the likelihood are estimated by the multivariate regression generalization of the model just considered, and then ML estimates of other parameters of the joint distribution of $X_1, \ldots, X_K$ are found by transformation, as discussed in Chapter 6.

REFERENCES

Aitkin, M., and Rubin, D. B. (1985). Estimation and hypothesis testing in finite mixture models, *J. Roy. Statist. Soc.* **B47**, 67–75.

Aitkin, M., and Wilson, G. T. (1980). Mixture models, outliers, and the EM algorithm, *Technometrics* **22**, 325–331.

Anderson, T. W. (1958). *An Introduction to Multivariate Statistical Analysis*. New York: Wiley.

Brown, D. T. (1959). A note on approximations to discrete probability distributions, *Inf. Control* **2**, 386–392.

Day, N. E. (1969). Estimating the components of a mixture of normal distributions, *Biometrika* **56**, 464–474.

Dempster, A. P., Laird, N. M., and Rubin, D. B. (1977). Maximum likelihood estimation from incomplete data via the EM algorithm (with discussion), *J. Roy. Statist. Soc.* **B39**, 1–38.

Dempster, A. P., Laird, N. M., and Rubin, D. B. (1980). Iteratively reweighted least squares for linear regression when errors are normal/independent distributed. *Multivariate Analysis* **V**, 35–37.

Krzanowski, W. J. (1980). Mixtures of continuous and categorical variables in discriminant analysis, *Biometrics* **36**, 493–499.

Krzanowski, W. J. (1982). Mixtures of continuous and categorical variables in discriminant analysis: a hypothesis-testing approach, *Biometrics* **38**, 991–1002.

Little, R. J. A. (1986). Robust estimation of the mean and covariance matrix from data with missing values, paper given at the Joint Statistical Meetings, Chicago, August 1986.

Little, R. J. A., and Schluchter, M. D. (1985). Maximum likelihood estimation for mixed continuous and categorical data with missing values, *Biometrika* **72**, 497–512.

Olkin, I., and Tate, R. F. (1961). Multivariate correlation models with mixed discrete and continuous variables, *Ann. Math. Statist.* **32**, 448–465.

Pettitt, A. N. (1985). Re-weighted least squares estimation with censored and grouped data: An application of the EM algorithm. *J. Roy. Statist. Soc.* **B47**, 253–261.

Rubin, D. B. (1983). Iteratively reweighted least squares, *Encyclopedia of the Statistical Sciences, Vol* 4. New York; Wiley, pp. 272–275.

PROBLEMS

1. Prove properties (ii) and (iii) that follow the statement of the model in Section 10.2.1.
2. Show that (10.2) provides ML estimates of the parameters for the complete-data loglikelihood (10.1).

3. Using the factored likelihood methods of Chapter 6, derive ML estimates of the general location model for the special case of one fully observed categorical variable Y and one continuous variable X with missing values.

4. Suppose that in Problem 3 X is fully observed and Y has missing values. Show that ML estimates for the general location model cannot be found by factoring the likelihood because the parameters of the appropriate factorization are not distinct. Suggest an alternative model for which the factorization is distinct and display ML estimates for this model.

5. Using Bayes's theorem, show that (10.7) follows from the definition of the general location model.

6. Derive the expressions in Section 10.2.3 for the conditional expectations of $w_{im}x_{ij}$ and $x_{ij}x_{ik}$ given $x_{\text{obs},i}$, S_i, and $\theta^{(t)}$, from properties of the general location model.

7. A survey of 20 graduates of a university class five years after graduation yielded the following results for the variables sex (1 = male, 2 = female), race (1 = white, 2 = other), and annual income, measured on a log scale (– denotes missing):

Case	1	2	3	4	5	6	7	8	9	10	11	12	13	14	15	16	17	18	19	20
Sex	1	1	1	2	2	2	2	2	2	2	2	1	1	2	2	1	1	1	2	2
Race	1	1	1	1	1	1	1	1	1	1	1	2	2	2	2	—	—	—	—	—
Income	25	46	31	5	16	26	8	10	2	—	—	20	29	—	32	—	—	38	15	—

(a) Compute ML estimates for the general location model applied to these data, based on complete cases only.

(b) Develop explicit formulas for the E and M steps (10.3)–(10.6) for these data, and carry out three steps of the EM algorithm, starting from estimates found in (a).

8. Repeat (b) of Problem 7, with the restriction that the variables race and sex are independent.

9. Derive the loglikelihood of the data in Problem 7 for the models of Problems 7 and 8, and hence derive the likelihood ratio chi-squared statistic for testing independence of race and sex. Note that the sample size is too small for this statistic to be considered chi-squared for this illustrative data set. (For help, see Little and Schluchter, 1985.)

10. Derive the weighting functions (10.16) and (10.19) for the models of Example 10.4 and Example 10.5.

11. Derive the weighting functions (10.23) and (10.24) for the models of Section 10.5.3.

12. Derive the E step equations in Section 10.5.4.

CHAPTER 11

Nonignorable Missing-Data Models

11.1. INTRODUCTION

In Section 5.3 we introduced the partition $Y = (Y_{\text{obs}}, Y_{\text{mis}})$ of the complete data Y into observed values, Y_{obs}, and missing values, Y_{mis}, and the missing-data indicator matrix R that identifies the pattern of missing data. We formulated models in terms of a probability distribution for Y with density $f(Y|\theta)$, indexed by unknown vector parameter θ, and a probability distribution $f(R|Y, \psi)$ for R given Y indexed by a vector parameter ψ. The likelihood ignoring the missing-data mechanism was defined to be any function of θ proportional to $f(Y_{\text{obs}}|\theta)$:

$$L(\theta|Y_{\text{obs}}) \propto f(Y_{\text{obs}}|\theta), \tag{11.1}$$

where $f(Y_{\text{obs}}|\theta)$ is obtained by integrating Y_{mis} out of the density $f(Y|\theta) = f(Y_{\text{obs}}, Y_{\text{mis}}|\theta)$. The full likelihood was defined to be any function of θ and ψ proportional to $f(Y_{\text{obs}}, R|\theta, \psi)$:

$$L(\theta, \psi|R, Y_{\text{obs}}) \propto f(Y_{\text{obs}}, R|\theta, \psi), \tag{11.2}$$

where $f(Y_{\text{obs}}, R|\theta, \psi)$ is obtained by integrating Y_{mis} out of the density $f(R|Y_{\text{obs}}, Y_{\text{mis}}, \psi)f(Y_{\text{obs}}, Y_{\text{mis}}|\theta)$. Inference about θ based on (11.1) was shown to be equivalent to ML estimation based on (11.2) when (1) the missing data are MAR, that is, $f(R|Y_{\text{obs}}, Y_{\text{mis}}, \psi) = f(R|Y_{\text{obs}}, \psi)$ for all ψ and Y_{mis} evaluated at the observed values of R and Y_{obs}, and (2) the parameters θ and ψ are distinct, as defined in Section 5.3. Examples in Chapters 6, 7, 8, 9, and 10 have all concerned models with likelihoods of the form (11.1), and hence were based on the assumption that conditions (1) and (2) apply. In this chapter we discuss models where the data are *not* MAR. ML estimation requires a model for the missing-data mechanism and maximization of the full likelihood (11.2).

An important distinction arises here between models where the missing-data mechanism is nonignorable but *known*, in the sense that the distribution of R given $Y = (Y_{obs}, Y_{mis})$ depends on Y_{mis} but does not contain unknown parameters ψ, and models where the missing-data mechanism is nonignorable and unknown, with lack of knowledge reflected in the unknown parameter ψ. A simple example with a known nonignorable mechanism was the censored exponential sample leading to the likelihood (5.15); since values were missing when they fell beyond a *known* censoring point c, the distribution of R given Y was fully determined. Other examples of known nonignorable mechanisms are presented in Section 11.3. ML estimation in these examples can often be achieved by the EM algorithm, which is discussed in Section 11.2 for the general case of known or unknown nonignorable mechanisms.

Sections 11.4–11.6 concern nonignorable models where the missing-data mechanism is nonignorable and ψ is unknown. That is, nonresponse is considered to be related in some partially unknown way to the values of Y, even after adjusting for covariate information X known for respondents and nonrespondents. Most models in the literature of this type concern the case where nonresponse is confined to a single variable. For example, Y are income amounts in a survey of individuals, X represents a set of fully recorded variables, such as age, sex and education, and nonresponse to income questions among individuals with the same value of X is assumed to depend on income amounts, but the exact form of this dependence is unknown.

Two approaches to formulating nonignorable missing-data models can be distinguished. As in the theory of Section 5.3, we may write the joint distribution of R and Y in the form

$$f(R, Y|X, \theta, \psi) = f(Y|X, \theta) f(R|Y, X, \psi), \tag{11.3}$$

where the first component characterizes the distribution of Y given X in the population, and the second component models the incidence of response as a function of X *and* Y. Alternatively, we may write

$$f(R, Y|X, \xi, \omega) = f(Y|X, R, \xi) f(R|X, \omega), \tag{11.4}$$

where the first distribution characterizes the distribution of y_i given x_i in the strata defined by different patterns of response and the second distribution models the incidence of patterns of response as a function of X alone. Observe that when missing-data is confined to a single variable so that R_i takes the values 0 and 1, we usually have no data with which to estimate the distribution $f(Y|X, R = 0, \xi)$ in (11.4), since this distribution relates to nonrespondents. The form (11.4) of the model emphasizes a basic difficulty inherent with the data. To make progress, the distribution $f(Y|X, R = 0, \xi)$ for nonrespondents must be related to the corresponding distribution $f(Y|X, R = 1, \xi)$ for respondents. In Section 11.5 this relationship is achieved by a Bayesian prior relating the parameters of the two distributions.

The formulation (11.3) is used with the models discussed in Sections 11.3 and 11.4. We shall see that in some cases the parameters of the models can be estimated without the explicit inclusion of prior information relating respondents and nonrespondents, unlike models based on (11.4). However, this property is deceptive, since there still is an implicit specification of prior information relating respondents and nonrespondents. Consequently, sensitivity to model specification is an equally serious scientific problem for *both* versions, (11.3) and (11.4), of the model, and in many applications it is prudent to calculate estimates for a variety of missing-data models, rather than to rely exclusively on one model.

11.2. LIKELIHOOD THEORY FOR NONIGNORABLE MODELS

Likelihood theory for θ and ψ based on (11.2) parallels that for θ when the nonresponse mechanism is ignorable, as discussed in Chapters 5, 6, and 7. In particular, ML estimates $(\hat{\theta}, \hat{\psi})$ are obtained by maximizing (11.2), and the inverse of the information matrix obtained by differentiating the loglikelihood twice with respect to (θ, ψ) supplies an estimated covariance matrix for the parameters, provided this matrix exists and samples are large enough to yield a quadratic loglikelihood in the region of $(\hat{\theta}, \hat{\psi})$.

Explicit ML estimates can be derived in special situations, such as Example 5.8. Often, however, iterative techniques are required to maximize the likelihood, as discussed in Section 7.1, for ignorable nonresponse. In particular, the EM algorithm has the following form for nonignorable models: (1) find initial estimates $\theta^{(0)}$, $\psi^{(0)}$ of (θ, ψ); (2) at iteration t, given current estimates $(\theta^{(t)}, \psi^{(t)})$ of (θ, ψ), the E step calculates

$$Q(\theta, \psi | \theta^{(t)}, \psi^{(t)}) = \int l(\theta, \psi | Y_{\text{obs}}, Y_{\text{mis}}, R) f(Y_{\text{mis}} | Y_{\text{obs}}, R, \theta = \theta^{(t)}, \psi = \psi^{(t)})\, dY_{\text{mis}},$$

where $l(\theta, \psi | Y_{\text{obs}}, Y_{\text{mis}}, R)$ is the complete-data loglikelihood and $f(Y_{\text{mis}} | Y_{\text{obs}}, R, \theta, \psi)$ is the density of the conditional distribution of the missing data given the observed values, θ and ψ. The M step finds $\theta^{(t+1)}$, $\psi^{(t+1)}$ to maximize Q:

$$Q(\theta^{(t+1)}, \psi^{(t+1)} | \theta^{(t)}, \psi^{(t)}) \geq Q(\theta, \psi | \theta^{(t)}, \psi^{(t)}) \qquad \text{for all } \theta, \psi.$$

Then $\theta^{(t+1)}$, $\psi^{(t+1)}$ replace $\theta^{(t)}$, $\psi^{(t)}$ in the next iteration of the algorithm. By theory analogous to that in Section 7.3, each iteration of this algorithm increases $L(\theta, \psi | Y_{\text{obs}}, R)$, and under rather general conditions the algorithm converges to a stationary value of the likelihood.

11.3. MODELS WITH KNOWN NONIGNORABLE MISSING-DATA MECHANISMS: GROUPED AND ROUNDED DATA

ML estimation for data where some observations are grouped into categories can be achieved by the EM algorithm, although standard algorithms can be adopted, as discussed in the monograph by Kulldorff (1961). The following three examples illustrate the use of the EM algorithm.

EXAMPLE 11.1. *Grouped Exponential Sample.* Suppose the hypothetical complete data are a random sample $(y_1, \ldots, y_n)$ from the exponential distribution with mean θ. In fact, values of Y are available for $m < n$ observations. The remaining $n - m$ observations are grouped into J categories, such that the jth category contains values of Y lying between a_j and b_j, and r_j observations are included in this category. This formulation includes censored data, where $a_j > 0$ and $b_j = \infty$, as well as situations where $m = 0$ and all the data are in grouped form.

The binary response indicator R of Section 11.1 needs to be expanded to a variable with $J + 1$ outcomes for this example. Specifically, let $R_i = 1$ if y_i is observed, and $R_i = j + 1$ if y_i falls in the jth nonresponse category, that is, lies between a_j and $b_j (j = 1, \ldots, J)$.

The hypothetical complete data belong to the regular exponential family with complete-data sufficient statistics $\sum_{i=1}^{n} y_i$. Hence the E step of the EM algorithm consists in calculating at iteration t

$$E\left(\sum_{i=1}^{n} y_i | Y_{\text{obs}}, R, \theta = \theta^{(t)}\right) = \sum_{i=1}^{m} y_i + \sum_{j=1}^{J} r_j \hat{y}_j^{(t)},$$

where the predicted values are given by

$$\begin{aligned} \hat{y}_j^{(t)} &= E(y | a_j \leqslant y < b_j; \theta^{(t)}) \\ &= \int_{a_j}^{b_j} y \exp\left(-\frac{y}{\theta^{(t)}}\right) dy \Big/ \int_{a_j}^{b_j} \exp\left(-\frac{y}{\theta^{(t)}}\right) dy, \end{aligned}$$

from the definition of the exponential distribution. Integrating by parts gives

$$\hat{y}_j^{(t)} = \theta^{(t)} + \frac{b_j e^{-b_j/\theta^{(t)}} - a_j e^{-a_j/\theta^{(t)}}}{e^{-b_j/\theta^{(t)}} - e^{-a_j/\theta^{(t)}}}.$$

The M step calculates

$$\theta^{(t+1)} = n^{-1}\left(\sum_{i=1}^{m} y_i + \sum_{j=1}^{J} r_j \hat{y}_j^{(t)}\right).$$

The predicted value for an observation censored at a_j is obtained by setting $b_j = \infty$, yielding

$$\hat{y}_j^{(t)} = \theta^{(t)} + a_j.$$

If all the $n - m$ grouped observations are censored, then an explicit ML estimate can be derived. The E and M steps can be combined to give

$$\theta^{(t+1)} = n^{-1}\left(\sum_{i=1}^{m} y_i + \sum_{j=1}^{J} r_j(\theta^{(t)} + a_j)\right).$$

Setting $\theta^{(t)} = \theta^{(t+1)} = \hat{\theta}$ and solving for $\hat{\theta}$ gives the solution

$$\hat{\theta} = m^{-1}\left(\sum_{i=1}^{m} y_i + \sum_{j=1}^{J} r_j a_j\right).$$

In particular, if $a_j = c$ for all j, that is, the observations have a common censoring point, then

$$\hat{\theta} = m^{-1}\left(\sum_{j=1}^{m} y_i + (n - m)c\right),$$

which is the estimate derived directly in Example 5.14.

EXAMPLE 11.2. *Grouped Normal Data with Covariates.* Suppose that data on an outcome variable Y are grouped with the same structure as Example 11.1, but now the hypothetical complete Y values are independent normal with a linear regression on fully observed covariates $X_1, X_2, \ldots, X_p$. That is, the value y_i for unit i is normal with mean $\beta_0 + \sum_{k=1}^{p} \beta_k x_{ik}$ and constant variance σ^2. The complete-data sufficient statistics are $\sum y_i$, $\sum y_i x_{ik}$ $(k = 1, \ldots, p)$ and $\sum y_i^2$. Hence the E step of the EM algorithm computes

$$E\left(\sum_{i=1}^{n} y_i \mid Y_{\text{obs}}, R, \theta = \theta^{(t)}\right) = \sum_{i=1}^{m} y_i + \sum_{i=m+1}^{n} \hat{y}_i^{(t)}$$

$$E\left(\sum_{i=1}^{n} y_i x_{ik} \mid Y_{\text{obs}}, R, \theta = \theta^{(t)}\right) = \sum_{i=1}^{m} y_i x_{ik} + \sum_{i=m+1}^{n} \hat{y}_i^{(t)} x_{ik}, \qquad k = 1, 2, \ldots, p,$$

$$E\left(\sum_{i=1}^{n} y_i^2 \mid Y_{\text{obs}}, R, \theta = \theta^{(t)}\right) = \sum_{i=1}^{m} y_i^2 + \sum_{i=m+1}^{n} \hat{y}_i^{(t)2} + \hat{s}_i^{(t)2},$$

where $\theta = (\beta_0, \beta_1, \ldots, \beta_p, \sigma^2)$, $\theta^{(t)} = (\beta_0^{(t)}, \beta_1^{(t)}, \ldots, \beta_p^{(t)}, \sigma^{(t)2})$ is the current estimate of θ, $\hat{y}_i^{(t)} = \mu_i^{(t)} + \sigma^{(t)}\delta_i^{(t)}$, $\hat{s}_i^{(t)2} = \sigma^{(t)2}(1 - \gamma_i^{(t)})$, $\mu_i^{(t)} = \beta_0^{(t)} + \sum_{k=1}^{p} \beta_k^{(t)} x_{ik}$ and $\delta_i^{(t)}$ and $\gamma_i^{(t)}$ are corrections for the nonignorable nonresponse. The latter take the form

$$\delta_i^{(t)} = -\frac{\phi(d_i^{(t)}) - \phi(c_i^{(t)})}{\Phi(d_i^{(t)}) - \Phi(c_i^{(t)})},$$

$$\gamma_i^{(t)} = \delta_i^{(t)2} + \frac{d_i^{(t)}\phi(d_i^{(t)}) - c_i^{(t)}\phi(c_i^{(t)})}{\Phi(d_i^{(t)}) - \Phi(c_i^{(t)})},$$

where ϕ and Φ are the standard normal density and cumulative distribution functions, and for units i in the jth category ($R_i = j + 1$, or equivalently, $a_j < y_i \leqslant b_j$),

$$c_i^{(t)} = (a_j - \mu_i^{(t)})/\sigma^{(t)} \qquad \text{and} \qquad d_i^{(t)} = (b_j - \mu_i^{(t)})/\sigma^{(t)}.$$

The M step consists in calculating the regression of Y on $X_1, \ldots, X_p$, using the expected values of the complete-data sufficient statistics found in the E step. This model is applied to a regression of log(blood lead) using grouped data in Hasselblad, Stead and Galke (1980).

EXAMPLE 11.3. *Censored Normal Data with Covariates (Tobit Model).* An important special case of the previous example occurs when positive values of Y are fully recorded but negative values are censored, that is, can lie anywhere in the interval $(-\infty, 0)$. In the notation of Example 11.2, all observed y_i are positive, $J = 1$, $a_1 = -\infty$, and $b_1 = 0$. We have for censored cases $c_i^{(t)} = -\infty$, $d_i^{(t)} = -\mu_i^{(t)}/\sigma^{(t)}$, $\delta_i^{(t)} = -\phi(d_i^{(t)})/\Phi(d_i^{(t)})$, and $\gamma_i^{(t)} = \delta_i^{(t)}(\delta_i^{(t)} + \mu_i^{(t)}/\sigma^{(t)})$. Hence

$$\hat{y}_i^{(t)} = E(y_i|\theta^{(t)}, x_i, y_i \leqslant 0) = \mu_i^{(t)} - \sigma^{(t)}\lambda(-\mu_i^{(t)}/\sigma^{(t)}),$$

where $\lambda(z) = \phi(z)/\Phi(z)$ (the inverse of the so-called Mills ratio), and $-\sigma^{(t)}\lambda(-\mu_i^{(t)}/\sigma^{(t)})$ is the correction for censoring. Substituting ML estimates of the parameters yields the predicted values

$$\hat{y}_i = E(y_i|\hat{\theta}, x_i, y_i \leqslant 0) = \hat{\mu}_i - \hat{\sigma}\lambda(-\hat{\mu}_i/\hat{\sigma}) \tag{11.5}$$

for censored cases, where $\hat{\mu}_i = \hat{\beta}_0 + \sum \hat{\beta}_k x_{ik}$. This model is sometimes called the Tobit model in the econometric literature (see Amemiya, 1984), after an earlier econometric application (Tobin, 1958).

11.4. STOCHASTIC CENSORING MODELS

11.4.1. Maximum Likelihood Estimation for Stochastic Censoring Models

An interesting extension of the model of Example 11.3 concerns an incompletely observed variable (Y_1) that again has a linear regression on covariates and is observed if and only if the value of another completely unobserved variable

(Y_2) exceeds a threshold (say zero). A common specification is given in the following example:

EXAMPLE 11.4. *Bivariate Normal Stochastic Censoring Model.* Suppose Y_1 is incompletely observed, Y_2 is never observed, p covariates X are fully observed, and for unit i, $f(Y|X, \theta)$ is specified by

$$\begin{pmatrix} y_{i1} \\ y_{i2} \end{pmatrix} \overset{\text{indep}}{\sim} N_2\left[\begin{pmatrix} x_i\beta_1 \\ x_i\beta_2 \end{pmatrix}, \begin{pmatrix} \sigma_1^2 & \rho\sigma_1 \\ \rho\sigma_1 & 1 \end{pmatrix}\right], \tag{11.6}$$

where $x_i = (x_{i0}, x_{i1}, \ldots, x_{ip})$ are the constant term $x_{i0} \equiv 1$, and predictors $x_{i1}, \ldots, x_{ip}$ for case i, β_1 and β_2 are $(p+1) \times 1$ vectors of regression coefficients, some of which may be set to zero a priori, and $N_2(a, b)$ denotes the bivariate normal distribution with mean a, covariance matrix b. Further, $f(R|X, Y, \psi)$ is specified by the degenerate distribution

$$\begin{aligned} R_{i1} &= \begin{cases} 1, & \text{if } y_{i2} > 0, \\ 0, & \text{if } y_{i2} \leqslant 0, \end{cases} \\ R_{i2} &\equiv 0, \end{aligned} \tag{11.7}$$

where R_{ij} is a response indicator for y_{ij} with $R_{ij} = 1$ when y_{ij} is observed and $R_{ij} = 0$ when y_{ij} is missing.

Another version of the factorization (11.3) is relevant for this model, where Y_2 is integrated out of the model and R refers only to missingness of Y_1. Let R denote the set of values $\{R_{i1}\}$, $\theta = (\beta_1, \sigma_1^2)$, and $\psi = (\beta_1, \beta_2, \rho)$. From (11.6) and (11.7), the distribution of R given X, Y_1 is Bernoulli with probability of response for unit i

$$\begin{aligned} \Pr(R_i = 1|y_{i1}, x_i) &= \Pr(y_{i2} > 0|y_{i1}, x_i) \\ &= \Phi\left[\frac{x_i\beta_2 + \rho\sigma_1^{-1}(y_{i1} - x_i\beta_1)}{\sqrt{1-\rho^2}}\right]. \end{aligned} \tag{11.8}$$

When $\rho \neq 0$, this probability is a monotonic function of the values of Y_1, which are sometimes missing. Hence by the theory of Chapter 7, the missing-data mechanism is nonignorable. If, on the other hand, $\rho = 0$ and β_2 and (β_1, σ^2) are distinct, then the missing-data mechanism is ignorable, and ML estimates of (β_1, σ^2) are obtained by least squares linear regression based on the complete cases.

This model was introduced by Heckman (1976) to describe selection of women into the labor force. Amemiya (1984) calls it a Type II Tobit model; note that the Tobit model of Example 11.3 is obtained when $Y_1 \equiv \sigma_1 Y_2$.

Two estimation procedures have been proposed for this model, ML and the two-step method of Heckman (1976). ML estimation was originally achieved using the algorithm of Berndt et al. (1974). We describe the EM algorithm for the case where no constraints are placed on the coefficients β_1, β_2, with hypothetical complete data defined as cases with both Y_1 and Y_2 completely observed. The complete-data sufficient statistics are then $\{\sum_i y_{i1}x_{ij}, \sum_i y_{i2}x_{ij}, \sum_i y_{i1}y_{i2}, \sum_i y_{i1}^2\}$ for $j = 0, 1, \ldots, p$. Since $\{x_{ij}\}$ are fully observed, the E step consists of replacing missing values of y_{i1}, y_{i2}, $y_{i1}y_{i2}$ and y_{i1}^2 by their expectations given the parameters and the observed data. Properties of the bivariate normal distribution yield

$$\begin{aligned}
E(y_{i2}|y_{i2} \leqslant 0) &= \mu_{i2} - \lambda(-\mu_{i2}),\\
E(y_{i2}|y_{i2} > 0) &= \mu_{i2} + \lambda(\mu_{i2}),\\
E(y_{i1}|y_{i2} \leqslant 0) &= \mu_{i1} - \rho\sigma_1\lambda(-\mu_{i2}),\\
E(y_{i2}^2|y_{i2} \leqslant 0) &= 1 + \mu_{i2}^2 - \mu_{i2}\lambda(-\mu_{i2}),\\
E(y_{i2}^2|y_{i2} > 0) &= 1 + \mu_{i2}^2 + \mu_{i2}\lambda(\mu_{i2}),\\
E(y_{i1}^2|y_{i2} \leqslant 0) &= \mu_{i1}^2 + \sigma_1^2 - \rho\sigma_1\lambda(-\mu_{i2})(2\mu_{i1} - \rho\sigma_1\mu_{i2}),\\
E(y_{i1}y_{i2}|y_{i2} \leqslant 0) &= \mu_{i1}(\mu_{i2} - \lambda(-\mu_{i2})) + \rho\sigma_1.
\end{aligned}$$

In these expressions, (a) $\lambda(\cdot)$ is defined as in Example 11.3, (b) conditioning on x_i and the parameters is implicit, (c) $\mu_{i1} = x_i\beta_1$, $\mu_{i2} = x_i\beta_2$, and (d) $y_{i2} \leqslant 0$ applies to cases with Y_1 and Y_2 missing, $y_{i2} > 0$ to cases with Y_1 observed, Y_2 missing. Current values of the parameters are substituted to yield estimates for the E step.

The M step consists of the following substeps performed with complete-data sufficient statistics replaced by estimates from the E step:

1. Regress Y_2 on X, yielding coefficients $\hat{\beta}_2$ of the response equation.
2. Regress Y_1 on Y_2 and X, yielding coefficients $\hat{\delta}$ for Y_2 and $\hat{\beta}_1^*$ for X, and residual variance $\hat{\sigma}_{1\cdot 2}^2$.
3. Set $\hat{\beta}_1 = \hat{\beta}_1^* + \hat{\delta}\hat{\beta}_2$, $\hat{\sigma}_1^2 = \hat{\sigma}_{1\cdot 2}^2 + \hat{\delta}^2$, and $\hat{\rho} = \hat{\delta}/\hat{\sigma}_1$.

When a priori constraints are imposed on the coefficients β_1 and β_2, the M step involves iterative calculations, so the simplicity of the algorithm is lost.

11.4.2. Sensitivity of ML Estimation to Normality

The model of Example 11.4 implies the following predicted values for missing cases:

$$E(y_{i1}|y_{i2} \leqslant 0, x_i, \hat{\theta}) = \hat{\mu}_{i1} - \hat{\rho}\hat{\sigma}_1\lambda(-\hat{\mu}_{i2}), \tag{11.9}$$

where $\hat{\mu}_{i1} = x_i\hat{\beta}_1$, $\hat{\mu}_{i2} = x_i\hat{\beta}_2$. Note that the censoring correction $-\hat{\rho}\hat{\sigma}_1\lambda(-\hat{\mu}_{i2})$

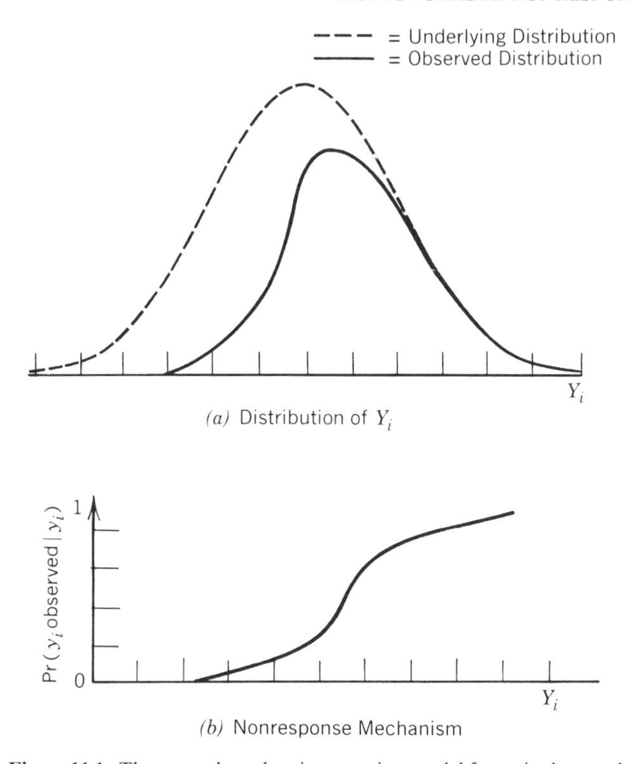

Figure 11.1. The normal stochastic censoring model for a single sample.

depends on the estimated correlation $\hat{\rho}$, a quantity that does not arise in the predictions (11.5) for the pure censoring model of Example 11.3. Thus, even though Y_1 and Y_2 are never jointly observed, their correlation ρ must be estimated.

Two assumptions of the model yield information on ρ that is exploited in ML estimation: (1) any a priori restriction placed on the coefficients β_1 and β_2 and (2) the assumption of normality of y_{i1} given x_i in the unrestricted population. To see the role of the latter assumption, consider the model in the absence of covariates, where $x_i \equiv 1$, the constant term. Figure 11.1*a* displays the model distributions of Y_1 in the unrestricted and the respondent populations, with the latter scaled so that the area underneath it is the fraction of respondents. By assumption, the unrestricted distribution of Y_1 is normal. The respondent distribution is skewed by the stochastic censoring, unless $\rho = 0$. Given a sample of respondents, ρ is estimated to account for the lack of normality in the sample. In other words, it fills in the nonrespondents to make the unrestricted sample look as normal as possible. Clearly, this procedure relies totally on the untestable assumption that the values of Y_1 in the unrestricted population *are* normal. In the absence of knowledge about this distribution, an equally plausible assumption might be that nonrespondents have the same skewed distribution as that observed for respondents in Figure 11.1. If this hypothesis is in fact true, the correction for selectivity in the normal model would add bias rather than eliminate it. The following example illustrates this possibility.

EXAMPLE 11.5. *Income Nonresponse in the Current Population Survey.* Lillard, Smith, and Welch (1982, 1986) apply the model of Example 11.4 to income nonresponse in four rounds of the Current Population Survey Income Supplement, conducted in 1970, 1975, 1976, and 1980. In 1980 their sample consisted of 32, 879 employed white civilian males aged 16–65 who reported receipt (but not necessarily amount) of W = wages and salary earnings and who were not self-employed. Of these individuals, 27,909 reported the value of W and 4970 did not. In the notation of Example 11.4, Y_1 is defined to equal $(W^{\gamma-1})/\gamma$, where γ is a power transformation of the kind proposed in Box and Cox (1964). The predictors X were chosen as

- Constant term
- Education (5 dummy variables; Sch 8, Sch 9–11, Sch 12, Sch 13–15, Sch 16+)
- Years of market experience (4 linear splines, Exp 0–5, Exp 5–10, Exp 10–20, Exp 20+)
- Probability of being in first year of market experience (Prob 1)
- Region (South or other),
- Child of household head (1 = yes, 0 = no)
- Other relative of household head or member of secondary family (yes, no)
- Personal interview (1 = yes, 0 = no)
- Year in survey (1 or 2)

The last four variables were omitted from the earnings equation; that is, their coefficients in the vector β_1 were set equal to zero. The variables education, years of market experience, and region were omitted from the response equation; that is, their coefficients in the vector β_2 were set to zero.

Most empirical studies model the logarithm of earnings, the transformation obtained by letting $\gamma \to 0$. Table 11.1 shows estimates of the regression coefficients β_1 for this transformation of earnings, calculated (1) by ordinary least squares (OLS) on the respondents, a procedure that effectively assumes ignorable nonresponse ($\rho = 0$); and (2) by ML for the model of Example 11.4. For the ML procedure $\hat{\rho} = -0.6812$, implying positive

Table 11.1 Estimates for the Regression of ln(Earnings) on Covariates: 1980 CPS

		Coefficient (β_1)	
Variable	OLS on Respondents	ML for Selection Model	Two-Step Method
Constant	9.5013 (.0039)	9.6816 (.0051)	10.0373 (.0173)
Sch 8	.2954 (.0245)	.2661 (.0202)	.2615 (.0241)
Sch 9–11	.3870 (.0206)	.3692 (.0169)	.3718 (.0203)
Sch 12	.6881 (.0188)	.6516 (.0158)	.6713 (.0185)
Sch 13–15	.7986 (.0201)	.7694 (.0176)	.8096 (.0198)
Sch 16+	1.0519 (.0199)	1.0445 (.0178)	1.0418 (.0195)
Exp 0–5	− .0225 (.0119)	− .0294 (.0111)	− .0425 (.0117)
Exp 5–10	.0534 (.0038)	.0557 (.0039)	.0561 (.0037)
Exp 10–20	.0024 (.0016)	.0240 (.0016)	.0448 (.0017)
Exp 20+	− .0052 (.0008)	− .0036 (.0008)	− .0033 (.0008)
Prob 1	−1.8136 (.1075)	−1.7301 (.0945)	−1.5311 (.1059)
South	− .0654 (.0087)	− .0649 (.0085)	− .0893 (.0086)
$\rho = \text{Corr}(y_1, y_2 \mid x)$	0	− .6842	
N	27909	32879	32879

Table 11.2 The Maximized Loglikelihood as a Function of γ, with Associated Values of $\hat{\rho}$

γ	Maximized Loglikelihood	$\hat{\rho}$
0	−300,613.4	−0.6812
0.45	−298,169.7	−0.6524
1.0	−300,563.1	0.8569

Source: Lillard, Smith, and Welch (1982).

corrections $-\hat{\rho}\hat{\sigma}_1\lambda(-\hat{\mu}_{2i})$ for nonignorability to the income amounts of nonrespondents (Eq. 11.9). The regression coefficients from OLS and ML in Table 11.1 are quite similar, but the difference in intercepts (9.68–9.50 = 0.18 on the log scale) between ML and OLS implies a roughly 20% increase in the predicted income amounts for nonignorable nonresponse, a substantial correction.

Lillard, Smith, and Welch (1982) fit the stochastic censoring model for a variety of other choices of γ. Table 11.2 shows the maximized loglikelihood for three values of γ, namely, 0 (the log model), 1 (the model for raw income amounts), and 0.45, the ML estimate for a random subsample of the data. The maximized loglikelihood for $\hat{\gamma} = 0.45$ is much larger than that for

$\gamma = 0$ or $\gamma = 1$, indicating that the normal selection models for raw and log-transformed earnings are not supported by the data.

Table 11.2 also shows values of $\hat{\rho}$ as a function of γ. Note that for $\gamma = 0$ and $\hat{\gamma} = 0.45$, $\hat{\rho}$ is negative, the distribution of respondent income residuals is left-skewed, and large income amounts for nonrespondents are needed to fill the right tail. On the other hand, when $\gamma = 1$, $\hat{\rho}$ is negative, the distribution of respondent residuals is right-skewed, and small income amounts for nonrespondents are needed to fill the left tail. Thus the table reflects sensitivity of the correction to skewness in the transformed-income respondent residuals.

Lillard, Smith, and Welch's best-fitting model, $\hat{\gamma} = 0.45$, predicts large income amounts for nonrespondents, in fact, 73% larger on average than imputations supplied by the Census Bureau, which uses a hot deck method that assumes ignorable nonresponse. However, as Rubin (1983) notes, this large adjustment is founded on the normal assumption for the population residuals from the $\gamma = 0.45$ model. It is quite plausible that nonresponse is ignorable and the unrestricted residuals follow the same (skewed) distribution as that in the respondent sample. Indeed, comparisons of Census Bureau imputations with IRS income amounts from matched CPS/IRS files do *not* indicate substantial underestimation (David et al., 1986).

11.4.3. Heckman's Two-Step Fitting Method

The original fitting procedure proposed by Heckman (1976) is not ML, and in fact is much easier to compute than ML, a characteristic that has led to its widespread application. It relies heavily on untestable assumptions about a priori zeros in the regression coefficient vectors β_1 and β_2, and may yield very misleading results if these assumptions are incorrect. Hence its application requires care.

In describing the method, it is convenient to use a notation that distinguishes the predictors of Y_1 from those of Y_2. Let X_2 denote the subset of covariates X assumed to predict Y_2 (and hence response), and let β_2^* denote the corresponding subvector of nonzero values of β_2. Similarly, let X_1 denote the covariates assumed to predict Y_1, and let β_1^* be the corresponding subvector of nonzero values of β_1. Then the model, (11.6) and (11.7), implies that

$$\Pr(R_{i1} = 1 | x_i) = \Phi(x_i \beta_2) \equiv \Phi(x_{i2} \beta_2^*) \tag{11.10}$$

and

$$\begin{aligned} E(y_{i1} | R_{i1} = 1, x_i) &= x_i \beta_1 + \rho \sigma_1 \lambda(x_i \beta_2) \\ &\equiv x_{i1} \beta_1^* + \rho \sigma_1 \lambda(x_{i2} \beta_2^*), \end{aligned} \tag{11.11}$$

where x_{i1} and x_{i2} denote values of X_1 and X_2 for unit i. Hence consistent estimates of β_1^* and β_2^* are obtained by (1) estimating β_2^* by $\tilde{\beta}_2^*$, from probit regression of R_{i1} on x_{i2} using the whole sample; and (2) estimating β_1^* and $\rho\sigma_1$ by OLS regression of y_{i1} on x_{i1} and $\tilde{\lambda}_i = \lambda(x_{i2}\tilde{\beta}_2^*)$, using the respondent sample. This is Heckman's two-step method. Slightly more efficient variants of the method can be formulated that replace OLS by generalized weighted least squares at the second step.

The two-step method does not work in the absence of covariates that predict response, since if x_{i2} is constant for all i then $\lambda(x_{i2}\beta_2^*)$ is also constant, and hence is confounded with the constant term in (11.11). If X_2 varies across units, but $X_1 = X$ and X_2 is a subset of the variables X_1, then the parameters β_1^* and $\rho\sigma_1$ in (11.1) are identified only through the nonlinearity of the transformation λ. For the method to work in practice, variables are needed in X_2 that are good predictors of response and do not appear in X_1, that is, are not associated with Y_1 when other covariates are controlled. For example, in Example 11.5 the variable *child of household head* is assumed to affect response but not to be associated with income. However, it seems hard to believe this lack of association with much conviction, particularly since *child of household head* could be acting as a proxy for income predictors not included in X_1. If this variable is associated with income, then the correction for nonignorable nonresponse from the two-step method may be quite spurious.

The last column of Table 11.1 shows the results of fitting the two-step method to the data of Example 11.5. The most striking difference between this method and ML is the larger intercept (10.0373 instead of 9.0816), which implies imputed amounts that are nearly 60% higher than those from OLS—not a very plausible correction. Lillard, Smith, and Welch (1982) also found considerable instability in the two-step estimates for different CPS samples. Little (1985a) attempts to explain this instability and provides further discussion of the assumptions of the method.

Variants of the normal stochastic censoring model that replace the probit model for response by uniform (Olsen, 1980) and logistic models (Olsen, 1980; Greenlees, Reece, and Zieschang, 1982) have also been considered, the last of these references in the context of CPS income nonresponse. Estimates for these models are also necessarily highly sensitive to the choice of a priori zeros in β_1 and β_2. (See Problem 11.6).

11.5. A PREDICTIVE BAYESIAN APPROACH TO NONRESPONSE BIAS

In this approach to nonresponse bias, proposed by Rubin (1977), we model the distribution of nonrespondents and respondents with separate parameters as in (11.4), and relate the parameters by a suitable Bayesian prior distribution. Nonignorable nonresponse is indexed by parameters that are fixed a priori, and the remaining distributional parameters are estimated by their Bayes posterior distributions. The influence of nonresponse is assessed by a probability interval based on the predictive distribution of the hypothetical complete-data statistic given the responding values. The basic idea is contained in the following simple example. The more general case given by Rubin is presented in Example 11.7.

EXAMPLE 11.6. *Sensitivity of the Sample Mean to Nonignorable Nonresponse.* In a simple random sample of n units on a variable Y, a proportion $p = (n - m)/n$ are nonrespondents. The summary statistic of interest is the average value of Y in the sample, which can be expressed as

$$\bar{y} = (1 - p)\bar{y}_{\text{R}} + p\bar{y}_{\text{NR}},$$

where $\bar{y}_{\text{R}}$ is the observed mean of respondents and $\bar{y}_{\text{NR}}$ is the unobserved mean for nonrespondents. Suppose that values of Y for respondents are normally distributed with mean α_{R} and variance σ_{R}^2, and for nonrespondents are normally distributed with mean α_{NR} and variance σ_{NR}^2. For simplicity we assume initially that the variances are equal, that is, $\sigma_{\text{R}}^2 = \sigma_{\text{NR}}^2 = \sigma^2$, and σ^2 is known. Also the respondent and nonrespondent values are independent, given α_{R}, α_{NR}, and σ^2. Hence

$$(\bar{y}_{\text{R}} | \alpha_{\text{R}}, \sigma^2) \sim N(\alpha_{\text{R}}, \sigma^2/(n - np)), \qquad (\bar{y}_{\text{NR}} | \alpha_{\text{NR}}, \sigma^2) \sim N(\alpha_{\text{NR}}, \sigma^2/(np)),$$

where $N(\alpha, \sigma^2)$ is the normal distribution with mean α, variance σ^2. Subjective prior notions about the similarity of respondents and nonrespondents are formalized by specifying a prior distribution for α_{NR} given α_{R} and σ^2. For the application discussed by Rubin, the normal prior

$$(\alpha_{\text{NR}} | \alpha_{\text{R}}, \sigma^2) \sim N(\alpha_{\text{R}}, \psi_2^2 \alpha_{\text{R}}^2),$$

for a selected value of ψ_2, was considered appropriate. The mean of this distribution, α_{R}, implies that it is as likely for the nonrespondent mean to be above or below the respondent mean. The quantity ψ_2 measures the subjective coefficient of variation for the nonrespondents' mean about the respondent mean. In particular, it implies that the investigator is 95% sure that the nonrespondent mean will fall in the interval

$$\alpha_{\text{R}}(1 \pm 1.96\psi_2).$$

If $\psi_2 = 0$, then the distributions of Y for respondents and nonrespondents are equal, and the missing data mechanism is ignorable.

The impact of nonresponse is assessed by the predictive distribution of $\bar{y}$ given $\bar{y}_{\text{R}}$ and ψ_2^2, which can be readily shown by applying Bayes's Theorem to be normal with mean $\bar{y}_{\text{R}}$ and variance

$$\begin{aligned}\text{Var}(\bar{y} | \bar{y}_{\text{R}}, \sigma^2) = p^2 \,\text{Var}(\bar{y}_{\text{NR}} | \bar{y}_{\text{R}}, \sigma^2) &= p^2 \,\text{Var}(\alpha_{\text{R}} | \bar{y}_{\text{R}}, \sigma^2) \\ &+ p^2 E\{\text{Var}(\bar{y}_{\text{NR}} | \alpha_{\text{NR}}) | \bar{y}_{\text{R}}, \sigma^2\} + p^2 E\{\text{Var}(\alpha_{\text{NR}} | \alpha_{\text{R}}) | \bar{y}_{\text{R}}, \sigma^2\}.\end{aligned}$$

The last term in this equation is independent of the sample size and reflects the uncertainty in the nonignorable component of the model. Assuming a flat prior for α_R and substituting for the prior variance of α_{NR} given α_R and σ^2, we obtain

$$\begin{aligned}\operatorname{Var}(\bar{y}|\bar{y}_R,\sigma^2) &= p^2\{\sigma^2/(np) + \sigma^2/(n-np) + \psi_2^2(\bar{y}_R^2 + \sigma^2/(n-np))\}\\ &= p^2\{\sigma^2/(np-np^2) + \psi_2^2(\bar{y}_R^2 + \sigma^2/(n-np))\}.\end{aligned}$$

Hence a Bayesian 95% probability interval for $\bar{y}$ given $\bar{y}_R$ and σ^2 is

$$\begin{aligned}\bar{y}_R \pm 1.96\sqrt{\operatorname{Var}(\bar{y}|\bar{y}_R,\sigma^2)} = \bar{y}_R &\pm 1.96p\{\sigma^2/(np-np^2)\\ &+ \psi_2^2(\bar{y}_R^2 + \sigma^2/(n-np))\}^{1/2}.\end{aligned}$$

This equation corresponds to Eq. (2.1) of Rubin's paper with $\bar{X}_{NR} = \bar{X}_R$, $\theta_1 = 0$.

Note that for large n the width of this interval is approximately $4p\psi_2\bar{y}_R$, and represents the uncertainty introduced by the nonignorable component of the model. Expressed as a proportion of the mean, it is obtained by multiplying the prior parameter ψ_2 by four times the proportion of missing values.

EXAMPLE 11.7. *Sensitivity of the Sample Mean to Nonignorable Nonresponse, in the Presence of Covariates.* A significant development of the previous example is to include q covariates $X = (X_1, \ldots, X_q)$, available for all units in the sample. If Y is linearly related to X in both respondent and nonrespondent populations, we have for the expected values

$$E(\bar{y}_R|\alpha_R,\beta_R,\sigma^2,X) = \alpha_R + \bar{X}_R\beta_R,\quad \operatorname{Var}(\bar{y}_R|\alpha_R,\beta_R,\sigma^2,X) = \sigma^2/(n-np)$$

$$E(\bar{y}_{NR}|\alpha_{NR},\beta_{NR},\sigma^2,X) = \alpha_{NR} + \bar{X}_{NR}\beta_{NR},\quad \operatorname{Var}(\bar{y}_{NR}|\alpha_{NR},\beta_{NR},\sigma^2,X) = \sigma^2/(np),$$

where $\bar{X}_R$ and $\bar{X}_{NR}$ are the observed respondent and nonrespondent means of X in the sample, and α_R, α_{NR}, β_R, β_{NR}, and σ^2 are unknown parameters. To relate the nonrespondent parameters to those of the respondents we assume that conditional on the respondent parameters, the nonrespondent parameters have priors

$$\begin{aligned}\beta_{NR} &\sim N_q(\beta_R, \psi_1^2\beta_R\beta_R^T)\\ \phi_{NR} &\sim N(\phi_R, \psi_2^2\phi_R^2),\end{aligned}$$

where $N_q(a, B)$ is the q-variate normal distribution with mean a and covariance matrix B, and $\phi_{NR} = \alpha_{NR} + \bar{X}_R\beta_{NR}$, $\phi_R = \alpha_R + \bar{X}_R\beta_R$ are parameters representing the Y means at $X = \bar{X}_R$ in the nonrespondent and respondent populations. The parameter ψ_1 measures a priori uncertainty about the regression coefficients. Writing $\beta_R^{(j)}$ and $\beta_{NR}^{(j)}$ for the jth components of β_R and β_{NR}, respectively, the prior implies that the investigator is 95% sure that $\beta_{NR}^{(j)}$

will fall in the interval

$$\beta_R^{(j)}(1 \pm 1.96\psi_1),$$

for all j. The parameter ψ_2 measures uncertainty in the adjusted mean and corresponds to ψ_2 in Example 11.6 for the case of no covariates. The nonresponse mechanism is ignorable for likelihood-based inferences if $\psi_1 = \psi_2 = 0$.

Assuming a flat prior distribution for ϕ_R and β_R, a Bayesian 95% probability interval for $\bar{y}$ given the data takes the form

$$\bar{y}_R[1 + h_0 \pm 1.96(\psi_1^2 h_1^2 + \psi_2^2 h_2^2 + h_3^2)^{1/2}]$$

where

$$h_0 = p(\bar{X}_{NR} - \bar{X}_R)b_R/\bar{y}_R$$

$$h_1^2 = h_0^2 + (p^2 s_R^2/\bar{y}_R^2)(\bar{X}_{NR} - \bar{X}_R)S_{xx}^{-1}(\bar{X}_{NR} - \bar{X}_R)^T$$

$$h_2^2 = p^2\{1 + s_R^2/[\bar{y}_R^2 n(1 - p)]\}$$

and

$$h_3^2 = p^2(s_R^2/\bar{y}_R^2)[1/(np - np^2) + (\bar{X}_{NR} - \bar{X}_R)S_{xx}^{-1}(\bar{X}_{NR} - \bar{X}_R)]^T.$$

Here b_R and s_R^2 are the slopes and residual variance from the least squares regression of Y_R on X_R, and S_{xx} is the sum-of-squares and cross-products matrix of X for respondents. In particular, the predicted mean is

$$\bar{y}_R(1 + h_0) = \bar{y}_R + p(\bar{X}_{NR} - \bar{X}_R)b_R;$$

the last term represents the covariance adjustment for differences in the means of X in the nonrespondent and respondent samples. The width of the interval

$$3.92\bar{y}_R(\psi_1^2 h_1^2 + \psi_2^2 h_2^2 + h_3^2)^{1/2}$$

involves three components. The first, $\psi_1^2 h_1^2$, is the relative variance due to uncertainty about the equality of the slopes of Y on X in the respondent and nonrespondent groups. The term $\psi_2^2 h_2^2$ reflects uncertainty about the equality of the Y means for respondents and nonrespondents at $X = \bar{X}_R$. The term h_3^2 represents the uncertainty introduced by nonresponse that is present even when the respondent and nonrespondent distributions are equal, that is, when $\psi_1 = \psi_2 = 0$ so that the nonresponse mechanism is ignorable.

It is instructive to examine these expressions when the responding sample size tends to infinity. The component h_3^2 tends to zero, and the width of the interval becomes approximately $4p\bar{y}_R\{\psi_2^2 + \psi_1^2[(\bar{X}_{NR} - \bar{X}_R)b_R/\bar{y}_R]^2\}^{1/2}$, compared with $4p\bar{y}_R\psi_2$ in the previous example. The fact that the interval is apparently larger when covariates are introduced seems contradictory, since the covariance adjustment should reduce uncertainty in the prediction of $\bar{y}$. However, the subjective parameter ψ_2 is *not* the same in the two examples.

The covariate adjustment should reduce the differences between the conditional means in the nonrespondent and respondent population at $\bar{X}_R$, and this should be reflected in the prior distributions by a smaller value of ψ_2 in this example than in Example 11.6. An unfortunate (but realistic) characteristic of the model is that it does not assess the improvement in prediction achieved by the introduction of covariates, since this depends on the relative sizes of ψ_2 in Example 11.6 and ψ_1 and ψ_2 in this example, quantities that are assessed without looking at the data.

EXAMPLE 11.8. *An Application of Example 11.7.* Rubin illustrates the method with data from a survey of 660 schools, 472 of which filled out a compensatory reading questionnaire consisting of 80 items. Twenty-one dependent variables (Y's) and 35 background variables (X's) were selected.

The dependent variables in the study measured characteristics of compensatory reading in the form of frequency with which they were present, and were scaled to lie between zero (never) and one (always). The restricted scale complicates interpretations for variables with means at the extreme ends of the scale; consequently, we focus attention on seven variables in the middle of the range. The dependent variables chosen here are as follows:

- 17B: Compensatory reading carried out during school hours released from other classwork
- 18A: Compensatory reading carried out during time released from social studies, science, and/or foreign language
- 18B: Compensatory reading carried out during time released from mathematics
- 23A: Frequency of organizing compensatory reading class into groups by reading grade level
- 23C: Frequency of organizing compensatory reading class into groups by shared interests
- 32A: Compensatory reading teaches textbooks other than basal readers
- 32D: Compensatory reading teaches teacher-prepared materials

The background variables X in the study described the school and the socioeconomic status and achievement of the students.

Table 11.3 presents summary statistics $\bar{y}_R$, R^2 (the squared multiple correlation between Y and X for respondents), h_0, h_1, h_2, and h_3. The values of h_0 represent the proportional adjustments to the means based on the regression on X. These are generally small, although for variable 18B the mean is increased by 6%, suggesting that nonrespondents carry out compensatory reading during time released from mathematics more often than respondents.

Table 11.3 Summary Statistics for Seven Dependent Variables from School Survey

Variable[a]	$\bar{y}_R$	R^2	h_0	h_1	h_2	h_3
17B	0.39322	0.54763	0.02096	0.02647	0.28549	0.02786
18A	0.23132	0.54378	0.00080	0.02300	0.28615	0.03961
18B	0.09067	0.53405	0.06331	0.07749	0.28974	0.07701
23A	0.65796	0.54942	0.00814	0.01020	0.28494	0.01060
23C	0.45240	0.59361	0.00368	0.00679	0.28492	0.00983
32A	0.33968	0.66787	0.00394	0.00394	0.28487	0.00596
32D	0.41489	0.57689	0.00130	0.00334	0.28487	0.00531

[a] For a description of the variables and the statistics, see text.

The values of h_1, h_2, and h_3 determine the contributions $\psi_1^2 h_1^2$, $\psi_2^2 h_2^2$, and h_3^2 to the squared width of the probability interval. The width of the 95% interval expressed as a percentage of the mean is tabulated in Table 11.4 as a function of ψ_1, and ψ_2. The low values of h_1 imply that uncertainty about equality of the slopes of the regressions for respondents and nonrespondents, modeled by the quantity ψ_1, has a negligible impact on the interval. The values of h_2 are only marginally greater than the proportion of missing values, $p = 0.2848$. Thus the contribution to the interval width of uncertainty about equality of the adjusted means in the respondent and nonrespondent populations is represented by $4h_2\psi_2 \simeq 4p\psi_2 = 1.14\psi_2$.

The quantity ψ_2 has a major impact on the interval widths. For example, the effect of increasing the value of ψ_2 from 0 to 0.1 is to treble the interval widths in variables 23A and 23C and to increase the interval widths in variables 32A and 32D by a factor of 5. On the other hand, for variables 17B, 18A, and in particular 18B, the component attributable to residual variance from the regression, h_3, is more pronounced, although the other component is still nonnegligible for $\psi_2 \geqslant 0.1$. The example illustrates dramatically the potential impact of nonresponse bias, and the extent to which it is dependent on quantities (such as ψ_2) that generally cannot be reliably estimated from the data. The only satisfactory resolution of the problem of nonignorable non-response is to obtain follow-up information for nonrespondents, as discussed in Section 12.6 below.

11.6. NONIGNORABLE MODELS FOR CATEGORICAL DATA

Two types of nonignorable models for incomplete categorical data have been considered. Pregibon (1977), Little (1982), and Nordheim (1984) introduce prior odds of response for categories of the table that modify the likelihood. Hierarchical loglinear models for the joint distribution of the categorical

Table 11.4 Widths of Subjective 95 Percent Intervals of $\bar{y}$, as Percentages of $\bar{y}_R$

	$\psi_1 = 0$				$\psi_1 = 0.4$			
Variable	$\psi_2 = 0$	$\psi_2 = 0.1$	$\psi_2 = 0.2$	$\psi_2 = .4$	$\psi_2 = 0$	$\psi_2 = 0.1$	$\psi_2 = 0.2$	$\psi_2 = .4$
17B	5.6	8.0	12.7	23.7	6.0	8.3	12.9	23.6
18A	7.9	9.8	13.9	24.2	8.1	9.9	14.0	24.3
18B	15.4	16.5	19.3	27.8	16.6	17.6	20.2	28.5
23A	2.1	6.1	11.6	22.9	2.3	6.1	11.6	22.9
23C	2.0	6.0	11.6	22.9	2.0	6.1	11.6	22.9
32A	1.2	5.8	11.5	22.8	1.2	5.8	11.5	22.8
32D	1.1	5.8	11.4	22.8	1.1	5.8	11.4	22.8

Table 11.5 A 2 × 2 Contingency Table with One Partially Classified Margin

Fully Classified ($R = 1$)

		Y_2 = 1	Y_2 = 2	
Y_1	1	$m_{11} = 100$	$m_{12} = 20$	$m_{1+} = 120$
	2	$m_{21} = 30$	$m_{22} = 50$	$m_{2+} = 80$
		$m_{+1} = 130$	$m_{+2} = 70$	$m = 200$

Partially Classified ($R = 0$)

		Y_2 = 1	Y_2 = 2	
Y_1	1	$r_{11} = ?$	$r_{12} = ?$	$r_1 = 40$
	2	$r_{21} = ?$	$r_{22} = ?$	$r_2 = 60$
				$r = 100$

variables and indicator variables for nonresponse are considered by Baker and Laird (1985), Fay (1986), and Little (1985b). We consider the latter approach here, as it is closer in spirit to the contingency tables models discussed in Chapter 9. Unlike those models, the nonignorable models discussed here involve subtle issues of estimability which we do not address in detail here. Attention is confined to a two-way contingency table with one supplemental margin, to convey basic ideas.

EXAMPLE 11.9. *Two-Way Contingency Table with One Supplemental Margin.* Suppose data are as in Example 9.1, with n observations on two categorical variables, Y_1 with levels $j = 1, \ldots, J$ and Y_2 with levels $k = 1, \ldots, K$, m completely classified units that form a two-way contingency table $\{m_{jk}\}$, and $r = n - m$ units classified by Y_1 but not by Y_2 that form a supplemental margin $\{r_j\}$. For illustration we shall fit models to the data set in Table 11.5, with $J = K = 2$.

Now define R to take the value 1 if Y_2 is observed, 0 if Y_2 is missing. Suppose that for fixed n, hypothetical complete observations have a multinomial distribution over the $J \times K \times 2$ table formed by Y_1, Y_2, and R. Let $\pi_{jk} = \Pr(Y_1 = j, Y_2 = k)$ and $\phi_{jk} = \Pr(R = 1 \mid Y_1 = j, Y_2 = k)$, so that $\Pr(Y_1 = j, Y_2 = k, R = 1) = \pi_{jk}\phi_{jk}$ and $\Pr(Y_1 = j, Y_2 = k, R = 0) = \pi_{jk}(1 - \phi_{jk})$. This model has $2JK - 1$ parameters, and the data have $JK + J - 1$ degrees of freedom to estimate them: JK from the fully classified data, J from the supplementary margin, less one for the constraint that the probabilities sum to one. Hence there are $2JK - 1 - (JK + J - 1) = J(K - 1)$ too many parameters in the model. We seek to reduce the number of parameters by placing hierarchical loglinear model restrictions on the cell probabilities. (Note that the loglinear models in Section 9.4 concerned the joint distribution of the Y's; here we are modeling the joint distribution of both the Y's and the response indicator, R.)

All the hierarchical models that include the main effects of Y_1, Y_2, and R are displayed in Table 11.6. The first column describes the model using the notation introduced in Section 9.4. The next three columns give the number of parameters in the model, the number of degrees of freedom for testing the

Table 11.6 Models for a Two-Way Table with One Supplemental Margin

	Degrees of Freedom			Lack of Fit		Example from Table 11.5, Estimated Cell Prob × 100			
Model	Model	Lack of Fit	Inestimable	x^2	df	π_{11}	π_{12}	π_{21}	π_{22}
(1) $\{Y_1 Y_2 R\}$	$2JK - 1$	0	$J(K-1)$	—	—	—	—	—	—
(2) $\{Y_1 Y_2, Y_1 R, Y_2 R\}$	$JK + J + K - 2$	0	$K - 1$	—	—	—	—	—	—
(3) $\{Y_1 Y_2, Y_1 R\}$	$JK + J - 1$	0	0	0	0	44.4	8.9	17.5	29.2
(4) $\{Y_1 Y_2, Y_2 R\}$	$JK + K - 1$	$\max(J-K, 0)$	$\max(K-J, 0)$	0	0	39.4	14.0	11.8	34.9
(5) $\{Y_1 Y_2, R\}$	JK	$J - 1$	0	10.75	1	44.4	8.9	17.5	29.2
(6) $\{Y_1 R, Y_2 R\}$	$2(J+K) - 3$	$(J-1)(K-1)$	$K - 1$	44.99	1	—	—	—	—
(7) $\{Y_1 R, Y_2\}$	$2J + K - 2$	$(J-1)(K-1)$	0	44.99	1	34.7	18.7	30.3	16.3
(8) $\{Y_1, Y_2 R\}$	$2K + J - 2$	$(J-1)K$	$K - 1$	55.74	2	—	—	—	—
(9) $\{Y_1, Y_2, R\}$	$J + K - 1$	$(J-1)K$		55.74	2	34.7	18.7	30.3	16.3

fit of the model, and the number of parameters in the model that are inestimable, in the sense that they do not appear in the likelihood. These quantities satisfy the relationship:

$$\text{df(model)} + \text{df(lack of fit)} - \text{df(inestimable)} = JK + J - 1,$$

the degrees of freedom in the data. The remaining six columns show fits to the data in Table 11.5—the likelihood ratio chi-squared statistic for lack of fit, its associated degrees of freedom, and estimates of the cell probabilities ($\times 100$).

The following properties of the models in Table 11.6 merit some discussion:

1. *Inestimability.* The models $\{Y_1 Y_2 R\}$, $\{Y_1 Y_2, Y_1 R, Y_2 R\}$, $\{Y_1 R, Y_2 R\}$, $\{Y_1, Y_2 R\}$, and, if $K > J$, $\{Y_1 Y_2, Y_2 R\}$ have inestimable parameters. Additional information is needed to estimate the cell probabilities for these models, so estimated probabilities are not given in the table.

Note that two of these models, $\{Y_1 R, Y_2 R\}$ and $\{Y_1, Y_2 R\}$, are inestimable, even though they have fewer parameters than degrees of freedom $JK + J - 1$ in the data. For example, consider the model for conditional independence of Y_1 and Y_2 given R, namely, $\{Y_1 R, Y_2 R\}$. The model has $2J + 2K - 3$ parameters—one for the marginal probability of response, $J + K - 2$ for the conditional distribution of Y_1 and Y_2 given $R = 1$, and $J + K - 2$ for the conditional distribution of Y_1 and Y_2 given $R = 0$. The latter two distributions both have $JK - 1$ probabilities, less $(J - 1)(K - 1)$ degrees of freedom since Y_1 and Y_2 are independent, given R. The incomplete-data likelihood factorizes into three components with distinct parameters, corresponding to the marginal distribution of R, the conditional distribution of Y_1 and Y_2 given $R = 1$, and the conditional distribution of Y_1 given $R = 0$. These three components provide estimates of $1 + (J + K - 2) + (J - 1) = 2J + K - 2$ parameters; the remaining $K - 1$ parameters in the model, corresponding to the distribution of Y_2 given $R = 0$, are inestimable. This leaves $(JK + J - 1) - (2J + K - 2) = (J - 1)(K - 1)$ degrees of freedom in the data, which correspond to lack of fit of the conditional independence assumption of Y_1 and Y_2 given $R = 1$.

2. *Ignorability.* The models $\{Y_1 Y_2, Y_1 R\}$ and $\{Y_2, Y_1 R\}$ are ignorable since missingness depends only on Y_1, which is fully observed. These models can be fitted using the methods of Chapter 9. The models $\{Y_1 Y_2, R\}$ and $\{Y_1, Y_2, R\}$ are also ignorable, since they assume that nonresponse is independent of Y_1 and Y_2, that is, that the data are MCAR. They yield the same estimates of $\{\pi_{jk}\}$ as their MAR counterparts, $\{Y_1 Y_2, Y_1 R\}$ and $\{Y_1, Y_2, Y_1 R\}$, respectively.

3. *Lack of Fit.* The lack-of-fit chi-squared for $\{Y_1 Y_2, R\}$ is based on a test of independence of Y_1 and R, using the $Y_1 \times R$ 2-way margin. The lack-of-fit

chi-squared for $\{Y_1 R, Y_2\}$ is based on a test of independence of Y_1 and Y_2, using the fully classified data. The lack-of-fit chi-squared for $\{Y_1, Y_2, R\}$ is found by summing the chi-squared statistics for $\{Y_1 Y_2, R\}$ and $\{Y_1 R, Y_2\}$.

4. *Estimation.* The ML estimate of π_{jk} for $\{Y_1 Y_2, Y_1 R\}$ or $\{Y_1 Y_2, R\}$ is $\hat{\pi}_{jk} = (m_{jk} + \hat{r}_{jk})/(m + r)$, where $\hat{r}_{jk} = (m_{jk}/m_j{}^{+})r_j$ is a filled-in count [cf. Eq. (9.3)]. One can view this estimate as arising from distributing the partially classified counts $\{r_j\}$ into the table to match the *row* distributions $\{m_{jk}/m_{j+}\}$ of the fully observed data, as in Examples 9.1 and 9.2.

Only one of the five nonignorable models in Table 11.6 can be fitted to data without additional prior information, namely, $\{Y_1 Y_2, Y_2 R\}$, which can be estimated if $K \leqslant J$. The model supposes that response to Y_2 depends on the value of Y_2 but not on the value of Y_1. The ML estimates of $\{\pi_{jk}\}$ for this model also have the form $\hat{\pi}_{jk} = (m_{jk} + \hat{r}^*_{jk})/(m + r)$, but now the filled-in values $\hat{r}^*_{jk}$ are such that $\hat{r}^*_{jk}/\hat{r}^*_{+k} = m_{jk}/m_{+k}$, that is, they match the *column* distributions of the fully classified data. These constraints, together with the constraints $\sum_k \hat{r}^*_{jk} = r_j$ for all j, yield $JK - K + J$ linear equations for the JK unknowns $\hat{r}^*_{jk}$. When $K > J$ there are fewer equations than parameters, and hence a priori constraints are required to define uniquely $\{\hat{r}^*_{jk}\}$ (and hence $\hat{\pi}_{jk}$). When $K < J$ there are more equations than parameters, and the ML estimates $\hat{r}^*_{jk}$ cannot satisfy the constraints exactly; the EM algorithm can be used to calculate $\{\hat{r}^*_{jk}\}$ in such cases. [See, for example, Baker and Laird (1985).] When $K = J$, the JK linear equations can be solved directly, yielding estimates without resorting to the EM algorithm iterations. In particular, for $J = K = 2$ we obtain the following equations for $\hat{r}^*_{11}$, $\hat{r}^*_{12}$, $\hat{r}^*_{21}$, and $\hat{r}^*_{22}$:

$$\hat{r}^*_{21} = \hat{r}^*_{11} m_{21}/m_{11}; \quad \hat{r}^*_{22} = \hat{r}^*_{12} m_{22}/m_{12}; \quad \hat{r}^*_{11} + \hat{r}^*_{12} = r_1; \quad \hat{r}^*_{21} + \hat{r}^*_{22} = r_2.$$

Solving yields $\hat{r}^*_{11} = (r_2 - r_1 m_{22}/m_{12})(m_{21}/m_{11} - m_{22}/m_{12})^{-1}$, and so on. For the data in Table 11.5 we obtain

$$\hat{r}^*_{11} = 200/11, \qquad \hat{r}^*_{12} = 240/11, \qquad \hat{r}^*_{21} = 60/11, \qquad \hat{r}^*_{22} = 600/11,$$

which yield the estimates of $\{\pi_{jk}\}$ in row (4) of Table 11.6.

The estimates obtained from solving these linear equations can be negative, and hence not ML. Baker and Laird (1985) show that to yield nonnegative estimates $\{\hat{r}^*_{jk}\}$, the marginal column odds $\{r_j/r_l\}$ must lie between the smallest and largest values of the column odds $\{m_{jk}/m_{lk}\}$, $k = 1, \ldots, K$. In our example, $r_1/r_2 = 40/60$ lies between $m_{11}/m_{21} = 100/30$ and $m_{12}/m_{22} = 20/50$, so this condition is satisfied. If this condition is not satisfied, then the estimates need to be modified to ensure that $\hat{r}^*_{jk} \geqslant 0$ for all j, k. Details are given in Baker and Laird (1985).

5. *Choice between Models.* It is important to note that in our example both the models $\{Y_1, Y_2, Y_1 R\}$ and $\{Y_1, Y_2, Y_2 R\}$ yield perfect fits to the data

with no degrees of freedom for fit. Thus it is not possible to choose between the estimates of $\{\pi_{jk}\}$ they supply, except by a-priori reasoning about which nonresponse mechanism is more plausible for the data set at hand.

The ideas of this example are generalized to a two-way table with two supplementary margins in Little (1985). In that case indicators R_1 and R_2 are introduced for response to Y_1 and Y_2, and models for the four-way table of Y_1, Y_2, R_1, and R_2 are considered. Higher-order tables can also be considered, at least in principle.

REFERENCES

Amemiya, T. (1984). Tobit models: a survey, *J. Econometrics* **24**, 3–61.

Baker, S., and Laird, N. (1985). Categorical response subject to nonresponse, Department of Biostatistics, Harvard School of Public Health, Boston, MA.

Berndt, E. B., Hall, B., Hall, R., and Hausman, J. A. (1974). Estimation and inference in nonlinear structural models, *Ann. Econ. Soc. Meas.* **3**, 653–665.

Box, G. E. P., and Cox, D. R. (1964). An analysis of transformations, *J. Roy. Statist. Soc.* **B26**, 211–252.

David, M. H., Little, R. J. A., Samuhel, M. E., and Triest, R. K. (1986). Alternative methods for CPS income imputation, *J. Am. Statist. Assoc.* **81**, 29–41.

Fay, R. E. (1986). Causal models for patterns of nonresponse, *J. Am. Statist. Assoc.* **81**, 354–365.

Greenlees, W. S., Reece, J. S., and Zieschang, K. D. (1982). Imputation of missing values when the probability of response depends on the variable being imputed, *J. Am. Statist. Assoc.* **77**, 251–261.

Hasselblad, V., Stead, A. G., and Galke, W. (1980). Analysis of coarsely grouped data from the lognormal distribution, *J. Am. Statist. Assoc.* **75**, 771–778.

Heckman, J. (1976). The common structure of statistical models of truncation, sample selection and limited dependent variables, and a simple estimator for such models, *Ann. Econ. Soc. Meas.* **5**, 475–492.

Kulldorff, G. (1961). *Contributions to the Theory of Estimation from Grouped and Partially Grouped Samples*. Stockholm: Almquist and Wiksell and New York: Wiley.

Lillard, L., Smith, J. P., and Welch, F. (1982). What do we really know about wages: The importance of nonreporting and census imputation, The Rand Corporation, Santa Monica, CA.

Lillard, L., Smith, J. P., and Welch, F. (1986). What do we really know about wages? The importance of nonreporting and Census imputation, *Journal of Political Economy*, **94**, 489–506.

Little, R. J. A. (1982). Models for nonresponse in sample surveys, *J. Am. Statist. Assoc.* **77**, 237–250.

Little, R. J. A. (1985a). A note about models for selectivity bias, *Econometrica* **53**, 1469–1474.

Little, R. J. A. (1985b). Nonresponse adjustments in longitudinal surveys: models for categorical data, *Bulletin of the International Statistical Institute*, **15.1**, 1–15.

Nordheim, E. V. (1984). Inference from nonrandomly missing data: An example from a genetic study on Turner's Syndrome, *J. Am. Statist. Assoc.* **79**, 772–780.

Olsen, R. J. (1980). A least squares correction for selectivity bias, *Econometrica* **48**, 1815–1820.

Pregibon, D. (1977). Typical survey data: estimation and imputation, *Survey Methodol.* **2**, 79–102.

Rubin, D. B. (1977). Formalizing subjective notions about the effect of nonrespondents in sample surveys, *J. Am. Statist. Assoc.* **72**, 538–543.

Rubin, D. B. (1983). Imputing Income in the CPS, in *The Measurement of Labor Cost* (Jack Triplett, Ed.). Chicago: University of Chicago Press.

Tobin, J. (1958). Estimation of relationships for limited dependent variables, *Econometrica* **26**, 24–36.

PROBLEMS

1. Carry out the integrations needed to derive the E step in Example 11.1.
2. Derive the expressions for the E step in Example 11.2. Also display the M step for this example explicitly.
3. Derive the expressions for the E step in Example 11.4.
4. Justify the M step computations given at the end of Example 11.4. In particular, why is the estimate of σ_1^2 not given simply from the regression of Y_1 on X?
5. Summarize the ML and two-step fitting methods for estimating parameters of the stochastic censoring model of Example 11.4, noting similarities and differences.
6. Consider the selection model of Example 11.4 when $x_i = (x_{i1}, z_i)$ where z_i is a single binary variable predictive of selection but with coefficient zero in the regression on Y_1 on X. The following gives the means of y_{i1} given x_i, classified by z_i and by whether y_{i1} is observed ($R_{i1} = 1$) or missing ($R_{i1} = 0$).

		z_i	
		1	0
R_{i1}	1	$x_i\beta_1 + \rho\sigma_1\lambda(\gamma_{i1})$	$x_i\beta_1 + \rho\sigma_1\lambda(\gamma_{io})$
	0	$x_i\beta_1 + \rho\sigma_1\lambda(-\gamma_{i1})$	$x_i\beta_1 + \rho\sigma_1\lambda(-\gamma_{io})$
	ALL	$x_i\beta_1$	$x_i\beta_1$

In the table, $\lambda(\cdot)$ is defined as in Example 11.3, and γ_{ij} is the mean of Y_2 for units with values $(x_i, z_i = j)$ of the covariates.

(a) Derive the expressions in the table.

(b) By considering the difference in means of Y_1 between responding and nonresponding cases for several values of γ_{ij}, show that the model implies that the means in the table have an approximately additive structure. (See Little, 1985a, for more details.)

7. Suppose that for the model of Example 11.4 a random subsample of nonrespondents to Y_1 are followed up, and values of Y_1 are obtained. Write down the loglikelihood for the resulting data and describe the E and M steps of the EM algorithm.

8. Show in the context of Example 11.7 that the posterior distribution of $\bar{y}$ given $\sigma^2 = s_{\mathrm{R}}^2$ is normal with mean $\bar{y}_{\mathrm{R}}(1 + h_0)$ and variance $\bar{y}_{\mathrm{R}}^2(\psi_1^2 h_1^2 + \psi_2^2 h_2^2 + h_3^2)$. What is the posterior mean and variance of variable 32D when $\psi_1 = \psi_2 = .5$?

9. For suitable parameterizations of the models, write down factored likelihoods for the models $\{Y_1 Y_2, Y_1 R\}$, $\{Y_1 Y_2, Y_2 R\}$, $\{Y_1 R, Y_2 R\}$, and $\{Y_1, Y_2 R\}$ in Example 11.8. State for each model which parameters (if any) are inestimable.

10. Verify the five sets of estimated cell probabilities in Table 11.6.

11. Redo Table 11.6 for the data in Table 11.5 with r_1 and r_2 multiplied by a factor of 10.

CHAPTER 12

The Model-Based Approach to Survey Nonresponse

12.1. BAYESIAN THEORY WITH COMPLETE RESPONSE

In Chapter 4 we treated survey nonresponse from the quasi-randomization viewpoint, where the values of variables (Y, Z) were treated as fixed and inferences were based on the known sampling distribution $f(I|Z)$ and the modeled response distribution $f(R|I, Y, Z)$. An alternative inferential approach is to specify a model for the item variables Y and use the model-based methodology discussed in Chapters 5–11. For inference about finite population quantities, a Bayesian approach where prior distributions are specified for unknown parameters in the model is more natural than a strict likelihood approach. Consequently, we adopt a fully Bayesian perspective here like that in Section 11.5. A more complete development is given in Rubin (1987, ch. 2).

Potential data from a survey with no nonresponse can be represented as in Figure 12.1, with rows representing units and columns variables. The survey design variables Z and the sample indicator variable I are known for all units in the population, and the item variables Y are recorded for the n sampled units with $I_i = 1$. The analysis of complete survey data might be regarded as an incomplete-data problem for the monotone pattern in Figure 12.1. The objective is to draw inferences about the nonsampled values of Y.

The Bayesian modeling approach to such data treats I and Y as realizations of random variables with joint distribution

$$f(Y, I|Z) = f(Y|Z)f(I|Y, Z). \tag{12.1}$$

Write $Y = (Y_{\text{inc}}, Y_{\text{exc}})$, where Y_{inc} is the set of Y values included in the sample, and Y_{exc} the set excluded from the sample. The data observed in the absence of nonresponse are then Y_{inc} and I, and, of course, Z. Inferences for a population quantity, such as the population mean of Y, $\bar{Y}$, are obtained from its distribu-

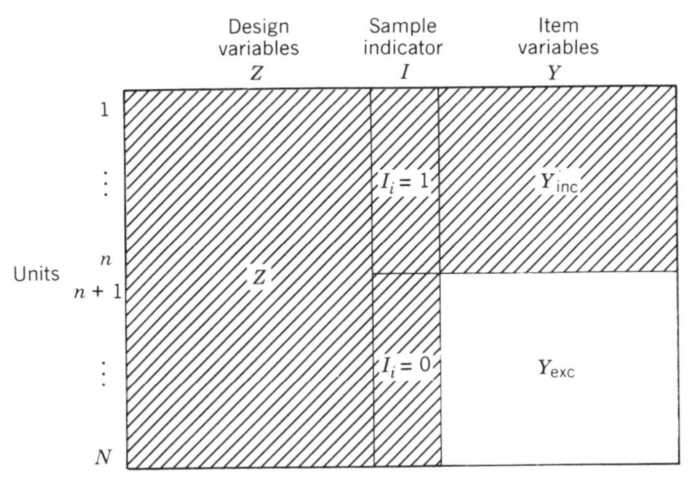

Figure 12.1. The data from a sample survey without nonresponse.

tion given the observed values, Y_{inc}, I, and Z. This distribution follows from the observed values of Y_{inc}, I, and Z and the distribution of the excluded Y values, Y_{exc}, given Y_{inc}, I, and Z:

$$f(Y_{exc} | Y_{inc}, I, Z) = \frac{f(Y, I | Z)}{\int f(Y, I | Z) \, dY_{exc}}. \tag{12.2}$$

A characteristic of this formulation is that the posterior distribution of Y_{exc} in (12.2) is explicitly conditional on I in addition to Y_{inc} and Z. Often Bayesian modelers ignore this extra conditioning and base inference on the distribution

$$f(Y_{exc} | Y_{inc}, Z) = \frac{f(Y | Z)}{\int f(Y | Z) \, dY_{exc}}. \tag{12.3}$$

Following Rubin (1976, 1978, 1987) and Little (1982) we say that the sampling mechanism is ignorable if the distributions for Y_{exc} given by (12.2) and (12.3) are identical. A sufficient condition for the sampling mechanism to be ignorable is that the distribution of I given Y, Z does not depend on Y_{exc}:

$$f(I | Y, Z) = f(I | Y_{inc}, Z). \tag{12.4}$$

If (12.4) holds, then the right side of (12.2) is

$$\frac{f(Y|Z) f(I | Y, Z)}{\int f(Y|Z) f(I | Y, Z) \, dY_{exc}} = \frac{f(Y|Z) f(I | Y_{inc}, Z)}{\int f(Y|Z) f(I | Y_{inc}, Z) \, dY_{exc}} = \frac{f(Y|Z)}{\int f(Y|Z) \, dY_{exc}},$$

which implies that (12.2) and (12.3) are identical.

Note that if the selection of units represented by the sampling mechanism, $f(I | Y, X)$, is achieved by probability sampling, then the function $f(I | Y_{inc}, Y_{exc}, Z)$

is known and does not depend on Y_{inc} or Y_{exc}. Thus any probability sampling mechanism is ignorable, and its probability distribution plays no direct role in Bayesian inferences. Other forms of the sampling mechanism may not be ignorable, and then inference based on (12.3) may be subject to bias. In this case, the full model, (12.1), is then hard to specify unless exclusion from the sample is determined by a *known* mechanism, such as censoring with known censoring points. Thus probability sampling plays an important role in the context of superpopulation modeling even though the sampling distribution is not used as the basis of inference. This point has been noted in the literature (see, for example, Rubin, 1976, 1978; Scott, 1977) but needs to be emphasized, since it provides the crucial argument against the belief that the modeling approach precludes the need to select units by probability sampling. Furthermore, even with ignorable sampling mechanisms the sample design does affect the impact of specification errors and thus may influence the choice of model indirectly.

EXAMPLE 12.1. *Stratified Random Sampling with Complete Response.* To illustrate the preceding theory, let Z be a variable indicating J strata in the population, such that $z_i = j$ if unit i belongs to stratum j, $j = 1, \ldots, J$. Let Y be the survey item measured for units in the sample. The distribution $f(Y|Z)$ is specified as

$$f(Y|Z) = \int f(Y|Z,\theta) f(\theta|Z)\, d\theta,$$

where $\theta = \{(\mu_j, \sigma_j^2), j = 1, \ldots, J\}$ are intermediate parameters in the model, with reference prior

$$f(\theta) = \prod_{j=1}^{J} \sigma_j^{-2},$$

and $f(Y|Z,\theta) = \Pi_{i=1}^{N} f(y_i|z_i,\theta)$, where for units in stratum j,

$$f(y_i|z_i = j, \theta) = (2\pi\sigma_j^2)^{-1/2} \exp[-(y_i - \mu_j)^2/2\sigma_j^2],$$

the normal distribution with mean μ_j and variance σ_j^2. The strategy of using an intermediate parameter θ that yields conditional independence of the units is common in Bayesian modeling.

The distribution $f(I|Y, Z)$ corresponds to stratified random sampling of n_j out of N_j units in stratum j. That is, $f(I|Y, Z)$ is constant for all samples $I = (I_1, \ldots, I_N)^{\mathrm{T}}$ that have n_j units in stratum j for $j = 1, \ldots, J$ and is zero otherwise. Since this distribution does not depend on Y_{exc}, the sampling mechanism is ignorable, and inference about excluded values can be based on the distribution (12.3). In particular, inference about the population mean $\bar{Y}$ is based on $f(\bar{Y}|Z, Y_{\text{inc}})$.

Let $\bar{y}_j$ and s_j^2 denote the sample mean and variance in stratum j. In large samples, the posterior distribution of μ_j is normal with mean $\bar{y}_j$ and variance s_j^2/n_j. Now the population mean $\bar{Y}_j$ in stratum j has the form

$$\bar{Y}_j = \left(\sum_{i \in \text{inc}} y_{ij} + \sum_{i \in \text{exc}} y_{ij}\right)/N_j,$$

the posterior mean of $\bar{Y}_j$ is

$$\begin{aligned} E(\bar{Y}_j | Y_{\text{inc}}, Z) &= \left(\sum_{i \in \text{inc}} y_{ij} + \sum_{i \in \text{exc}} E(y_{ij} | Y_{\text{inc}}, Z)\right)/N_j \\ &= \bar{y}_j, \end{aligned}$$

since for excluded units, $E(y_{ij}| Y_{\text{inc}}, Z) = E[E(y_{ij}|\mu_j, Y_{\text{inc}}, Z)| Y_{\text{inc}}, Z] = E(\mu_j| Y_{\text{inc}}, Z) = \bar{y}_j$. It can also be shown that the posterior variance of $\bar{Y}_j$ is

$$\text{Var}(\bar{Y}_j| Y_{\text{inc}}, Z) = (1 - n_j/N_j)s_j^2/n_j.$$

Thus the finite population correction $(1 - n_j/N_j)$ appears in the estimate of precision for the population quantity $\bar{Y}_j$ in the same way as it arises in stratified random sampling in randomization theory. In this and other examples, inferences about parameters (μ_j) differ from inferences about their finite population analogs $(\bar{Y}_j)$ by finite population corrections, which can be ignored if the proportions sampled (n_j/N_j) are small. In large samples, then, the posterior distribution of $\bar{Y}$ is normal with mean

$$E(\bar{Y} | Y_{\text{inc}}, Z) = \sum_{j=1}^{J} P_j \bar{y}_j \tag{12.5}$$

and variance

$$\text{Var}(\bar{Y} | Y_{\text{inc}}, Z) = \sum_{j=1}^{J} P_j^2 (n_j^{-1} - N_j^{-1}) s_j^2, \tag{12.6}$$

where $P_j = N_j/N$. Note that from the randomization perspective of Chapter 4, (12.5) is the stratified mean usually used to estimate $\bar{Y}$, and (12.6) is the standard estimate of variance in repeated sampling (Cochran, 1977). Thus Bayesian probability intervals based on (12.5) and (12.6) are equivalent to the confidence intervals obtained from randomization theory.

12.2. BAYESIAN MODELS FOR SURVEY DATA WITH NONRESPONSE

It is notationally convenient to divide the survey variables into two groups U and Y, where U are observed for all sampled items and Y are subject to

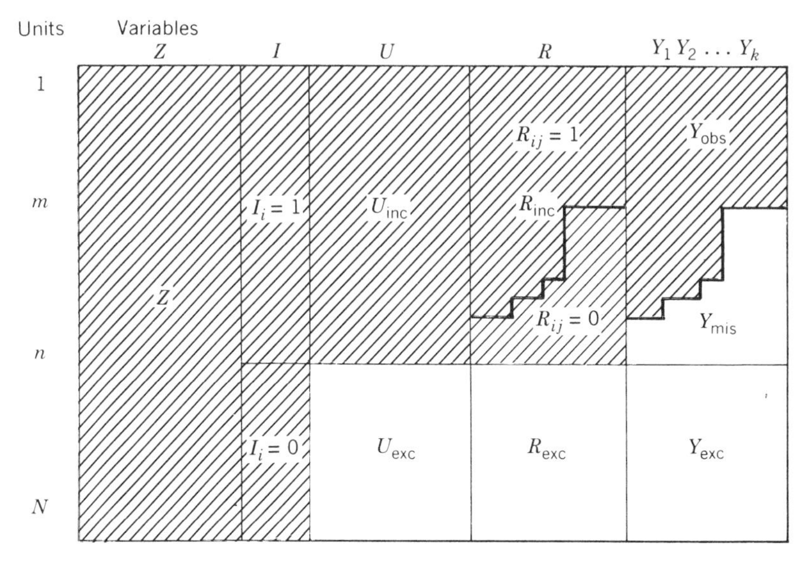

Figure 12.2. The data from a sample survey with monotone nonresponse.

nonresponse. The response pattern for Y is described by the response indicator matrix $R = (R_{ij})$, where $R_{ij} = 1$ if y_{ij} is recorded for unit i and $R_{ij} = 0$ otherwise. The data are represented schematically in Figure 12.2 as in Little (1982). Values of U, R, Y included in the sample are denoted by U_{inc}, R_{inc}, Y_{inc} and the excluded items by U_{exc}, R_{exc}, Y_{exc}, respectively. The included items of Y_{inc} are further divided into observed values Y_{obs} and missing values Y_{mis}. Shaded areas in the diagram represent the data $(Z, I, U_{inc}, R_{inc}, Y_{obs})$ and values of the sample and response indicators are shown as blocks of 1's and 0's. The units are arranged in rows so that the first m units are observed on all variables, the next $n - m$ units are sampled but incomplete, and the remaining $N - n$ are not sampled.

The diagram illustrates a special case of item nonresponse where the observed values of Y have a *monotone* pattern, as discussed in Chapter 6. In such cases the whole data set is also monotone, with Z and I completely observed; Z and I more observed than U and R; and Z, I, U, R more observed than Y. As discussed in Chapter 6, it is easier to devise efficient methods for handling missing values for monotone patterns than for more general patterns, although the theory presented here is completely general. A special case of a monotone pattern is *unit nonresponse* where all partially observed variables are missing for the same units and there are no complete item variables U.

A full Bayesian model specifies a joint distribution of I, U, R, Y given Z. The distribution may be specified as a product of conditional distributions of the form

$$f(Y, R, U, I|Z) = f(Y, U|Z)f(R|Y, U, Z)f(I|Y, R, U, Z). \tag{12.7}$$

The first factor on the right side of (12.7) is entirely analogous to the first factor on the right side of (12.1) with Y replaced by (Y, U). The last factor in (12.7) is analogous to the second factor in (12.1) with Y replaced by (Y, U) and the additional conditioning on R, which appears because in the presence of possible nonresponse, sampling decisions can depend on response patterns. Because of this extra conditioning when nonresponse is possible, the sampling mechanism is said to be ignorable if $f(I|Y, R, U, Z)$ depends only on observed values $(Y_{obs}, R_{inc}, U_{inc}, Z)$:

$$f(I|Y, R, U, Z) = f(I|Y_{obs}, R_{inc}, U_{inc}, Z). \tag{12.8}$$

Finally, the middle factor in (12.7) is new and represents the nonresponse mechanism, the conditional distribution of R given (Y, U, Z). The factorization given by (12.7) has the same general form as the factorization given by (11.3) for nonignorable nonresponse with an extra factor reflecting the sampling mechanism.

Inferences for population quantities now follow from the observed values $(Y_{obs}, R_{inc}, U_{inc}, I, Z)$ and the conditional distribution of the unobserved values $(Y_{exc}, Y_{mis}, U_{exc})$ given the observed values:

$$f(Y_{exc}, Y_{mis}, U_{exc}|Y_{obs}, R_{inc}, U_{inc}, I, Z) = \frac{\int f(Y, R, U, I|Z)\, dR_{exc}}{\iiiint f(Y, R, U, I|Z)\, dY_{exc}\, dY_{mis}\, dR_{exc}\, dU_{exc}}. \tag{12.9}$$

Notice that the posterior distribution in (12.9) is explicitly conditional on I and R_{inc} in addition to (Y_{obs}, Y_{inc}, Z). Often Bayesian modelers even when faced with nonresponse ignore this extra conditioning and base inferences on the distribution

$$f(Y_{exc}, Y_{mis}, U_{exc}|Y_{obs}, U_{inc}, Z) = \frac{f(Y, U|Z)}{\iiint f(Y, U, Z)\, dY_{exc}\, dY_{mis}\, dU_{exc}}. \tag{12.10}$$

Following Rubin (1976, 1987) and the definition in Section 12.1. we say that the sampling and nonresponse mechanisms are jointly ignorable if the distributions (12.9) and (12.10) are identical. A sufficient condition for ignorability of the nonresponse mechanism when the sampling mechanism is ignorable is that the distribution of R given Y, U, Z does not depend on the unobserved values Y_{exc}, Y_{mis}, U_{exc}:

$$f(R|Y, U, Z) = f(R|Y_{obs}, U_{inc}, Z).$$

If this equation and (12.8) hold,

$$f(Y, R, U, I|Z) = f(Y, U|Z)f(I|Y_{\text{obs}}, R_{\text{inc}}, U_{\text{inc}}, Z)f(R|Y_{\text{obs}}, U_{\text{inc}}, Z),$$

and thus the right side of (12.9) equals the right side of (12.10).

Detailed conditions for when the sampling and nonresponse mechanisms can be ignored are given in Little (1982). Paraphrasing the conditions, we can say that given probability sampling, the sampling and nonresponse mechanism can be ignored if the response distribution does not depend on the values of items that are missing for some units. In particular, the incidence of response is assumed to be independent of the missing variable within subclasses defined by the values of design variables Z and completely observed item variables, U_{inc}. These conditions are nearly the same as the MAR and distinct parameter assumptions of previous chapters in the context of likelihood-based parametric inference; Rubin (1987, ch. 2) discusses the correspondence.

In the next section we describe methods based on model distributions of the form (12.9) that are based on the assumption that the response and sampling mechanisms are ignorable. Nearly all analytic procedures for handling nonresponse in sample survey practice effectively make this assumption.

12.3. METHODS BASED ON IGNORABLE NONRESPONSE MODELS

A series of simple examples is now used to illustrate the application of the Bayesian theory of Section 12.2. Some results agree with standard quasi-randomization estimates and standard errors from Chapter 4, but easily derived extensions of these standard results indicate the inherent flexibility of the modeling approach for deriving estimates and standard errors in non-standard situations. All examples assume simple random sampling of a finite population where Y is subject to nonresponse but U is not. Since no example posits the existence of design variables Z, the appearance of Z is suppressed in all expressions; however, extensions to stratified samples are straightforward. All examples assume that nonresponse is ignorable but possibly dependent on the values of the completely recorded survey items U_{inc}.

EXAMPLE 12.2. *Adjustment Cell Models with Known Adjustment Cell Counts.* Suppose a simple random sample of n units is taken from a population of N units, and a particular survey item Y is subject to nonresponse that corresponds to random subsampling of the sampled values within adjustment cells formed from a variable U, recorded for all units in the sample. Let N_j and $\bar{Y}_j$, respectively, denote the population size and the population mean of Y in adjustment cell $U = j$. Our objective is to estimate the overall mean

$$\bar{Y} = \sum P_j \bar{Y}_j,$$

where $P_j = N_j/N$.

Suppose we specify that values of Y within adjustment cell j are iid normal with mean μ_j and variance σ_j^2 and that μ_j and $\ln \sigma_j^2$ have locally uniform priors as in Example 12.1. Assuming large samples and known $\{N_j\}$, the posterior distribution of $\bar{Y}$ given the respondent data can be shown (by arguments similar to those in Example 12.1) to be normal with mean

$$E(\bar{Y} \mid Y_{\text{obs}}, U_{\text{inc}}, \{N_j\}) = \sum P_j \bar{y}_{jR}, \tag{12.11}$$

and variance

$$\text{Var}(\bar{Y} \mid Y_{\text{obs}}, U_{\text{inc}}, \{N_j\}) = \sum P_j^2 (m_j^{-1} - N_j^{-1}) s_{jR}^2, \tag{12.12}$$

where m_j is the respondent sample size and $\bar{y}_{jR}$ and s_{jR}^2 are the sample mean and variance of the respondent Y values in cell j. Observe that (12.11) is the poststratified estimator, and (12.12) is its sampling variance from quasi-randomization theory (4.15) and (4.16). Hence these expressions yield Bayesian probability intervals identical to confidence intervals from the frequentist approach in Section 4.4.3.

EXAMPLE 12.3. *Adjustment Cell Models with Unknown Adjustment Cell Counts.* Suppose we have the setup of the previous example, but the adjustment cell totals $\{N_j\}$ are unknown, as is usual when U is a surveyed item. Bayesian inference for $\bar{Y}$ is based on its posterior distribution given Y_{obs}, which is obtained from the posterior distribution of $\bar{Y}$ given Y_{obs} and $\{N_j\}$ in Example 12.2 by integrating over the posterior distribution of $\{N_j\}$ given $Y_{\text{obs}}, U_{\text{inc}}$:

$$f(\bar{Y} \mid Y_{\text{obs}}, U_{\text{inc}}) = \int f(\bar{Y} \mid Y_{\text{obs}}, U_{\text{inc}}, \{N_j\}) f(\{N_j\} \mid Y_{\text{obs}}, U_{\text{inc}}) d\{N_j\}.$$

In particular, the mean and variance of this distribution are

$$\begin{aligned} E(\bar{Y} \mid Y_{\text{obs}}, U_{\text{inc}}) &= E[E(\bar{Y} \mid Y_{\text{obs}}, U_{\text{inc}}, \{N_j\} \mid Y_{\text{obs}}, U_{\text{inc}})] \\ &= \sum E(P_j \mid Y_{\text{obs}}, U_{\text{inc}}) \bar{y}_{jR} \end{aligned} \tag{12.13}$$

and

$$\begin{aligned} \text{Var}(\bar{Y} \mid Y_{\text{obs}}, U_{\text{inc}}) &= E[\text{Var}(\bar{Y} \mid \{N_j\}, Y_{\text{obs}}, U_{\text{inc}}) \mid Y_{\text{obs}}, U_{\text{inc}}] \\ &\quad + \text{Var}[E(\bar{Y} \mid \{N_j\}, Y_{\text{obs}}, U_{\text{inc}}) \mid Y_{\text{obs}}, U_{\text{inc}}] \\ &= \sum [E(P_j^2 \mid Y_{\text{obs}}, U_{\text{inc}})/m_j - E(P_j \mid Y_{\text{obs}}, U_{\text{inc}})/N] s_{jR}^2 \\ &\quad + \text{Var}(\textstyle\sum P_j \bar{y}_{jR} \mid Y_{\text{obs}}, U_{\text{inc}}), \end{aligned} \tag{12.14}$$

where the summations are over adjustment cells, $j = 1, \ldots, J$. Suppose that units are distributed over adjustment cells as iid multinomial random variables with index 1 and probabilities $(\theta_1, \ldots, \theta_J)$, with a locally uniform prior on $(\theta_1, \ldots, \theta_J)$. Then

$$E(\theta_j | Y_{obs}, U_{inc}) = p_j,$$
$$\text{Var}(\theta_j | Y_{obs}, U_{inc}) = p_j(1 - p_j)/n,$$
$$\text{Cov}(\theta_j, \theta_k | Y_{obs}, U_{inc}) = -p_j p_k / n$$

where $p_j = n_j/n$, the sample proportion in cell j. Also

$$E(P_j | \{\theta_j\}, Y_{obs}, U_{inc}) = [n_j + (N - n)\theta_j]/N,$$
$$\text{Var}(P_j | \{\theta_j\}, Y_{obs}, U_{inc}) = (N - n)\theta_j(1 - \theta_j)/N^2,$$
$$\text{Cov}(P_j, P_k | \{\theta_j\}, Y_{obs}, U_{inc}) = -(N - n)\theta_j\theta_k/N^2.$$

Hence

$$\begin{aligned} E(P_j | Y_{obs}, U_{inc}) &= E[E(P_j | \{\theta_j\}, Y_{obs}, U_{inc}) | Y_{obs}, U_{inc}] \\ &= [n_j + (N - n)p_j]/N \\ &= p_j; \end{aligned}$$

$$\begin{aligned} \text{Var}(P_j | Y_{obs}, U_{inc}) &= E[\text{Var}(P_j | \{\theta_j\}, Y_{obs}, U_{inc}] \\ &\quad + \text{Var}[E(P_j | \{\theta_j\}, Y_{obs}, U_{inc}) | Y_{obs}, U_{inc}] \\ &= (1 - 1/N)(1 - n/N)p_j(1 - p_j)/n; \end{aligned}$$

$$\begin{aligned} &\text{Cov}(P_j, P_k | Y_{obs}, U_{inc}) \\ &\quad = E[\text{Cov}(P_j, P_k | \{\theta_j\}, Y_{obs}, U_{inc} | Y_{obs}, U_{inc}] \\ &\qquad + \text{Cov}[E(P_j | \{\theta_j\}, Y_{obs}, U_{inc}), E(P_k(\{\theta_j\}, Y_{obs}, U_{inc}) | Y_{obs}, U_{inc}] \\ &\quad = -(1 - 1/N)(1 - n/N)p_j p_k/n. \end{aligned}$$

Substituting these expressions in (12.13) and (12.14), we find

$$E(\bar{Y} | Y_{obs}, U_{inc}) = \sum_{j=1}^{J} p_j \bar{y}_{jR} = \bar{y}_{wc},$$

the weighting cell estimator (4.10) and

$$\begin{aligned} \text{Var}(\bar{Y} | Y_{obs}, U_{inc}) &= \sum_{j=1}^{J} p_j^2 [1 - m_j/(Np_j)] s_{jR}^2/m_j \\ &\quad + n^{-1}(1 - 1/N)(1 - n/N) \sum_{j=1}^{J} [p_j(\bar{y}_{jR} - \bar{y}_{wc})^2 \\ &\quad + p_j(1 - p_j)s_{jR}^2/m_j], \end{aligned}$$

after some algebra. This expression is approximately equal to the estimated mean squared error of the weighting cell estimator given after (4.12).

The flexibility of the modeling approach for obtaining estimates and standard errors becomes apparent when the adjustment cells are formed by the joint levels of two or more factors. Suppose

$$\bar{Y} = \sum_{j=1}^{J} \sum_{k=1}^{K} P_{jk} \bar{Y}_{jk},$$

where the suffix (jk) labels the adjustment cell formed by the levels of two classifying variables, $U_1 = j$ and $U_2 = k$. The model

$$(y_i | u_{i1} = j, u_{i2} = k, \theta) \sim N(\mu_{jk}, \sigma_{jk}^2)$$

for the responding and nonresponding units in cell (jk), with $\theta = \{\mu_{jk}, \sigma_{jk}^2\colon j = 1, \ldots, J, k = 1, \ldots, K\}$ and locally uniform priors on μ_{jk} and $\ln \sigma_{jk}^2$, leads to weighting cell or poststratified estimators, as before. If the number of responding units m_{jk} is small for certain cells, then it may be preferable to fit a more parsimonious model for the cell means. For example, an additive model

$$(y_i | u_{i1} = j, u_{i2} = k, \theta) \sim N(\mu + \alpha_j + \beta_k; \sigma_{jk}^2)$$

where now $\theta = \{\mu, \alpha_j, \beta_k\colon j = 1, \ldots, J, k = 1, \ldots, K\}$ may provide an appropriate fit to the data. This model may be fitted by least squares, yielding predictions for the nonresponding values of Y. Also, a pooled estimate of variance can be obtained by fitting a model that assumes the variance σ_{jk}^2 is the same for all values of j and/or k.

Another modeling strategy, which may reduce the added variance of the weighting cell estimator $\bar{y}_{wc}$, over the poststratified estimator (12.11), is to model the cell proportions P_{jk}. For example, we might be prepared to assume that the classifying factors U_1 and U_2 are independent, leading to the estimate (ignoring finite population corrections)

$$\hat{P}_{jk} = n_{j+} n_{+k} / n^2$$

of P_{jk}, where $n_{j+} = \sum_{k=1}^{K} n_{jk}$, $n_{+k} = \sum_{j=1}^{J} n_{jk}$.

The flexibility of the modeling approach in smoothing the cell means, variances, and proportions is clearly useful when adjustment cells are determined by crossing three or more factors, a situation where the poststratified and weighting cell estimators are limited by the requirement of retaining a sufficient number of responding units in each cell with a nonrespondent. This flexibility also extends to the use of interval-scaled covariates to predict the values of missing items, as the next example indicates.

EXAMPLE 12.4. *Regression Imputation.* More generally, suppose that a simple random sample of size n is drawn of $(y_i, u_{i1}, \ldots, u_{iK})$, where the K variables $U_1, \ldots, U_K$ are recorded for all units in the sample, and the incidence of response is independent of Y, conditional on $U_1, \ldots, U_K$. We assume that

$$(y_i | u_{i1}, \ldots, u_{iK}, \theta) \sim (\mu_i, \sigma^2 v_i), \quad (12.15)$$

$$\mu_i = \beta_0 + \sum_j \beta_j \mu_{ij},$$

$v_i = v(u_{i1}, \ldots, u_{iK})$ is a known function characterizing the heterogeneity of variance, and the parameters $\theta = (\beta_0, \ldots, \beta_K, \ln \sigma^2)$ are assumed to have locally uniform priors. Nonresponding values of Y are then estimated by their posterior means

$$E(y_i | Y_{\text{obs}}, U_{\text{inc}}) = \hat{\beta}_0 + \sum_j \hat{\beta}_j u_{ij},$$

where $(\hat{\beta}_0, \hat{\beta}_1, \ldots, \hat{\beta}_K)$ are estimated coefficients found by weighted least squares, weighting unit i by v_i^{-1}.

Special cases of estimators based on this model include the estimators of the previous example, obtained from (12.15) with $v_i = 1$ and $U_1, \ldots, U_K$ representing dummy variables for the adjustment cells. They also include the ratio estimator

$$y_i = (\bar{y}_R / \bar{u}_R) u_i$$

of missing y's, where $\bar{y}_R$ and $\bar{u}_R$ are the means of Y and a single variable U for responding units. This estimator is obtained by setting $K = 1$, $\beta_0 = 0$, and $v_i = u_{i1}$ in (12.15). If $K = 1$ and $v_i = 1$, we obtain the regression estimator of missing y_i's:

$$\hat{y}_i = \bar{y}_R + \hat{\beta}(u_i - \bar{u}_R). \quad (12.16)$$

This estimator also occurs in the randomization theory in the context of double sampling (Cochran, 1977, ch. 12), where U is a variable recorded in a large initial sample, and the variable Y is recorded for a randomly chosen subset of the originally sampled units.

In a strict application of model-based methods, the regression model (12.15) is used not only to estimate the missing values, but also to derive inferences for population quantities. An important aspect of this process is that models are chosen to "conform" to the sample design, in the sense that they are not sensitive to specification error. For recent discussions of this important topic for completely recorded data, see, for example, Royall and Herson (1973), or Hansen, Madow, and Tepping (1982) and its discussions, especially those by Little and Rubin; also see Rubin (1985).

A more restricted use of the model is simply to supply imputed values, with estimation for population quantities from the completed data being carried out using randomization-based methods. The model-based approach to imputing missing values is, of course, not limited to linear models of the form (12.15). For a binary Y, for example, a logistic regression is usually preferable.

For categorical Y and $U_1, \ldots, U_K$, the data form a partially classified contingency table, with responding units classified by $Y, U_1, \ldots, U_K$ and the nonresponding units classified by $U_1, \ldots, U_K$ only. Loglinear models for contingency tables may be fitted to these data, as discussed in Chapter 9. These models provide estimates $\hat{p}(c|U_1, \ldots, U_K)$ of the conditional probability $p(c|U_1, \ldots, U_K)$ of a nonrespondent having the category $Y = c$, given values $U_1, \ldots, U_K$ of the covariates. Imputation can be achieved by assigning a partially classified unit to cell c with probability $\hat{p}(c|U_1, \ldots, U_K)$, a procedure closely related to the EM algorithm for ML estimation discussed in Chapter 6.

Since one imputed value cannot possibly represent the uncertainty about which value to impute for the missing item, if each missing item is replaced with an imputed value, some adjustment needs to be made in the analysis for inferences to be generally valid. One adjustment approach is multiple imputation.

12.4. MULTIPLE IMPUTATION

Multiple imputation refers to the procedure of replacing each missing value by a vector of $M \geqslant 2$ imputed values. The M values are ordered in the sense that M completed data sets can be created from the vectors of imputations; replacing each missing value by the first component in its vector of imputations creates the first completed data set, replacing each missing value by the second component in its vector creates the second completed data set, and so on. Standard complete-data methods are used to analyze each data set. When the M sets of imputations are repeated random draws under one model for nonresponse, the M complete-data inferences can be combined to form one inference that properly reflects uncertainty due to nonresponse under that model. When the imputations are from two or more models for nonresponse, the combined inferences under the models can be contrasted across models to display the sensitivity of inference to models for nonresponse, a particularly critical activity when nonresponse is nonignorable.

Multiple imputation was first proposed in Rubin (1978), although the idea appears in Rubin (1977). A comprehensive treatment is given in Rubin (1987), and other references include Rubin (1986), Herzog and Rubin (1983), Li (1985), Schenker (1985), and Rubin and Schenker (1986). The method has potential for application in a variety of contexts. It appears particularly promising in complex surveys with standard complete-data analyses that are difficult to modify analytically in the presence of nonresponse. Here we provide a brief overview of multiple imputation and illustrate its use.

As already indicated in Chapters 2–4, the practice of imputing for missing values is very common. Single imputation has the obvious practical advantage of allowing standard complete-data methods of analysis to be used. Imputation also has an advantage in many contexts in which the data collector (e.g., the Census Bureau) and the data analyst (e.g., a university social scientist) are different individuals, because the data collector may have access to more and better information about nonrespondents than the data analyst. For example, in some cases, information protected by confidentiality constraints (e.g., zip codes of dwelling units) may be available to help impute missing values (e.g., annual incomes). The obvious disadvantage of single imputation is that imputing a single value treats that value as known, and thus without special adjustments, single imputation cannot reflect sampling variability under one model for nonresponse or uncertainty about the correct model for nonresponse.

Multiple imputation shares both advantages of single imputation and rectifies both disadvantages. Specifically, when the M imputations are repetitions under one model for nonresponse, the resulting M complete-data analyses can be easily combined to create an inference that validly reflects sampling variability because of the missing values, and when the multiple imputations are from more than one model, uncertainty about the correct model is displayed by the variation in valid inferences across the models. The only disadvantage of multiple imputation over single imputation is that it takes more work to create the imputations and analyze the results. The extra work in analyzing the data, however, is really quite modest in today's computing environments, since it basically involves performing the same task M times instead of once.

Multiple imputations ideally should be drawn according to the following protocol. For each model being considered, the M imputations of Y_{mis} are M repetitions from the posterior predictive distribution of Y_{mis}, each repetition corresponding to an independent drawing of the parameters and missing values. In practice, implicit models can often be used in place of explicit models. Both types of models are models illustrated in Herzog and Rubin (1983), where repeated imputations are created using (1) an explicit regression model and (2) an implicit model, which is a modification of the Census Bureau's hot deck.

The analysis of a multiply imputed data set is quite direct. First, each data set completed by imputation is analyzed using the same complete-data method that would be used in the absence of nonresponse. Let $\hat{\theta}_l$, W_l, $l = 1, \ldots, M$ be M complete-data estimates and their associated variances for an estimated θ, calculated from M repeated imputations under one model. For instance, in the context of Example 12.1, $\hat{\theta}_l$ is given by the right side of (12.5), calculated using the lth set of imputed values for Y_{mis}, and W_l is given by the right side of (12.6), also calculated using the lth set of imputed values. The combined estimate is

$$\bar{\theta}_M = \sum_{l=1}^{M} \frac{\hat{\theta}_l}{M}. \tag{12.17}$$

The variability associated with this estimate has two components: the average within-imputation variance,

$$\overline{W}_M = \sum_{l=1}^{M} \frac{\hat{W}_l}{M}, \tag{12.18}$$

and the between-imputation component,

$$B_M = \frac{\sum(\hat{\theta}_l - \bar{\theta}_M)^2}{M - 1} \tag{12.19}$$

[with vector θ, $(\cdot)^2$ replaced by $(\cdot)^T(\cdot)$.]. The total variability associated with $\bar{\theta}_M$ is

$$T_M = \overline{W}_M + \frac{M + 1}{M} B_M, \tag{12.20}$$

where $(M + 1)/M$ is an adjustment for finite M. With scalar θ, the reference distribution for interval estimates and significance tests is a t distribution,

$$(\theta - \bar{\theta}_M) T_M^{-1/2} \sim t_\nu, \tag{12.21}$$

where the degrees of freedom,

$$\nu = (M - 1)\left[1 + \frac{1}{M + 1} \frac{\overline{W}_M}{B_M}\right]^2, \tag{12.22}$$

is based on a Satterthwaite approximation (Rubin and Schenker, 1986; Rubin, 1987). It is interesting to note that $\overline{W}_M/B_M$ estimates the quantity $(1 - \gamma)/\gamma$, where γ is the fraction of information about θ missing due to nonresponse. Observed and missing information are defined in Section 7.5.

For θ with r components, significance levels for null values of θ can be obtained from M repeated complete-data estimates $\hat{\theta}_l$ and variance–covariance matrices $\bar{U}_l$, using multivariate analogs of (12.17)–(12.21). Less precise p values can be obtained directly from M repeated complete-data significance levels. Details may be found in Rubin (1987).

Although multiple imputation is most directly motivated from the Bayesian perspective, the resultant inferences can be shown to possess good sampling properties. For example, Rubin and Schenker (1986) show that in many cases interval estimates created using only two imputations provide randomization-based coverages close to their nominal levels.

EXAMPLE 12.5. *Multiple Imputation Inferences for Stratified Random Samples (Examples 12.1 and 12.2 continued).* To illustrate multiple imputation, consider inference for a population mean $\bar{Y}$ from a stratified random sample, using the model in Example 12.1. With complete data, inference for $\bar{Y}$ would be made using the statement

$$\left(\bar{Y} - \sum_{j=1}^{J} P_j \bar{y}_j\right) \sim N\left[0, \left(\sum_{j=1}^{J} P_j^2 (n_j^{-1} - N_j^{-1}) s_j^2\right)\right]. \tag{12.23}$$

Now suppose that only m_j of the n_j units in stratum j are respondents. With multiple imputation, each of the $\sum_j (n_j - m_j)$ missing units would have M imputations, thereby creating M completed data sets and M values of the stratum means and variances, say, $\bar{y}_{j(l)}$ and $s_{j(l)}^2$, $l = 1, \ldots, M$. From (12.17) and (12.23), the multiple imputation estimate of $\bar{Y}$ is the average of the M complete-data estimates of $\bar{Y}$,

$$\hat{\bar{Y}} = \sum_{l=1}^{M} \left[\sum_{j=1}^{J} P_j \bar{y}_{j(l)}\right] \Big/ M. \tag{12.24}$$

From (12.19)–(12.20) and (12.23), the variability associated with $\hat{\bar{Y}}$ is the sum of the two components displayed in (12.25):

$$\sum_{l=1}^{M} \left[\sum_{j=1}^{J} P_j^2 (n_j^{-1} - N_j^{-1}) s_{j(l)}^2\right] \Big/ M + \frac{M+1}{M} \sum_{l=1}^{M} \sum_{j=1}^{J} (P_j \bar{y}_{j(l)} - \hat{\bar{Y}})^2 / (M-1). \tag{12.25}$$

From (12.21) and (12.22) resulting inferences for $\bar{Y}$ follow from the statement that $(\bar{Y} - \hat{\bar{Y}})$ is distributed as t with center zero, squared scale given by (12.25), and degrees of freedom given by (12.22).

EXAMPLE 12.6. *Creating Multiple Imputations for Stratified Random Samples with Ignorable Nonresponse* (*Examples 12.5 continued*). Since the multiple imputations are drawn from a predictive distribution, an intuitive method for creating such imputations is the hot deck, which draws the non-respondents' values at random from the respondents' values in the same stratum. Arguments in Rubin (1979) and Herzog and Rubin (1983) can be used to show that for infinite M, the resultant multiple imputation estimator given by (12.24) equals the poststratified estimator of Example 12.2 given by the right side of (12.11); however, the resultant multiple imputation variance given by (12.25) is less than the variance of the poststratified estimator given by the right side of (12.12). The source of this problem is that hot deck imputation does not reflect uncertainty about the stratum parameters. Simple extensions of the hot deck do reflect such uncertainty and therefore with large M yield not only the poststratified estimator but the correct associated variance.

First consider a method based on an implicit model, which is called the *approximate Bayesian Bootstrap* by Rubin and Schenker (1986). For $l = 1, \ldots,$

M carry out the following steps independently: For each stratum, first create n_j possible values of Y by drawing n_j values at random with replacement from the m_j observed values of Y in stratum j, and second, draw the $n_j - m_j$ missing values of Y at random with replacement from these n_j values. Results in Rubin and Schenker (1986) or Rubin (1987) can be used to show that this method is proper for large M in the sense that it will yield the poststratified estimator and its associated variance in this case.

Creating multiple imputations using the explicit normal model of Example 12.2, in which the Y-values within stratum j are iid normal with mean μ_j and variance σ_j^2 and the prior distribution on $(\mu_j, \ln \sigma_j)$ is locally uniform, is also proper in this sense. This method is called *fully normal* imputation by Rubin and Schenker (1986), and is defined by M independent applications of the following two steps: first, for each stratum draw (μ_j, σ_j^2) from their joint posterior distribution, and second draw the $n_j - m_j$ missing values of Y as iid normal with mean and variance given by the drawn values of μ_j and σ_j^2.

12.5. NONIGNORABLE NONRESPONSE

The model-based approaches to survey nonresponse we have considered thus far have been based on an assumption of ignorable nonresponse, in the sense that respondents and nonrespondents with the same values of recorded variables do not differ systematically on the values of variables missing for the nonrespondents; that is, the missing values are assumed to be MAR. By using nonignorable models for nonresponse, such as illustrated in Chapter 11, the model-based approach can be applied to surveys with nonresponse suspected of being nonignorable. In fact, Examples 11.5 and 11.7 use nonignorable models on data from actual surveys.

An important issue that arises with nonignorable nonresponse is whether the objective of the statistical analysis is (1) to provide a single valid inference or (2) to display sensitivity of conclusions to nonignorable nonresponse by providing a variety of inferences each of which is valid under an assumed model for nonresponse. Obviously, the objective of obtaining one valid inference is generally very demanding, since in common practice there is no hard evidence from which to specify one correct model for nonresponse. Consequently, although less satisfying than providing one valid inference, displaying sensitivity through a variety of conditionally valid inferences is a much more realistic goal. Such a display of sensitivity was conducted in Example 11.7. Multiple imputation, with imputed values drawn from different models for nonresponse, provides a convenient display of sensitivity of inference to those models.

Table 12.1 The Sample Cross-classified by Response and by an Adjustment Cell Variable Known Only for Respondents[a]

		Response indicator R		
		$R_i = 1$	$R_i = 0$	Total
Adjustment cell	1	m_1	$(n_1 - m_1)$	(n_1)
	⋮	⋮	⋮	⋮
	j	m_j	$(n_j - m_j)$	(n_j)
	⋮	⋮	⋮	⋮
	J	m_J	$(n_J - m_J)$	(n_J)
	Total	m	$n - m$	n

[a] Quantities in parentheses are not observed.

EXAMPLE 12.7. *A Nonignorable Adjustment Cell Model.* Suppose that as in Example 12.3, adjustment cells are formed so that in cell j the distribution of Y is the same for nonrespondents and for respondents and is normal with mean μ_j and variance σ_j^2. Unlike Example 12.3, however, the variable defining adjustment cells is recorded only for respondents in the sample. For example, suppose that nonresponse to a survey is ignorable within income categories and income is recorded for respondents but not for nonrespondents. Then we know the number of respondents (m_j) in cell j but do not know the number of nonrespondents ($n_j - m_j$); see table 12.1. We wish to estimate the population mean

$$\bar{Y} = \sum_{j=1}^{J} P_j \bar{Y}_j,$$

where, as before, P_j and $\bar{Y}_j$ are the population proportion and mean in cell j. If the sample sizes n_j in each cell j were known, we could estimate $\bar{Y}$ by the weighting cell estimator

$$E(\bar{Y} \mid Y_{\text{obs}}, R_{\text{inc}}, \{n_j\}) = \sum_{j=1}^{J} \frac{n_j \bar{y}_{jR}}{n} = \sum_{j=1}^{J} p_j \bar{y}_{jR} = \bar{y}_{\text{wc}}.$$

Instead we estimate $\bar{Y}$ by the posterior mean

$$E(\bar{Y} \mid Y_{\text{obs}}, R_{\text{inc}}) = \sum_{j=1}^{J} \frac{\hat{n}_j \bar{y}_{jR}}{n}, \qquad (12.26)$$

where $\hat{n}_j = E(n_j \mid Y_{\text{obs}}, R_{\text{inc}})$ is the estimated count in cell j. The estimation of n_j is very sensitive to prior specifications, as can be seen from Table 12.1 by noting that the problem is to distribute the $n - m$ nonrespondents into the adjustment cells. The assumption of ignorability corresponds to independence of the rows and columns of this table and leads to estimating the population mean by $\bar{y}_R$, the respondents' mean in the sample.

Let ψ_j be the a priori probability of response in cell j, and suppose (unrealistically) that this quantity is known. Assuming a multinomial distribution for the counts in the table, and flat priors on the cell probabilities, we obtain

$$E(n_j|\{\psi_j\}, Y_{\text{obs}}, R_{\text{inc}}) = nm_j\psi_j^{-1}\left(\sum_{k=1}^{J} m_k\psi_k^{-1}\right). \tag{12.27}$$

Hence one strategy is to specify plausible values for the response probabilities and then calculate the estimate of $\bar{Y}$ obtained by substituting (12.27) in (12.26). In fact, the response probabilities only affect the estimate of Y through their relative values, so $J - 1$ rather than J quantities need to be specified. Choosing various values for the $\{\psi_j\}$ and calculating the corresponding estimates displays sensitivity of estimation to prior specification of the nonresponse mechanism.

EXAMPLE 12.8. *Nonignorable Models for Partially Classified Contingency Tables.* Nonignorable models for categorical data are considered in Section 12.6. Little (1982) presents ML estimates for a nonignorable model related to a model discussed by Pregibon (1977) for a problem involving continuous and categorical variables. We give here a special case of this model where nonresponse is confined to a single categorical variable.

Suppose we have m units classified by categorical variables $Y, U_1, \ldots, U_K$ and $n - m$ units classified by $U_1, \ldots, U_K$ only. Let $p(c|U_1, \ldots, U_K)$ denote the conditional probability of a randomly selected unit in the population having the value $y = c$, given values $U_1, \ldots, U_K$ of the covariates. For an ignorable nonresponse mechanism, the E step of the EM algorithm involves distributing a proportion $p^{(t)}(c|U_1, \ldots, U_K)$ of the subset of the $n - m$ partially classified units with values $U_1, \ldots, U_K$ into the cell $y = c$, where $p^{(t)}$ denotes the estimate of p at the tth iteration (Chapter 9 provides details). Thus the odds of assigning a unit to $y = c$ rather than $y = c'$ are

$$p^{(t)}(c|U_1, \ldots, U_K)/p^{(t)}(c'|U_1, \ldots, U_K). \tag{12.28}$$

Suppose now that we assume that a priori a nonresponding unit has odds $\pi(c, c')$ of belonging to category $y = c$ rather than $y = c'$. The posterior odds are given by Bayes's theorem as the prior odds times (12.28). Thus sensitivity of estimation to nonignorable nonresponse can be easily displayed as a function of the prior odds when estimation is by the EM algorithm or through the use of multiple imputations under various assumptions about the prior odds.

12.6. NONIGNORABLE NONRESPONSE WITH FOLLOW-UPS

The only way to reduce sensitivity of inference to nonignorable nonresponse is to reduce nonresponse or accumulate information about how nonrespondents differ from respondents on the outcome variables under investigation. There exists a rather extensive applied statistical literature on methods for reducing initial nonresponse in surveys. The three volumes produced by the National Academy of Sciences Panel on Incomplete Data provide an excellent pathway into the literature (Madow, Nisselson, and Olkin, 1983; Madow, Olkin, and Rubin, 1983; Madow and Olkin, 1983). The most direct method for accumulating information on nonrespondents is to follow up at least some of them to obtain the desired information. Even if only a few nonrespondents are followed up, these can be exceedingly helpful in reducing sensitivity of inference, as the following simulation experiment illustrates.

EXAMPLE 12.9. *Decreased Sensitivity of Inference with Follow-ups.* Glynn, Laird, and Rubin (1986) performed a series of simulations using normal and lognormal data, which can be used to study the decreased sensitivity of inference when follow-up data are obtained from nonrespondents. For the normal data, a sample of 400 standard normal deviates was drawn from an essentially infinite population, the logistic nonresponse mechanism $\Pr(R_i = 1 | y_i) = \exp(1 + y_i)/[1 + \exp(1 + y_i)]$ was applied to create 101 nonrespondents. Then, various fractions of the 101 nonrespondents were randomly sampled to create follow-up data among nonrespondents. The resultant data consisted of (y_i, R_i) for respondents and follow-up nonrespondents, but only R_i for nonfollow-up nonrespondents.

Two models were used to analyze the data. First, a Bayesian model was used like that in Example 11.5, with the initial respondents' data assumed to be $N(\mu_1, \sigma_1^2)$ and the nonrespondents' data assumed to be $N(\mu_0, \sigma_0^2)$ with the prior distribution on $(\mu_1, \mu_0, \ln \sigma_1, \ln \sigma_0)$ proportional to a constant. In this model the population Y_i values are modeled as a mixture of two normal populations where the mixing proportion is unknown. Also, the data were analyzed under the correct normal/logistic response selection model:

$$(y_i | \mu, \sigma^2) \sim N(\mu, \sigma^2),$$

$$\text{logit}[\Pr(R_i = 1 | y_i, \alpha_0, \alpha_1)] = \alpha_0 + \alpha_1 y_i,$$

where the prior distribution on $(\mu, \alpha_0, \alpha_1, \ln \sigma)$ was proportional to a constant. This model is similar to the stochastic selection models considered in Section 11.4.

The entire simulation was repeated with a different data set, with 400 lognormal values (exponentiated standard normal deviates) and 88 nonrespondents created using the nonignorable logistic response mechanism

Table 12.2 Sample Moments of Generated Data[a]

	Normal Data			Lognormal Data		
	N	Sample Mean	Sample Standard Deviation	N	Sample Mean	Sample Standard Deviation
Respondents	299	0.150	0.982	312	1.857	2.236
Nonrespondents	101	−0.591	0.835	88	0.724	0.571
Total	400	−0.037	1.000	400	1.608	2.047
Population values		0.0	1.0		1.649	2.161

[a] Normal data are sampled from the normal (0, 1) distribution; lognormal data are the exponentiated normal values. Response is determined by a logistic response function: $\Pr(R = 1 | y) = \exp(\alpha_0 + \alpha_1 y)/[1 + \exp(\alpha_0 + \alpha_1 y)]$, where $(\alpha_0, \alpha_1) = (1, 1)$ for normal data and $(0, 1)$ for lognormal data.

Table 12.3 Estimates of Population Mean Using Respondent Data of Table 12.2 and Follow-up Data from Some Nonrespondents

Normal Data			Lognormal Data		
Number of Follow-ups	Mixture Model	Selection Model	Number of Follow-ups	Mixture Model	Selection Model
11	−0.010	−0.009	9	1.58	0.934
24	−0.025	−0.029	21	1.60	1.030
28	−0.006	−0.008	25	1.61	1.054
101	−0.037	−0.037	88	1.61	1.605

$\Pr(R_i = 1 | y_i) = \exp(y_i)/[1 + \exp(y_i)]$. Again various fractions of the nonrespondents were randomly sampled to create follow-up data among the nonrespondents. The same two models used to analyze the normal data were applied to the lognormal data. Note that, whereas for the normal data the selection model was correct and the mixture model incorrect, for the lognormal data both models are incorrect.

Table 12.2 summarizes the generated data, both normal and lognormal. Table 12.3 gives estimates of the population means for both models with both the normal and lognormal data. Several trends are readily apparent: First, the mixture model appears to be somewhat more robust than the selection model, doing as well as the selection model when the selection model is correct and doing better than the selection model when neither is correct. Second, the larger the fraction of follow-ups, the better the estimates under both models. Third, using the mixture model, even a few follow-ups yield reasonable estimates. Glynn, Laird, and Rubin (1986) use multiple imputation to draw inferences from survey data of retired men with follow-ups using an extension of the mixture model that includes covariates.

REFERENCES

Box, G. E. P., and Tiao, G. C. (1973). *Bayesian Inference in Statistical Analysis*. Reading, MA: Addison-Wesley.

Cochran, W. G. (1977). *Sampling Techniques*. New York: Wiley.

Glynn, R., Laird, N., and Rubin, D. B. (1986). Selection modelling versus mixture modelling with nonignorable nonresponse, *Proceedings of the 1985 Educational Testing Service Conference on Selection Modelling*.

Hansen, M. H., Madow, W. G., and Tepping, J. (1983). An evaluation of model-dependent and probability-sampling inferences in sample surveys, *J. Amer. Statist. Assoc.* **78**, 776–807.

Herzog, T., and Rubin, D. B. (1983). Using multiple imputations to handle nonresponse in sample surveys, *Incomplete Data in Sample Surveys, Volume 2: Theory and Bibliography*, New York: Academic Press, pp. 209–245.

Li, K. H. (1985). Hypothesis testing in multiple imputation—with emphasis on mixed-up frequencies in contingency tables, Ph.D. Thesis, The University of Chicago.

Little, R. J. A. (1982). Models for nonresponse in sample surveys, *J. Amer. Statist. Assoc.* **77**, 237–250.

Madow, W. G., Nisselson, H., and Olkin, I. (1983). *Incomplete Data in Sample Surveys, vol. 1: Report and case studies*. New York : Academic Press.

Madow, W. G., and Olkin, I. (1983). *Incomplete Data in Sample Surveys, vol. 3: Proceedings of the Symposium*, New York: Academic Press.

Madow, W. G., Olkin, I., and Rubin, D. B. (1983). *Incomplete Data in Sample Surveys, vol. 2: Theory and Bibliographies*. New York: Academic Press.

Pregibon, D. (1977). Typical survey data: estimation and imputation, *Survey Methodol.* **2**, 70–102.

Royall, R. M., and Herson, J. (1973). Robust estimation from finite populations, *J. Amer. Statist. Assoc.* **68**, 883–889.

Rubin, D. B. (1976). Inference and missing data, *Biometrika* **63**, 581–592.

Rubin, D. B. (1977). Formalizing subjective notions about the effect of nonrespondents in sample surveys, *J. Amer. Statist. Assoc.* **72**, 538–543.

Rubin, D. B. (1978). Multiple imputations in sample surveys—a phenomenological Bayesian approach to nonresponse, *Imputation and Editing of Faulty or Missing Survey Data*, U.S. Department of Commerce, pp. 1–23.

Rubin, D. B. (1979). Illustrating the use of multiple imputation to handle nonresponse in sample surveys, *Proceedings of the 1979 Meetings of the ISI-IASS*, Manila.

Rubin, D. B. (1985). The use of propensity scores in applied Bayesian inference. *Bayesian Statistics 2*. (Bernardo, J. M., De Groot, M. H., Lindley, D. V., and Smith, A. F. M. Eds). Amsterdam: North Holland, pp. 463–472.

Rubin, D. B. (1986). Statistical matching using file concatenation with adjusted weights and multiple imputations, *J. Business Econ. Statist.* **4**, 87–94.

Rubin, D. B. (1987). *Multiple Imputation for Nonresponse in Surveys*. New York: Wiley.

Rubin, D. B., and Schenker, N. (1986). Multiple imputation for interval estimation from simple random samples with ignorable nonresponse, *J. Amer. Statist. Assoc.* **81**, 366–374.

Scott, A. J. (1977). On the problem of randomization in survey sampling, *Sankhya* **C39**, 1–9.

PROBLEMS

1. Let $\bar{y}$ and s^2 be the sample mean and variance of y for a simple random sample of size n from a population of size N. Show that, under the normal model, the large sample posterior distribution of the population mean $\bar{Y}$ is normal with mean $\bar{y}$ and variance $(n^{-1} - N^{-1})s^2$. Hence infer (12.11) and (12.12).
2. Review Bayesian results on the multinomial distribution leading to the posterior moments of θ_j and θ_k in Example 12.3. See, for example, Box and Tiao (1973).
3. Fill in details in the argument leading to the mean and variance of $\bar{Y}$ in Example 12.3.
4. Show that the posterior variance of $\bar{Y}$ in Example 12.3 and the mean squared error of the weighting cell estimator in Example 4.2 yield the same answers in large samples.
5. Suppose in the context of Example 12.5 that multiple imputations are created using a hot deck where M and all n_j are large. Show that the resulting multiple imputation estimate of $\bar{Y}$, $\hat{\bar{Y}}$ in (12.24), equals (12.11) and that the associated variance given by (12.25) is less than (12.12).
6. Show that approximate Bayesian bootstrap and fully normal imputation methods of Example 12.6 give the poststratified estimator and associated variances for large M and n_j.

Author Index

Subject Index

Applied Probability and Statistics (Continued)

JOHNSON and KOTZ • Distributions in Statistics
Discrete Distributions
Continuous Univariate Distributions—1
Continuous Univariate Distributions—2
Continuous Multivariate Distributions

JUDGE, HILL, GRIFFITHS, LÜTKEPOHL and LEE • Introduction to the Theory and Practice of Econometrics

JUDGE, GRIFFITHS, HILL, LÜTKEPOHL and LEE • The Theory and Practice of Econometrics, *Second Edition*

KALBFLEISCH and PRENTICE • The Statistical Analysis of Failure Time Data

KISH • Statistical Design for Research

KISH • Survey Sampling

KUH, NEESE, and HOLLINGER • Structural Sensitivity in Econometric Models

KEENEY and RAIFFA • Decisions with Multiple Objectives

LAWLESS • Statistical Models and Methods for Lifetime Data

LEAMER • Specification Searches: Ad Hoc Inference with Nonexperimental Data

LEBART, MORINEAU, and WARWICK • Multivariate Descriptive Statistical Analysis: Correspondence Analysis and Related Techniques for Large Matrices

LINHART and ZUCCHINI • Model Selection

LITTLE and RUBIN • Statistical Analysis with Missing Data

McNEIL • Interactive Data Analysis

MAINDONALD • Statistical Computation

MANN, SCHAFER and SINGPURWALLA • Methods for Statistical Analysis of Reliability and Life Data

MARTZ and WALLER • Bayesian Reliability Analysis

MIKÉ and STANLEY • Statistics in Medical Research: Methods and Issues with Applications in Cancer Research

MILLER • Beyond ANOVA, Basics of Applied Statistics

MILLER • Survival Analysis

MILLER, EFRON, BROWN, and MOSES • Biostatistics Casebook

MONTGOMERY and PECK • Introduction to Linear Regression Analysis

NELSON • Applied Life Data Analysis

OSBORNE • Finite Algorithms in Optimization and Data Analysis

OTNES and ENOCHSON • Applied Time Series Analysis: Volume I, Basic Techniques

OTNES and ENOCHSON • Digital Time Series Analysis

PANKRATZ • Forecasting with Univariate Box-Jenkins Models: Concepts and Cases

PIELOU • Interpretation of Ecological Data: A Primer on Classification and Ordination

PLATEK, RAO, SARNDAL and SINGH • Small Area Statistics: An International Symposium

POLLOCK • The Algebra of Econometrics

PRENTER • Splines and Variational Methods

RAO and MITRA • Generalized Inverse of Matrices and Its Applications

RÉNYI • A Diary on Information Theory

RIPLEY • Spatial Statistics

RIPLEY • Stochastic Simulation

ROSS • Engineering Statistics

RUBIN • Multiple Imputation for Nonresponse in Surveys

RUBINSTEIN • Monte Carlo Optimization, Simulation, and Sensitivity of Queueing Networks

SCHUSS • Theory and Applications of Stochastic Differential Equations

SEAL • Survival Probabilities: The Goal of Risk Theory

(continued from front)